MAP OF BORNEO

SABAH

Kota Kinabalu
Kundasang 1
Papar 2
Beaufort 3
Labuan 4 Telupid
Sipitang 5 Sandakan
6 Tambunan
BRUNEI
Labi 7
Meligan 8 (Maliau Basin) 9
11 10 Long Pasia
12 Pulau Sebatik
13 Bario (Kelabit Highlands)
14 17

0   150
km

SARAWAK

EAST KALIMANTAN

15
Bau Kuching 18
16 21 20

WEST KALIMANTAN

22 Sangkulirang
KALIMANTAN 19

23

CENTRAL KALIMANTAN

24

SOUTH KALIMANTAN
25

ORCHIDS OF BORNEO
Vol. 4

Bulbophyllum anceps Rolfe *(Photo: C.L. Chan)*

ORCHIDS OF BORNEO

VOL. 4

J.J. WOOD

Series Editor
P.J. Cribb

Series Co-ordinators
C.L. Chan and A. Lamb

Line illustrations by
C.L. Chan, L. Gurr, F.L. Liew, J. Stone,
S. Stuart-Smith, J.J. Vermeulen and Yong Ket Hyun

Colour photographs by
T.J. Barkman, K. Barrett, C.L. Chan, J. Dransfield,
A. Lamb, A. Schuiteman, J. Stone and A. Vogel

The Sabah Society
Kota Kinabalu

in association with

2003

Published by

THE SABAH SOCIETY,
P.O. Box 10547,
88806 Kota Kinabalu, Sabah, Malaysia.
e-mail: sabsoc@po.jaring.my
Website: http://www.sabah.org.my/sabahsociety

in association with

THE ROYAL BOTANIC GARDENS,
Kew, Richmond,
Surrey TW9 3AB,
England.

Orchids of Borneo Vol. 4
    by J.J. Wood

Series Editor: P.J. Cribb
Series Co-ordinators: C.L. Chan and A. Lamb

First published 20 June 2003

Perpustakaan Negara Malaysia Cataloguing-in-Publication Data

Wood, J.J.
    Orchids of Borneo / J.J. Wood.
    Includes index
    Bibliography ; 239–242
    ISBN 967-99947-7-5 (v. 4)
    1. Orchids—Borneo. 2. Orchids—Borneo—Identification.
    I. Title.
    584.15095983

Printed in Malaysia.

CONTENTS

LIST OF COLOUR PLATES vi

ACKNOWLEDGEMENTS ix

ERRATA AND NOMENCLATURAL CHANGES IN
VOLUMES ONE, TWO AND THREE xi

CHAPTER 1
    REVISED CLASSIFICATION SYSTEM OF BORNEAN ORCHIDS 1

CHAPTER 2 – REVISED KEYS TO GENERA 3

CHAPTER 3 – DESCRIPTIONS AND FIGURES 19

REFERENCES 239

IDENTIFICATION LIST 243

HERBARIUM ABBREVIATIONS 255

COLOUR PLATES 257

INDEX TO ORCHID SCIENTIFIC NAMES 308

LIST OF COLOUR PLATES

(Taxon number in this account given in brackets)

Pl. 1   A.   *Abdominea minimiflora* (Hook.f.) J.J. Sm. (1)
B.   *Agrostophyllum glumaceum* Hook.f. (2)
C.   *Aphyllorchis montana* Rchb.f. (5)
D.   *Ania borneensis* (Rolfe) Senghas (3)
E.   *Ania ponggolensis* A. Lamb in H. Turner (4)

Pl. 2   A.   *Appendicula congesta* Ridl. (8)
B.   *Appendicula congesta* Ridl. (8)
C.   *Appendicula foliosa* Ames & C. Schweinf. (9)
D.   *Appendicula foliosa* Ames & C. Schweinf. (9)

Pl. 3   A.   *Appendicula calcarata* Ridl. (7)
B.   *Appendicula rostellata* J.J. Sm. (11)
C.   *Ascidieria longifolia* (Hook.f.) Seidenf. (12)
D.   *Brachypeza indusiata* (Rchb.f.) Garay (13)

Pl. 4   A.   *Bromheadia divaricata* Ames & C. Schweinf. (15)
B.   *Bulbophyllum lemniscatoides* Rolfe (18)
C.   *Bulbophyllum coriaceum* Ridl. (17)
D.   *Calanthe otuhanica* C.L. Chan & T.J. Barkman (19)
E.   *Cheirostylis spathulata* J.J. Sm. (20)

Pl. 5   A.   *Cleisostoma suaveolens* Blume (21)
B.   *Cordiglottis fulgens* (Ridl.) Garay (22)
C.   *Corybas serpentinus* J. Dransf. (24)
D.   *Corybas serpentinus* J. Dransf. (24)
E.   *Cymbidium kinabaluense* K.M. Wong & C.L. Chan (25)
F.   *Cymbidium kinabaluense* K.M. Wong & C.L. Chan (25)

Pl. 6   A.   *Cystorchis salmoneus* J.J. Wood (27)
B.   *Cystorchis saprophytica* J.J. Sm. (28)
C.   *Dendrobium bifarium* Lindl. (30)
D.   *Dendrobium aurantiflammeum* J.J. Wood (29)

Pl. 7   A.   *Dendrobium cinnabarinum* Rchb.f. (31)
B.   *Dendrobium cinnabarinum* Rchb.f. (31)
C.   *Dendrobium cinnabarinum* Rchb.f. (31)
D.   *Dendrobium cinnabarinum* Rchb.f. (31)
E.   *Dendrobium dearei* Rchb.f. (32)

Pl. 8   A.   *Dendrobium derryi* Ridl. (33)
B.   *Dendrobium derryi* Ridl. (33)

	C.	*Dendrobium derryi* Ridl. (33)
	D.	*Dendrobium hendersonii* A.D. Hawkes & A.H. Heller (35)
	E.	*Dendrobium kurashigei* Yukawa (36)

Pl. 9	A.	*Dendrobium lamrianum* C.L. Chan (37)
	B.	*Dendrobium lamrianum* C.L. Chan (37)
	C.	*Dendrobium metachilinum* Rchb.f. (39)
	D.	*Dendrobium parthenium* Rchb.f. (40)
	E.	*Dendrochilum alpinum* Carr (42)

Pl. 10	A.	*Dendrochilum alpinum* Carr (42)
	B.	*Dendrochilum joclemensii* Ames (46)
	C.	*Dendrochilum joclemensii* Ames (46)
	D.	*Dilochia rigida* (Ridl.) J.J. Wood (54)
	E.	*Diplocaulobium brevicolle* (J.J. Sm.) Kraenzl. (55)
	F.	*Diplocaulobium longicolle* (Lindl.) Kraenzl. (56)

Pl. 11	A.	*Dossinia marmorata* E. Morren (57)
	B.	*Dossinia marmorata* E. Morren (57)
	C.	*Entomophobia kinabaluensis* (Ames) de Vogel (58)
	D.	*Entomophobia kinabaluensis* (Ames) de Vogel (58)

Pl. 12	A.	*Epigeneium zebrinum* (J.J. Sm.) Summerh. (59)
	B.	*Eria cymbidifolia* Ridl. var. *cymbidifolia* (61)
	C.	*Eria saccifera* Hook.f. (63)

Pl. 13	A.	*Geesinkorchis alaticallosa* de Vogel (64)
	B.	*Geesinkorchis alaticallosa* de Vogel (64)
	C.	*Habenaria lobbii* Rchb.f. (67)
	D.	*Habenaria lobbii* Rchb.f. (67)

Pl. 14	A.	*Goodyera condensata* Ormerod & J.J. Wood (66)
	B.	*Hetaeria anomala* (Lindl.) Rchb.f. (68)
	C.	*Liparis anopheles* J.J. Wood (69)
	D.	*Liparis grandiflora* Ridl. (70)
	E.	*Liparis grandiflora* Ridl. (70)

Pl. 15	A.	*Liparis lacerata* Ridl. (71)
	B.	*Liparis lacerata* Ridl. (71)
	C.	*Pennilabium struthio* Carr (75)
	D.	*Oberonia patentifolia* Ames & C. Schweinf. (73)

Pl. 16	A.	*Peristylus hallieri* J.J. Sm. (76)
	B.	*Peristylus hallieri* J.J. Sm. (76)
	C.	*Phalaenopsis cochlearis* Holttum (77)
	D.	*Pholidota clemensii* Ames (78)
	E.	*Pholidota clemensii* Ames (78)

Pl. 17 A. *Pholidota sigmatochilus* (Rolfe) J.J. Sm. (80)
        B. *Pholidota sigmatochilus* (Rolfe) J.J. Sm. (80)
        C. *Pholidota ventricosa* (Blume) Rchb.f. (81)
        D. *Pholidota ventricosa* (Blume) Rchb.f. (81)

Pl. 18 A. *Pholidota ventricosa* (Blume) Rchb.f. (81)
        B. *Pholidota ventricosa* (Blume) Rchb.f. (81)
        C. *Podochilus marsupialis* Schuit. (82)
        D. *Pomatocalpa kunstleri* (Hook.f.) J.J. Sm. (83)

Pl. 19 A. *Pomatocalpa kunstleri* (Hook.f.) J.J. Sm. (83)
        B. *Porphyrodesme sarcanthoides* (J.J. Sm.) U.W. Mahyar (84)
        C. *Porphyrodesme sarcanthoides* (J.J. Sm.) U.W. Mahyar (84)
        D. *Sarcoglyphis potamophila* (Schltr.) Garay & W. Kittr. (85)

Pl. 20 A. *Schoenorchis buddleiflora* (Schltr. & J.J. Sm.) J.J. Sm. (86)
        B. *Schoenorchis endertii* (J.J. Sm.) E.A. Christenson & J.J. Wood (87)
        C. *Schoenorchis endertii* (J.J. Sm.) E.A. Christenson & J.J. Wood (87)
        D. *Smitinandia micrantha* (Lindl.) Holttum (88)
        E. *Smitinandia micrantha* (Lindl.) Holttum (88)

Pl. 21 A. *Spathoglottis kimballiana* Hook.f. (89)
        B. *Thrixspermum triangulare* Ames & C. Schweinf. (91)
        C. *Thrixspermum tortum* J.J. Sm. (90)

Pl. 22 A. *Trichoglottis jiewhoei* J.J. Wood, A. Lamb & C.L. Chan (92)
        B. *Trichoglottis jiewhoei* J.J. Wood, A. Lamb & C.L. Chan (92)

Pl. 23 A. *Trichoglottis kinabaluensis* Rolfe (93)
        B. *Trichoglottis kinabaluensis* Rolfe (93)
        C. *Trichoglottis magnicallosa* Ames & C. Schweinf. (94)

Pl. 24 A. *Trichoglottis tinekeae* Schuit. (95)
        B. *Trichoglottis tinekeae* Schuit. (95)
        C. *Trichotosia vestita* (Lindl.) Kraenzl. (96)
        D. *Vanilla havilandii* Rolfe (97)
        E. *Vanilla kinabaluensis* Carr (98)

Pl. 25 A. *Ventricularia borneensis* J.J. Wood (99)
        B. *Ventricularia borneensis* J.J. Wood (99)
        C. *Vrydagzynea grandis* Ames & C. Schweinf. (100)

ACKNOWLEDGEMENTS

I would like to acknowledge the continuing support for this project expressed by Professor Simon Owens, Keeper of the Herbarium at Kew. Datuk Chan Chew Lun, President of the Sabah Society and Co-ordinator of this project, has also unfailingly provided enthusiasm and support throughout. Thanks are also due to Dr. Phillip Cribb, Deputy Keeper of the Kew herbarium, for his work as series editor and for his constructive criticism and continuing commitment to the project.

Anthony Lamb, former Manager of the Agricultural Park and Orchid Centre at Tenom, has provided valued hospitality during stays in Sabah. This project owes much to his unrivalled knowledge of Bornean orchids in the field and enthusiasm.

Once again the collective skill of several artists have helped in accurately portraying the orchids described in this volume. I would like to thank C.L. Chan, Linda Gurr, Lucy F.L. Liew, Judi Stone, Susanna Stuart-Smith, Jaap Vermeulen and Yong Ket Hyun for their invaluable contribution.

The beautiful colour plates depict photographs kindly provided by Todd J. Barkman, Kath Barrett, C.L. Chan, John Dransfield, Tony Lamb, André Schuiteman, Jess Stone and Art Vogel.

I would like to express my thanks to Datuk Chan Chew Lun who, as ever, has maintained such a high standard of production throughout. Finally, the financial support magnanimously provided by Mr Tan Jiew Hoe (Singapore), without which this volume would not have been published, is gratefully acknowledged.

ERRATA AND NOMENCLATURAL CHANGES IN VOLUMES ONE, TWO AND THREE

ERRATA IN VOLUME ONE

Figure 8. The plant figured as *Bulbophyllum microglossum* Ridl. is *B. goebelianum* Kraenzl. according to L.A. Garay (pers. comm.).

Figure 31. O'Byrne (1998) comments that the plant figured as *Dendrobium singkawangense* J.J. Sm. is a closely related undescribed species.

Figure 81. The plant figured as *Pristiglottis hasseltii* (Blume) Cretz. & J.J. Sm. is *P. hydrocephala* J.J. Sm. according to André Schuiteman of the Rijksherbarium, Leiden University (pers. comm.).

Figure 87. E.A. Christenson (pers. comm.) disagrees with the transfer by Senghas of *Renanthera auyongii* Christenson to *Renantherella* as a subspecies of *R. histrionica* (Rchb.f.) Ridl. He believes *Renantherella* to be a poorly defined genus and that *Renanthera auyongii* is not closely related to *Renanthera histrionica* Rchb.f., the latter having differently coloured, resupinate flowers with a shorter spur.

Figure 92. Subsequent studies have shown that the plant figured as *Trichoglottis scapigera* Ridl. is *T. zollingeriana* (Kraenzl.) J.J.Sm.

Plate 16D. Christenson (2001) points out that the plant depicted is typical *Phalaenopsis corningiana* Rchb.f

ERRATA IN VOLUME TWO

Figure 1. The plant figured as *Bulbophyllum disjunctum* Ames & C. Schweinf. is *B. farinulentum* J.J. Sm. subsp. *farinulentum* (Vermeulen, 2002).

Figure 92. Vermeulen (pers. comm.) has confirmed that the plant figured as *Bulbophyllum macranthum* Lindl. is a closely related species.

NOMENCLATURAL CHANGES AFFECTING VOLUMES ONE AND THREE

VOLUME ONE

Figure 15. *Ceratochilus jiewhoei* J.J. Wood & Shim has been placed in a new genus, *Jejewoodia*, by the Polish botanist D.L. Szlachetko (1995). A second undescribed species is reported from Sabah.

Figure 78. Phalaenopsis violacea Witte. This species was originally described from Sumatra and has deep rose-pink or, less often, white or bluish flowers. Plants from Borneo, referred to *P. violacea* in the literature and horticultural trade, are commonly known as the "Lundu Orchid", the State flower of Sarawak. This Bornean *Phalaenopsis* has also been adopted as the state flower of Sarawak. Christenson & Whitten (1995), however, present a convincing argument using morphological and chemotaxonomic data for the recognition of two species. The Bornean plant is now recognised as *P. bellina* (Rchb.f.) Christenson, which was originally described as a variety of *P. violacea*. This has white, greenish white, yellow or, rarely, orange flowers with a characteristic intense deep purple blotch on the inner half of the lateral sepals. The perianth is never uniformly pigmented as in *P. violacea*. In *P. bellina* the petals are ovate and usually more than 1.3 cm broad, while in *P. violacea* they are elliptic and generally less than 0.7 cm broad. *Phalaenopsis bellina* has rather sickle-shaped lateral sepals, the tips of all three sepals forming an isosceles triangle. In *P. violacea* the lateral sepals are not sickle-shaped and the tips of all three form an equilateral triangle. The leaves of *P. bellina*, although variable, are generally broader than those of *P. violacea*. The flowers of *P. bellina* have a lemony fragrance, while those of *P. violacea* are spicy. Kaiser (1993) analysed the floral fragrances of both species and found that of *P. bellina* to be composed almost entirely of 64% geraniol and 32% linalool, while that of *P. violacea* was dominated by 55% elemicin and 27% cinnamyl alcohol. These differing fragrance compositions presumably play an important role in pollinator specificity.

Christenson & Whitten report that Bornean-type plants have recently been collected in Peninsular Malaysia. That these taxa are sympatric for part of their range reinforces their treatment as separate species rather than maintenance as subspecies.

VOLUME THREE

Figure 63. Dendrochilum papillilabium J.J. Wood is now treated at subspecific rank under *D. tenompokense* Carr (Wood, 2000).

Plate 3E. The plant figured as *Bulbophyllum polygaliflorum* J.J. Wood is *B. dracunculus* J.J. Verm.

CHAPTER 1

REVISED CLASSIFICATION SYSTEM OF BORNEAN ORCHIDS

(Updated from Dressler, 1990 & 1993; see also Pridgeon *et al.*, 1999).

Subfamily **APOSTASIOIDEAE**

Apostasia, Neuwiedia

Subfamily **CYPRIPEDIOIDEAE**

Paphiopedilum

Subfamily **ORCHIDOIDEAE**

Tribe Cranichideae
  Subtribe Goodyerinae — *Anoectochilus, Cheirostylis, Cystorchis, Dossinia, Erythrodes, Goodyera, Hetaeria, Hylophila, Kuhlhasseltia, Lepidogyne, Macodes, Myrmechis, Pristiglottis, Rhomboda, Vrydagzynea, Zeuxine*
  Subtribe Spiranthinae — *Spiranthes*

Tribe Diurideae
  Subtribe Acianthinae — *Corybas, Pantlingia*
  Subtribe Cryptostylidinae — *Cryptostylis*

Tribe Orchideae
  Subtribe Orchidinae — *Platanthera*
  Subtribe Habenariinae — *Habenaria, Peristylus*

Subfamily **VANILLOIDEAE**

Tribe Vanilleae
  Subtribe Galeolinae — *Cyrtosia, Erythrorchis, Galeola*
  Subtribe Vanillinae — *Vanilla*
  Subtribe Lecanorchidinae — *Lecanorchis*

Subfamily **EPIDENDROIDEAE**

Tribe Tropidieae — *Corymborkis, Tropidia*

Tribe Gastrodieae
  Subtribe Gastrodiinae — *Didymoplexiella, Didymoplexis, Gastrodia**
  Subtribe Epipogiinae — *Epipogium, Stereosandra*

Tribe Neottieae
  Subtribe Limodorinae — *Aphyllorchis*

Tribe Nervilieae — *Nervilia*

Tribe Malaxideae — *Hippeophyllum, Liparis, Malaxis, Oberonia*

---

* The genus *Neoclemensia* is now considered as congeneric with *Gastrodia*.

Tribe Cymbidieae
    Subtribe Bromheadiinae   *Bromheadia*
    Subtribe Eulophiinae   *Dipodium, Eulophia, Geodorum, Oeceoclades*
    Subtribe Thecostelinae   *Thecopus, Thecostele*
    Subtribe Collabiinae   *Chrysoglossum, Claderia, Collabium*
    Subtribe Cyrtopodiinae   *Cymbidium, Grammatophyllum, Pilophyllum,*
                                  Porphyroglottis
    Subtribe Acriopsidinae   *Acriopsis*
Tribe Arethuseae
    Subtribe Bletiinae   *Acanthephippium, Ania, Calanthe, Mischobulbum,*
                                  Nephelaphyllum, Pachystoma, Phaius, Plocoglottis,
                                  Spathoglottis, Tainia
    Subtribe Arundinae   *Arundina, Dilochia*
Tribe Epidendreae
    Subtribe Glomerinae   *Agrostophyllum*
    Subtribe Polystachyinae   *Polystachya*
Tribe Coelogyneae
    Subtribe Coelogyninae   *Chelonistele, Coelogyne, Dendrochilum,*
                                    Entomophobia, Geesinkorchis, Nabaluia, Pholidota
Tribe Podochileae
    Subtribe Eriinae   *Ascidieria, Ceratostylis, Eria, Porpax, Sarcostoma,*
                                    Trichotosia
    Subtribe Podochilinae   *Appendicula, Poaephyllum, Podochilus*
    Subtribe Thelasiinae   *Octarrhena, Phreatia, Thelasis*
    Subtribe Dendrobiinae   *Dendrobium, Diplocaulobium, Epigeneium,*
                                    Flickingeria
    Subtribe Bulbophyllinae   *Bulbophyllum, Trias*
Tribe Vandeae
    Subtribe Aeridinae   *Abdominea, Adenoncos, Aerides, Arachnis,*
                                    Ascocentrum, Ascochilopsis, Ascochilus,
                                    Biermannia, Bogoria, Brachypeza, Chamaeanthus,
                                    Chroniochilus, Cleisocentron, Cleisomeria,
                                    Cleisostoma, Cordiglottis, Dimorphorchis, Doritis,
                                    Dyakia, Gastrochilus, Grosourdya, Jejewoodia,
                                    Kingidium, Luisia, Macropodanthus, Malleola,
                                    Micropera, Microsaccus, Microtatorchis,
                                    Ornithochilus, Papilionanthe, Paraphalaenopsis,
                                    Pennilabium, Phalaenopsis, Pomatocalpa,
                                    Porphyrodesme, Porrorhachis, Pteroceras,
                                    Renanthera, Rhynchostylis, Robiquetia,
                                    Sarcoglyphis, Schoenorchis, Smitinandia,
                                    Spongiola, Staurochilus, Taeniophyllum,
                                    Thrixspermum, Trichoglottis, Tuberolabium, Vanda,
                                    Ventricularia

CHAPTER 2
REVISED KEYS TO GENERA

Since the publication of volume one of this series in 1994, changes in the status of certain genera and the recognition of several reinstated, new or additional genera have necessitated the production of revised generic keys. These changes are listed below:

Subfamily *Epidendroideae*, Tribe *Gastrodieae*, subtribe *Gastrodiinae*: *Neoclemensia* Carr. Paul Ormerod (pers. comm.) and I consider this monospecific genus to be conspecific with *Gastrodia*.

Subfamily *Epidendroideae*, Tribe *Podochileae*, subtribe *Bulbophyllinae*: *Bulbophyllum* is the largest orchid genus in Borneo, from where approximately 214 species have been recorded. The genus exhibits great polymorphy in both vegetative and floral structure, and considerable skill is required when determining taxa. Recently, Garay *et al.* (1994) have produced a revised overview of the *Bulbophyllum* alliance in which, among other things, the sections *Cirrhopetalum*, *Epicrianthes* and *Osyricera* are reinstated as genera. In addition, several new genera, viz. *Mastigion*, *Rhytionanthos* and *Synarmosepalum*, represented in Borneo, are also recognised. The generic and sectional delimitation within the alliance is still the subject of great controversy, however. Vermeulen (pers. comm.), for example, disagrees with much of what has been proposed. It is becoming clear that, only after further studies of the whole alliance, will the generic delimitations between *Bulbophyllum* and proposed "segregate genera" be clarified. In the meantime, these segregate genera are best treated within *Bulbophyllum* sensu lato. In the revised generic key provided in this volume I have indicated Garay's proposed genera under *Bulbophyllum* sensu lato (see couplets 70 to 75). Garay's concept, as it affects Bornean taxa, is summarised below.

Segregate genera of *Bulbophyllum* proposed by Garay *et al.* (1994):

Cirrhopetalum Lindl. Garay *et al.* (1994) reinstate *Cirrhopetalum* among the genera of the *Bulbophyllum* alliance on the primary characters exhibited by the type species *Bulbophyllum longiflorum* Thou.

Epicrianthes Blume. About ten species are recorded from Borneo, most of restricted distribution. Reinstated at generic level by Garay & Kittredge (1985).

Hapalochilus (Schltr.) Sengh. The centre of speciation lies in New Guinea. The sole Bornean species, *H. lohokii* (J.J. Verm. & Lamb) Garay, Hamer & Siegerist, represents the westernmost extension of range (see *Orchids of Borneo*, volume 3: fig. 77).

Mastigion Garay, Hamer & Siegerist. The Greek name *Mastigion* is a diminutive of *mastix*, a whip, in reference to the lateral sepals which resemble a small whip. *M. putidum* (Teijsm. & Binn.) Garay, Hamer & Siegerist is the sole representative in Borneo.

Osyricera Blume. This genus was created by Blume in 1825 based upon *O. crassifolia* Blume from Sumatra and Java. *O. osyriceroides* (J.J.Sm.) Garay, Hamer & Siegerist is the sole representative in Borneo.

Rhytionanthos Garay, Hamer & Siegerist. The generic name is derived from the Greek *rhytion*, a small drinking horn, and *anthos*, flower, in allusion to the shape of the united lateral sepals. *R. mirum* (J.J.Sm.) Garay, Hamer & Siegerist is the sole representative in Borneo.

Synarmosepalum Garay, Hamer & Siegerist. The generic name is derived from the Greek *synarmos*, to join together, in reference to the three fused sepals. The sole representative in Borneo is *S. heldiorum* (J.J.Verm.) Garay, Hamer & Siergerist.

Subfamily *Epidendroideae*, Tribe *Vandeae*, subtribe *Aeridinae*:

Ascochilus Ridl. This genus had not previously been recorded with certainty from Borneo until *A. emarginatus* (Blume) Schuit. was confirmed as occurring in Sarawak.

Ceratochilus jiewhoei J.J. Wood & Shim has been placed in a new genus, *Jejewoodia* by D.L. Szlachetko (1995). I agree *C. jiewhoei* is sufficiently distinct and that Szlachetko's proposal is a valid one.

Renantherella Ridl. Christenson (pers. comm.) considers this monospecific genus to be conspecific with *Renanthera*.

Ventricularia Garay. A recently described second species of this formerly monospecific genus from Peninsular Malaysia, viz. *V. borneensis* J.J. Wood, represents an eastward extension of range.

KEY TO GENERA
(EXCLUDING HETEROMYCOTROPHES AND TRIBE VANDEAE, SUBTRIBE AERIDINAE)

1. Flowers with 2 or 3 fertile anthers **2**
    Flowers with a single fertile anther **4**

2. ·Perianth segments similar, the lip never deeply saccate **3**
    Perianth segments very unequal, the lip deeply saccate, slipper- or pouch-shaped. Anthers 2, lateral. Staminode median, large and shield-shaped .... **Paphiopedilum**

3. Anthers 2, with or without a staminode. Inflorescence usually branched, curved and spreading, never erect .... **Apostasia**
    Anthers 3, staminode absent. Inflorescence simple, erect .... **Neuwiedia**

4. Anther erect or bending back, never short and operculate at apex of column. Leaves usually spirally arranged, convolute, not articulated at base **5**
    Anther eventually bending downward over column apex to become operculate, or operculate at column apex but not bending downward. Leaves distichous, usually articulate at base **33**

5. Plants exclusively terrestrial **6**
    Plants climbing **31**

6. Rostellum elongate, equalling the anther. Root-stem tuberoids (tubers) absent **7**
    Rostellum usually shorter than the anther. Root-stem tuberoids (tubers) present **26**

7. Stems tough and rigid. Leaves plicate **8**
    Stems weak and fleshy, often brittle, never tough and rigid. Leaves convolute or conduplicate **9**

8. Lip widest at apex. Column long. Inflorescence often branched .... **Corymborkis**
    Lip widest at base. Column short. Inflorescence simple .... **Tropidia**

9. Pollinia sectile. Roots scattered along rhizome **10**
    Pollinia not sectile. Roots in a close fascicle **25**

10. Flowers resupinate **11**
    Flowers non-resupinate **24**

11. Inner surface of lip hypochile with a bicarinate discoid keel along median line ... **Rhomboda**
    Bicarinate discoid keel absent **12**

12. Spur or saccate base of lip containing neither glands nor hairs (hairs may occur near mid-lobe only) **13**
    Lip hairy within or having papillae or glands on either side near base or in spur or sac **14**

13. Lip with spur which projects between lateral sepals .... **Erythrodes**
    Lip saccate, entirely enclosed by lateral sepals .... **Hylophila**

14. Lip hairy within .... **Goodyera**
    Lip otherwise **15**

15. Apex of lip not abruptly widened into a distinct spathulate or transverse, bilobed blade ... **16**
    Apex of lip abruptly widened into a distinct spathulate or transverse, bilobed blade **18**

16. Saccate base of lip with a transverse row of small calli. Plants robust, up to 100 cm tall ....
    .... **Lepidogyne**
    Saccate base of lip or spur containing stalked or sessile glands. Plants much smaller **17**

17. Hypochile swollen at base into twin lateral sacs each containing a sessile gland. Epichile with fleshy involute margins, forming a tube .... **Cystorchis**
    Hypochile otherwise, containing 2 stalked glands. Epichile otherwise .... **Vrydagzynea**

18. Claw of lip with a toothed or pectinate flange on either side .... **Anoectochilus**
    Claw of lip otherwise **19**

19. Leaves dark green with greenish-yellow, golden or pink median and secondary nerves ....
    .... **Dossinia**
    Leaves without such coloured secondary nerves, although median nerve sometimes coloured **20**

5

20. Sepals connate for half their length to form a swollen tube .... **Cheirostylis**
    Dorsal sepal and petals connivent, forming a hood, or free **21**

21. Dorsal sepal and petals connivent, forming a hood **22**
    Dorsal sepal and petals free **23**

22. Column without appendages .... **Kuhlhasseltia**
    Column with 2 narrow wings which project into the base of the lip .... **Pristiglottis**

23. Lip with a long claw. Stigmas on short processes. Inflorescence 1- to 2-flowered ....
    .... **Myrmechis**
    Lip with a short claw. Stigmas sessile. Inflorescence several-flowered .... **Zeuxine**

24. Lip and column twisted to one side .... **Macodes**
    Lip and column straight .... **Hetaeria**

25. Flowers resupinate, small, arranged spirally in a dense inflorescence .... **Spiranthes**
    Flowers non-resupinate, large, arranged in all directions in a lax inflorescence ....
    .... **Cryptostylis**

26. Lip without a spur **27**
    Lip spurred **28**

27. Inflorescence produced with the leaf. Lip orbicular. Column with a tooth-like process below
    .... **Pantlingia**
    Inflorescence produced before the leaf. Lip 3-lobed, the base embracing the column.
    Column without a tooth-like process .... **Nervilia**

28. Lip 2-spurred, tubular below. Flowers helmet-shaped .... **Corybas**
    Lip with 1 spur, not tubular below. Flowers otherwise **29**

29. Stigmas each on a stigmatophore extending from the column, free from hypochile ....
    .... **Habenaria**
    Stigmas not freely extending in front of column, sometimes connate or adpressed to lip
    hypochile **30**

30. Lip simple, strap-shaped. Spur cylindric, rather long, not swollen at apex. Stigmas joined to
    form a concave structure, free from hypochile .... **Platanthera**
    Lip 3-lobed. Spur short, usually globular, saccate or fusiform. Stigmas convex, cushion-like,
    connate with or adpressed to hypochile .... **Peristylus**

31. Leaves fleshy, never plicate. Stems fleshy. Pollinia soft and mealy, as monads .... **Vanilla**
    Leaves plicate, stem never fleshy, often rather tough, sometimes brittle. Pollinia 2, cleft .. **32**

32. Habit monopodial. Stems not distant on a creeping rhizome. Leaves distichous, imbricate,
    ensiform. Flowers pale yellowish with pink to crimson blotches .... **Dipodium**
    Habit sympodial. Stems placed distantly on a creeping rhizome. Leaves elliptic, neither
    distichous nor imbricate. Flowers green .... **Claderia**

33. Plants terrestrial **34**
    Plants epiphytic or lithophytic **61**

34. Pollinia 2 or 4, naked, i.e. without caudicles, viscidia and stipes usually absent **35**
    Pollinia 2 to 8, with caudicles (sometimes reduced), or a stipes **38**

35. Column-foot absent **36**
    Column-foot prominent **37**

36. Column long. Flowers usually resupinate. Lip without 2 large basal auricles, apex rarely
    pectinate .... **Liparis** (in part)
    Column short. Flowers non-resupinate. Lip with 2 large basal auricles, apex often pectinate
    .... **Malaxis**

37. Rhizomatous part of shoot (sometimes also the non-rhizomatous part) carrying one-noded
    pseudobulbs .... **Epigeneium** (in part, sometimes *E. kinabaluense*)
    Non-rhizomatous part of shoot consisting of several internodes, wholly or partly fleshy, with
    or without pseudobulbs .... **Dendrobium**
    (in part, some species in sections *Conostalix* & *Distichophyllum*)

38. Inflorescences numerous or not, borne along a slender leafy stem, lateral or terminal **39**
    Inflorescences never numerous, usually solitary, never borne along a slender, leafy stem,
    usually lateral, sometimes axillary or terminal **44**

39. Inflorescences lateral **40**
    Inflorescences terminal **42**

40. Pollinia 2. Inflorescences not numerous. Flowers *c.* 4 cm across .... **Cymbidium**
    (in part, *C. elongatum*)
    Pollina 6 or 8. Inflorescences numerous. Flowers much smaller **41**

41. Pollina 6. Leaf sheaths and flowers glabrous .... **Appendicula** (in part)
    Pollina 8. Leaf sheaths and flowers usually covered in reddish brown hispid hairs ....
    .... **Trichotosia** (in part)

42. Pollina 2 .... **Bromheadia** (in part, *B. borneensis, B. crassifolia, B. finlaysoniana*)
    Pollinia 8 .... **43**

43. Flowers large, up to 8 cm across (sometimes peloric). Petals much broader than sepals.
    Inflorescence usually unbranched. Floral bracts small, acute, persistent .... **Arundina**
    Flowers much smaller. Petals similar to sepals. Inflorescence branching. Floral bracts
    conspicuous, concave, deciduous .... **Dilochia**

44. Pollinia 2 .... **45**
    Pollinia 4 or 8 .... **47**

45. Plants densely covered in yellowish brown hairs. Flowers non-resupinate .... **Pilophyllum**
    Plants glabrous. Flowers resupinate **46**

46. Lip mobile. Column with 2 fleshy basal keels. Spur formed by column-foot .... **Chrysoglossum**
    Lip immobile. Column without basal keels. Mentum formed by column-foot, base of lateral sepals and base of lip .... **Collabium**

47. Pollinia 4 .... **48**
    Pollinia 8 .... **53**

48. Inflorescence arcuate, strongly decurved .... **Geodorum**
    Inflorescence otherwise **49**

49. Lip not spurred **50**
    Lip spurred **52**

50. Lip divided into a distinct, somewhat saccate hypochile and 2-lobed epichile. Pseudobulbs flattened, always 2-leaved .... **Geesinkorchis**
    Lip otherwise. Pseudobulbs never flattened, sometimes elongated into a leafy stem, 1 to many-leaved **51**

51. Lip convex, adnate to sides and apex of column-foot to form a sac, usually with an elastic hinge that springs when touched. Pseudobulbs often elongated into a leafy stem .... **Plocoglottis**
    Lip never convex, free or fused at base to base of column, without an elastic hinge. Pseudobulbs short, often enclosed in sheathing leaf-bases .... **Cymbidium**
    (in part, *C. borneense, C. ensifolium* subsp. *haematodes, C. lancifolium*)

52. Lip entire, or 3-lobed (mid-lobe not bilobulate) .... **Eulophia**
    (in part, *E. graminea, E. spectabilis*)
    Lip '4-lobed', mid-lobe bilobulate .... **Oeceoclades**

53. Pseudobulbs absent; replaced by a fleshy subterranean rhizome, swelling into a horizontal fusiform, sometimes V-shaped, tuber. Leaves usually several, often withering before inflorescence appears. Lip with a small basal pouch .... **Pachystoma**
    Plants pseudobulbous **54**

54. Lip spurred, or gibbous and partially adnate to and embracing column to form a tube **55**
    Lip not spurred **58**

55. Pseudobulbs always 1-leaved. Plants remaining green when damaged **56**
    Pseudobulbs 2- to several-leaved. Plants turning bluish-black when damaged. Lip spurred or gibbous **57**

56. Inflorescence lateral .... **Ania**
    Inflorescence terminal .... **Nephelaphyllum**

57. Column margins fused with the base of the lip over nearly their entire length. Lip normally spurred .... **Calanthe**
    Column margins fused with lip only at or near the base. Lip shortly spurred or gibbous .... **Phaius**

58. Pseudobulbs 1-leaved **59**
    Pseudobulbs 2- to several-leaved **60**

59. Leaf base cordate in mature plants, petiole absent .... **Mischobulbum**
    Leaf base ± decurrent along a petiole .... **Tainia**

60. Flowers urn-shaped, sepals fleshy, fused to form a swollen tube, free at the apices. Lip movably hinged to a column-foot, not clawed or callose .... **Acanthephippium**
    Flowers with free, usually spreading sepals. Lip not movably hinged, mid-lobe very narrowly clawed, with 2 ovoid, often pubescent basal calli. Column-foot absent .... **Spathoglottis**

61. Pollinia 2 or 4, naked, i.e. without caudicles **62**
    Pollinia 2 to 8, with distinct, though sometimes reduced, caudicles **76**

62. Column-foot absent. Leaves equitant, distichous, bilaterally flattened **63**
    Column-foot prominent. Leaves dorsiventral, or occasionally bilaterally flattened (in *Dendrobium* sections *Aporum, Oxystophyllum* and *Strongyle* only) **64**

63. Groups of leaves close together. Column short .... **Oberonia**
    Groups of leaves approximately 4 cm apart. Column long .... **Hippeophyllum**

64. Lip usually immobile, not hinged at base. Mentum often spur-like **65**
    Lip movably hinged to column-foot. Mentum saccate **68**

65. Rhizomatous part of shoot (sometimes also the non-rhizomatous part) bearing one-noded, 1- or 2-, rarely 3-leaved pseudobulbs **66**
    Non-rhizomatous part of shoot (when present) consisting of several internodes, with or without 1- to several-noded pseudobulbs. Flowers ephemeral or long-lasting **67**

66. Erect parts of shoot closely set, tufted, 15–25 cm high, consisting of a single internode tapering from a fleshy base into a slender neck, with 1 apical leaf and 1- to 2-flowered successive inflorescences. Flowers on very long pedicels, ephemeral. Sepals and petals narrowly caudate (in Bornean species) .... **Diplocaulobium**
    Erect parts of shoot spreading or suberect, never tufted, consisting of several internodes bearing one-noded, 1-, 2-, or rarely 3-leaved pseudobulbs. Inflorescence 1-to several-flowered. Flowers on shorter pedicels, long-lived. Sepals and petals narrowly elliptic .... **Epigeneium**

67. Stems superposed, the non-rhizomatous part of the shoot consisting of several quite long, thin internodes, the uppermost pseudobulbous and 1-leaved. Flowers always ephemeral ... . **Flickingeria**
    Stems not superposed; either 1) rhizomatous, 2) erect and many-noded, 3) erect and 1-noded or several-noded from a many-noded rhizome, or 4) rhizome absent, new stems of many nodes arising from base of old ones. Leaves 1 to many. Flowers long-lived or ephemeral .... **Dendrobium**

68.   Anther-cap with a prolongation in front, of varying shape, appearing rostrate, cornute or petaloid (deeply lacerate in Bornean *T. tothastes*). Column usually with insignificant stelidia. Sepals more or less forming a triangle in outline .... **Trias**
        Anther-cap otherwise. Column stelidia usually prominent. Sepals forming a variable outline **69**

69.   Column without a distinct foot; if swollen at base it does not form a mentum **70**
        Column with a distinct foot forming a mentum **71**

70.   Inflorescence one-flowered. Anther-cap ebullate ....
        .... **Bulbophyllum** (*Hapalochilus* sensu Garay *et al.*)
        Inflorescence racemose with a more or less fusiform rachis. Anther-cap bullate ....
        .... **Bulbophyllum** (*Osyricera* sensu Garay *et al.*)

71.   Dorsal sepal free from lateral sepals **72**
        All three sepals basally two thirds connate, urceolate. Petals linear ....
        .... **Bulbophyllum** (*Synarmosepalum* sensu Garay *et al.*)

72.   Column-foot naked, exposed between dorsal and lateral sepals, hence lateral sepals adnate to column-foot only near its apex. Lateral sepals more or less free at base then involute so as to become partially connivent along outer margins, with free, long-caudate apices, whip-like .... **Bulbophyllum** (*Mastigion* sensu Garay *et al.*)
        Column-foot never exposed between dorsal and lateral sepals, hence lateral sepals adnate to sides of column-foot. Not this combination of characters **73**

73.   Fully mature flowers with the lateral sepals twisted once to form a convex blade through conjoined outer margins .... **Bulbophyllum** (*Cirrhopetalum* sensu Garay *et al.*)
        Fully mature flowers with the lateral sepals not twisted, or if twisted once or more, the outer margins are neither conjoined nor form a convex blade **74**

74.   Lateral sepals involute, firmly united along both margins, horn- or pouch-like ....
        .... **Bulbophyllum** (*Rhytionanthos* sensu Garay *et al.*)
        Lateral sepals free or connate, never horn- or pouch-like **75**

75.   Plants caulescent or pendent. Rhizome stem-like, densely covered with whitish or silver greyish, imbricate sheaths. Petals digitate, either with stiff segments or segments provided with movable fleshy ornaments (palea) .. **Bulbophyllum** (*Epicrianthes* sensu Garay *et al.*)
        Plants rhizomatous, if caulescent or pendent, then without whitish or silver greyish sheaths. Petals entire, lobed or lacerate, with soft, pliable segments without movable ornaments .....
        .... **Bulbophyllum** (sensu Garay *et al.*)

76.   Stems slender, leafy, without pseudobulbs **77**
        Stems pseudobulbous, pseudobulbs sometimes small and entirely enclosed by imbricate leaf sheaths **90**

77.   Pollinia 2 or 4 .... **78**
        Pollinia 6 or 8 .... **80**

78. Pollinia 2. Flowers medium-sized, without mentum .... **Bromheadia** (in part)
     Pollinia 4. Flowers very small, with distinct mentum **79**

79. Stems slender, often branched, with many close, distichous leaves up to 1.2 cm long ....
     .... **Podochilus**
     Stems very short, tufted, with 1–2 linear leaves 6–12 cm long .... **Sarcostoma**

80. Pollinia 6 .... **Appendicula**
     Pollinia 8 .... **81**

81. Inflorescence terminal, usually globose, surrounded by bracts. Flowers white or yellow ....
     .... **Agrostophyllum**
     Inflorescence lateral, terminal or subterminal, never of globose heads. Flowers variously
     coloured **82**

82. Column-foot absent **83**
     Column with a short or long foot **85**

83. Leaves laterally compressed or terete, distichous. Flowers yellowish green .... **Octarrhena**
     Leaves dorsiventral, linear to linear-elliptic or strap-shaped **84**

84. Inflorescence and sepals white-tomentose. Flowers arranged in whorls, non-resupinate ....
     .... **Ascidieria**
     Inflorescence and sepals glabrous. Flowers not arranged in whorls, resupinate ....
     .... **Thelasis** (in part, e.g. *T. carinata, T. micrantha*)

85. Leaf sheaths covered with reddish brown, or rarely white, hispid hairs. Leaves never fleshy
     and subterete .... **Trichotosia**
     Leaf sheaths glabrous. Leaves sometimes fleshy and subterete **86**

86. Stems one-leaved .... **Ceratostylis**
     Stems few- to many-leaved **87**

87. Stems short, entirely enclosed by imbricate leaf sheaths. Inflorescence a densely flowered
     raceme with small bracts .... **Phreatia**
                         (in part, e.g. *P. amesii, P. densiflora, P. monticola, P. secunda*)
     Stems elongate, leafy throughout entire length **88**

88. Inflorescence terminal or subterminal, usually densely many-flowered, densely hirsute.
     Floral bracts small .... **Eria** (section *Mycaranthes*)
     Inflorescence axillary, few-flowered, glabrous **89**

89. Floral bracts large and often brightly coloured .... **Eria** (section *Cylindrolobus*)
     Floral bracts minute, green or brownish .... **Poaephyllum**

90. Pollinia 2 .... **91**
     Pollinia 4 or 8 .... **94**

91.   Lip joined at its base with an outgrowth from the column and with column-foot to form a tube at right angles to base of column .... **Thecostele**
      Lip otherwise **92**

92.   Flowers non-resupinate. Lip convex when viewed from above, scoop-shaped when viewed from below, hairy, bee-like. Column with large, curved stelidia. Habit similar to *Grammatophyllum* .... **Porphyroglottis**
      Flowers resupinate. Lip otherwise. Column lacking stelidia **93**

93.   Plants very large, with pseudobulbs up to 3 m or more long. Flowers up to 10 cm across. Sepals and petals up to 2.6 cm wide, with large irregular blotches. Stipes U-shaped .... **Grammatophylum**
      Plants much smaller. Flowers up to *c*. 5.7 cm across. Sepals and petals narrow, normally plain (blotched in *C. kinabaluense*). Stipes absent .... **Cymbidium**
                               (all epiphytic species except *C. lancifolium*)

94.   Pollinia 4 .... **95**
      Pollinia 8 .... **105**

95.   Inflorescence terminal **96**
      Inflorescence lateral **103**

96.   Flowers with a distinct column-foot, always non-resupinate .... **Polystachya**
      Flowers without a column-foot, resupinate, or, more rarely, non-resupinate **97**

97.   Pollinia attached to a stipes .... **Geesinkorchis** (in part)
      Stipes absent **98**

98.   Basal half of the narrow, saccate lip adnate to basal half of column. Apical half of lip separated by a transverse, high, fleshy callus .... **Entomophobia**
      Lip otherwise **99**

99.   Lip hypochile with long, slender lateral front lobes .... **Nabaluia**
      Lip hypochile without such lobes **100**

100. Column usually with lateral arms (stelidia) .... **Dendrochilum**
      Column without lateral arms **101**

101. Lip hypochile saccate, distinctly separate from epichile. Lip rarely 3-lobed .... **Pholidota**
      Lip hypochile, although often concave, not sharply distinct from epichile. Lip almost always 3-lobed **102**

102. Side-lobes of lip (when present) narrow, borne from front part of hypochile at right angles to the epichile. Hypochile narrow, saccate .... **Chelonistele**
      Side-lobes of lip broad, widening gradually from base of lip. Hypochile ± concave, broader and rarely saccate .... **Coelogyne**

103. Lip joined at its base with an outgrowth from the column and with column-foot to form a tube at right angles to base of column .... **Thecopus**
Lip otherwise **104**

104. Lateral sepals united into a synsepalum. Stipes long, linear .... **Acriopsis**
Lateral sepals free. Stipes absent .... **Cymbidium** (in part, *C. lancifolium* only)

105. Sepals connate to varying degrees, forming a tube. Pseudobulbs flattened .... **Porpax**
Sepals free **106**

106. Column with a prominent foot. Rachis usually hirsute or woolly. Pseudobulbs rarely flattened .... **Eria** (in part)
Column absent or short. Rachis glabrous. Pseudobulbs sometimes flattened **107**

107. Column with a short foot. Anther-cap horizontal on top of column, not beaked .... **Phreatia** (in part, e.g. *P. listrophora, P. sulcata*)
Column-foot absent. Anther-cap vertical behind column, beaked .... **Thelasis** (in part, e.g. *T. capitata, T. carnosa, T. variabilis*)

KEY TO HETEROMYCOTROPHIC GENERA
(LEAFLESS "SAPROPHYTIC" TERRESTRIALS LACKING CHLOROPHYLL)

1. Flowers with sepals and petals fused (connate) to a varying degree, often appearing campanulate and always resupinate **2**
Flowers with free, spreading or connivent sepals and petals, not appearing campanulate, or lateral sepals connate; resupinate or non-resupinate **4**

2. Column with long decurved arms (stelidia), foot absent .... **Didymoplexiella**
Column without long decurved arms, with a short foot **3**

3. Pollinia 4. Dorsal sepal and petals adnate to form a single trifid segment forming a shallow cup or tube with the partially connate lateral sepals. Stigma near column apex .... **Didymoplexis**
Pollinia 2. Sepals and petals connate to form a 5-lobed tube which is sometimes gibbous at the base, and which may be split between the lateral sepals. Stigma at base of column .... **Gastrodia** (including *Neoclemensia*)

4. Flowers always resupinate, lip lowermost **5**
Flowers non-resupinate or resupinate **10**

5. Stem branching, tough and wiry. Sepals and petals surrounded by a shallow denticulate calyculus (cup) .... **Lecanorchis**
Stem simple, slender or fleshy. Sepals and petals not encircled by a shallow denticulate calyculus (cup) **6**

6. Lip divided into a distinct hypochile and epichile. Hypochile with or without twin lateral sacs **7**
Lip not divided into a distinct hypochile and epichile **8**

7. Hypochile swollen at base into twin lateral sacs, each containing a globular sessile gland. Epichile with fleshy involute margins, tube-like. Flowers pink to reddish, tipped with white .... **Cystorchis** (*C. aphylla*, *C. salmoneus*, *C. saprophytica*)
    Hypochile without such sacs. Epichile 3-lobed. Flowers greenish white or creamy white and purple .... **Aphyllorchis**

8. Flowers large, reddish brown. Lip 3-lobed, saccate .... **Eulophia** (*E. zollingeri* only)
    Flowers small, white, or white flushed with purple at apex. Lip entire, with or without a spur **9**

9. Lip spurred, strap-shaped, margin not undulate .... **Platanthera** (*P. saprophytica* only)
    Lip not spurred, narrowly elliptic, margin undulate .... **Stereosandra**

10. Stem simple. Flowers non-resupinate. Lip with a short spur .... **Epipogium**
    Stem branching. Flowers resupinate or non-resupinate. Spur absent **11**

11. Stems long and climbing. Flowers resupinate. Fruits dry, dehiscent **12**
    Stems short, never climbing. Flowers non-resupinate. Sepals brownish mealy or blackish ramentaceous on reverse. Fruits succulent and indehiscent or dry and dehiscent **13**

12. Rachis and flowers furfuraceous-pubescent. Stems stout. Column stout, arcuate, clavate .... **Galeola**
    Rachis and flowers glabrous. Column slender, erect .... **Erythrorchis**

13. Plant robust, with several thick, fleshy stems borne from each rhizome. Sepals obtuse, concave, brownish mealy on reverse. Fruits succulent, indehiscent .... **Cyrtosia**
    Plant slender, with a single narrow, wiry stem borne from each rhizome. Sepals acute, reflexed (*T. saprophytica*), or lateral sepals connate (*T. connata*), blackish ramentaceous on reverse. Fruits dry, dehiscent .... **Tropidia** (*T. connata* and *T. saprophytica* only)

KEY TO GENERA OF TRIBE VANDEAE, SUBTRIBE AERIDINAE

1. Pollinia 4 .... **2**
    Pollinia 2 .... **28**

2. Pollinia more or less equal, globular, free from each other (*Group 1*) **3**
    Pollinia appearing as 2 pollen masses, each completely divided into either unequal, or more or less equal, semiglobular free halves (*Group 2*) **6**

3. Plants without leaves, or leaves reduced to minute brown scales. Stem minute. Roots terete or flattened, containing chlorophyll .... **Taeniophyllum**
    Plants with normal leaves. Roots lacking chlorophyll **4**

4. Large terrestrial, *Phalaenopsis*-like. Leaves radical. Inflorescences long, erect, many-flowered .... **Doritis**
    Small epiphytes. Leaves borne along a distinct stem. Inflorescence 1- to 4-flowered **5**

5.   Leaves dorsiventral. Inflorescence 1- to 4-flowered. Flowers green .... **Adenoncos**
      Leaves bilaterally flattened/compressed. Inflorescence 2-flowered. Flowers white ....
      .... **Microsaccus**

6   Flowers without a distinct column-foot **7**
      Flowers with a distinct, though sometimes short, column-foot **22**

7.   Leaves bilaterally flattened, distichous, with sheathing bases, resembling those of
      Microsaccus. Mid-lobe of lip expanded into a broadly oblong-elliptic, emarginate blade ...
      .... **Jejewoodia**
      Leaves otherwise **8**

8.   Lip not adnate to column, movable. Sepals and petals narrow, usually rather spathulate. Spur
      short and conical .... **Arachnis**
      Lip adnate to column, not movable **9**

9.   Spur with a distinct longitudinal internal median septum **10**
      Spur without a distinct longitudinal internal septum **13**

10.   Rostellum projection short or long, turned obliquely sideward and upward, supporting a thin
      linear stipes, sometimes to 9 times as long as diameter of pollinia .... **Micropera**
      Rostellum projection and stipes otherwise **11**

11.   Column with a raised fleshy, laterally compressed rostellum which sits on top of the
      clinandrium and has a longitudinal furrow along its edge into which the stipes and dorsally
      placed pollinia recline .... **Sarcoglyphis**
      Column without such a structure **12**

12.   Floral bracts, ovary and flowers densely pubescent. Floral bracts large, longer than flowers
      .... **Cleisomeria**
      Floral bracts, ovary and flowers glabrous. Floral bracts minute .... **Cleisostoma**

13.   Back wall of spur without calli and/or outgrowths **14**
      Back wall of spur ornamented with calli and/or outgrowths **19**

14.   Hypochile of lip globose-saccate, the side-lobes reduced to low, often fleshy edges of the
      sac, mid-lobe fan-shaped .. **Gastrochilus** (*G. patinatus* only, see also *Group 4*, couplet 44)
      Hypochile otherwise **15**

15.   Mid-lobe of lip distinctly pectinate-fringed. Stipes linear, about 4 times as long as diameter
      of pollinia .... **Ornithochilus**
      Mid-lobe otherwise. Stipes about twice as long as diameter of pollinia **16**

16.   Spur or sac separated from apical portion of lip by a fleshy transverse wall or ridge **17**
      Spur or sac not separated from apical portion of lip by a fleshy transverse wall or ridge .. **18**

17.   Flowers pale greenish yellow or cinnamon orange, spotted black. Rostellum projection
      large, narrow at base, rising in front of column .... **Abdominea**
      Flowers creamy white with lilac-pink patch on lip. Rostellum projection narrow, somewhat
      decurved .... **Smitinandia**

18. Inflorescence 2-flowered. Flowers lemon-yellow. Spur gibbous .... **Ventricularia**
     Inflorescence many-flowered. Not this combination of characters **19**

19. Lip as long as or longer than dorsal sepal. Flowers small, white to pink, bluish or mauve.
     Leaves narrowly lanceolate or terete .... **Schoenorchis**
     Lip much shorter than dorsal sepal. Flowers red or yellow, showy .... **Renanthera**
     (including *Renantherella*)

20. Lip with a tongue or valvate callus, often forked at the tip, projecting diagonally from deep
     inside the spur .... **Pomatocalpa**
     Lip with an often hairy ligulate tongue placed close to the spur entrance or at the base of the
     lip **21**

21. Inflorescence branched, scape long, several-flowered .... **Staurochilus**
     Inflorescence unbranched, scape short, often several close together, one- to few-flowered ....
     .... **Trichoglottis**

22. Flowers large, showy, dimorphic, the basal two always strongly scented, differently
     coloured from those above .... **Dimorphorchis**
     Flowers much smaller, not dimorphic **23**

23. Lip without a distinct spur or sac, but hypochile often somewhat concave. Flowers
     ephemeral **24**
     Lip with a distinct spur or sac. Flowers long-lasting **25**

24. Leaves terete .... **Cordiglottis**
     Leaves dorsiventral .... **Thrixspermum**

25. Spur or sac with a median longitudinal septum .... **Cleisostoma**
     Spur or sac without a longitudinal septum **26**

26. Lip epichile with two forward pointing teeth emerging from base .... **Kingidium**
     Lip epichile otherwise **27**

27. Stems very short. Inflorescence borne below the leaves. Flowers small, greenish yellow and
     white, marked with crimson on the lip. Lip deeply saccate .... **Bogoria**
     Stems long, usually pendent. Inflorescences axillary. Flowers translucent lavender-blue or
     dark lilac-pink. Lip distinctly spurred .... **Cleisocentron**

28. Pollinia sulcate or porate **29**
     Pollinia entire (*Group 5*) **46**

29. Pollinia sulcate, i.e. more or less, but not completely cleft or split (*Group 3*) **30**
     Pollinia porate (*Group 4*) **42**

30. Column-foot absent or very indistinct **31**
     Column-foot distinct, though sometimes short **35**

31. Flowers often large and showy, usually a few, well spaced on a raceme. Stipes short and broad, entire, shelf-like .... **Vanda**

Flowers small to medium-sized, crowded on to a usually densely many-flowered raceme or panicle. Stipes linear, spathulate, uncinnate, rarely hamate. Rostellum prominent, bifid or long and pointed **32**

32. Leaves linear, acute, fleshy. Inflorescence and flowers scarlet-red .... **Porphyrodesme**

Leaves broader, unequally bilobed. Inflorescence and flowers otherwise **33**

33. Plants small, stem and inflorescence less than 4 cm. Rachis very fleshy, clavate. Flowers borne in succession, minute .... **Ascochilopsis**

Plants larger. Rachis not clavate. Flowers not borne in succession, small to medium sized **34**

34. Stems short. Leaves borne close together, most often with many light-coloured nerves. Lip entire or obscurely 3-lobed, deeply saccate or with a short backward-pointing spur without interior ornaments .... **Rhynchostylis**

Stems rather long. Leaves distant, without pale nerves. Lip 3-lobed, spur often apically inflated and occasionally with callosities or scales within .... **Robiquetia**

35. Leaves terete, sometimes up to 165 cm long **36**

Leaves dorsiventral, much shorter **37**

36. Stems up to 2 m long. Lip spurred, ecallose. Column-foot entire .... **Papilionanthe**

Stems very short. Lip not spurred, with a conduplicate, plate-like callus situated at the junction of the mid- and side lobes. Column-foot 3-fingered .... **Paraphalaenopsis**

37. Spur or sac, if present, developed from the hypochile. Epichile dorsiventral **38**

Spur or sac borne centrally on lip. Epichile reduced, fleshy **39**

38. Spur absent, or rudimentary. Lip with a least one forward-pointing forked appendage. Flowers few, sometimes large and showy, distichous .... **Phalaenopsis**

Spur well developed. Forked appendages absent. Flowers many, facing in every direction, developing simultaneously .... **Aerides**

39. Lip very fleshy, nearly completely solid with a very narrow, almost hair-thin tube extending its full length inside, the entrance of which is a narrow basal slit .... **Ascochilus**

Lip otherwise **40**

40. Rostellum projection long, slender. Lip bent upward so as to make a right angle with column-foot, distinctly unguiculate. Flowers long-lasting, developing simultaneously .... **Macropodanthus**

Rostellum projection inconspicuous. Lip continuing the line of and usually flush with column-foot. Flowers usually ephemeral, usually developing successively, a few open at a time **41**

41. Column-foot longer than the column proper .... **Pteroceras**

Column-foot short, column proper elongate .... **Brachypeza**

17

42. Leaves terete. Lip neither saccate nor spurred .... **Luisia**

     Leaves dorsiventral. Lip saccate or spurred **43**

43. Lip mobile on a short but distinct column-foot. Spur or sac absent .... **Biermannia**

     Lip immobile. Column-foot absent. Spur or sac present **44**

44. Lip with a globose-saccate hypochile, the epichile separated from it by a transverse ridge connecting front edges of side-lobes .... **Gastrochilus** (see also *Group 2*, couplet 14)

     Lip with a rather long, cylindric or extinctoriform spur and a ligulate mid-lobe **45**

45. Spur without a prominent backwall callus. Rostellar projection broadly triangular. Stipes without apical appendages. Pollinator guides absent .... **Ascocentrum**

     Spur with a prominent backwall callus. Rostellar projection elongate, attenuate, sigmoid. Stipes with apical appendages. A pair of crimson pollinator guides present at throat of spur .... **Dyakia**

46. Column with a distinct foot. Lip movable **47**

     Column without a foot. Lip not movable **49**

47. Lip without a spur, side-lobes sometimes fimbriate .... **Chamaeanthus**

     Lip spurred or saccate **48**

48. Spur-like conical portion of lip more or less solid. Peduncle short, glabrous .. **Chroniochilus**

     Sac or spur thin-walled, without interior fleshiness. Peduncle longer, often prickly-hairy .... **Grosourdya**

49. Lip with a bristle or tooth inside near the apex. Floral bracts conspicuous, triangular, leafy .... **Microtatorchis**

     Lip without an apical bristle or tooth. Floral bracts not leafy **50**

50. Side-lobes of lip very large, often fringed .... **Pennilabium**

     Side-lobes of lip small, never fringed **51**

51. Mid-lobe of lip resembling a small spongy pouch, hollow above, solid towards apex .... **Spongiola**

     Mid-lobe of lip otherwise **52**

52. Lip not truly spurred, but with a spur-like tubular cavity. Lateral sepals adpressed to lip .... **Porrorhachis**

     Lip spurred. Lateral sepals not adpressed to lip **53**

53. Rachis slender, never thickened and sulcate, or clavate. Column hammer-shaped. Stipes linear-spathulate, much broadened at apex .... **Malleola**

     Rachis fleshy, sulcate, or clavate. Column short and stout. Stipes linear, much reduced .... **Tuberolabium**

CHAPTER 3

DESCRIPTIONS AND FIGURES

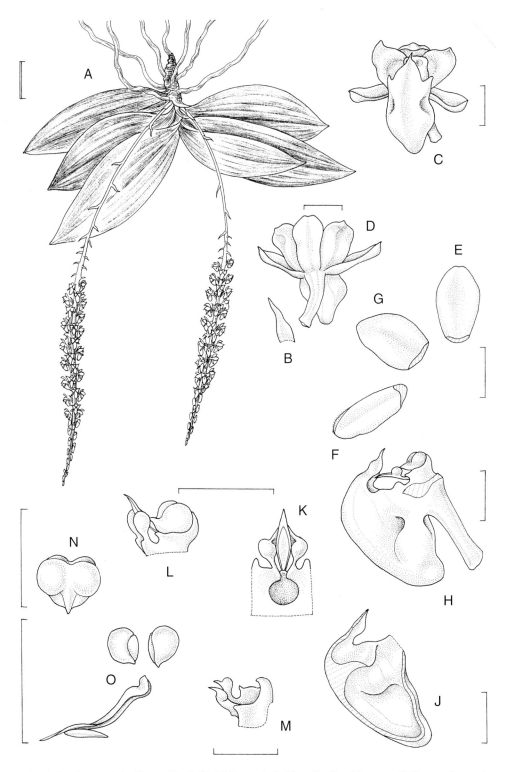

Figure 1. Abdominea minimiflora (Hook.f.) J.J.Sm. - A: habit. - B: floral bract. - C: flower, front view. - D: flower, back view. - E: dorsal sepal. - F: lateral sepal. - G: petal. - H: column, pedicel with ovary and lip, side view. - J: lip, longitudinal section. - K: column apex with anther-cap, front view. - L: column apex with anther-cap, side view. - M: column apex, anther-cap removed, side view. - N: anther-cap, back view. - O: pollinarium, pollinia detached. All drawn from *Lamb & Lohok* AL 368/85 by Susanna Stuart-Smith. Scale: single bar = 1 mm; double bar = 1 cm.

1. ABDOMINEA MINIMIFLORA (Hook.f.) J.J. Sm.

Abdominea minimiflora (*Hook.f.*). *J.J. Sm.* in Bull. Jard. Bot. Buitenzorg, ser. 2, 25: 98 (1917). Type: Peninsular Malaysia, Perak, *Scortechini* 635b (holotype K).

Saccolabium minimiflorum Hook.f., Fl. Brit. Ind. 6: 59 (1890).
Saccolabium cortinatum Ridl. in J. Bot. 36: 215 (1898). Type: Peninsular Malaysia, Selangor, near Kuala Lumpur, *Ridley* s.n. (holotype SING).
Abdominea micrantha J.J. Sm. in Bull. Jard. Bot. Buitenzorg, ser. 2, 14: 53 (1914). Type: Java, Goeah Gadjah unterhalb Tjikarang Saät, bei Kelapa Noenggal bei Buitenzorg, 400–500 m, Dec. 1912, *Backer* 5913 (holotype BO).
Schoenorchis minimiflora (Hook.f.) Ames, Orchidaceae 5: 241 (1915).
Schoenorchis philippinensis Ames, Orchidaceae 5: 241 (1915). Types: Philippines, Luzon, Nueva Viscaya Province, 15 Jan. 1913, *McGregor* Bur. Sci 20117 (syntype AMES); Mindanao, Lake Lanao, Camp Keithley, *M.S. Clemens* 1058 (syntype AMES); Bukidnon Subprovince, July 1913, *Escritor* Bur. Sci. 21396 (syntype AMES).

Epiphyte. Stem up to 1.5 cm long, erect to pendent, producing several elongate roots. **Leaves** 3–7, spreading; blade 2–5.5 × 0.7–2 cm, obliquely oblong-elliptic, attenuate below, unequally bidentate at apex, larger tooth falcate, *c.* 9-nerved; sheath 2–3 mm long, tubular, margin recurved. **Inflorescences** emerging from back of sheath, simple, rarely with a solitary branch, elongate, laxly many-flowered, *c.* 7 mm wide; peduncle 1–3.5 cm long, slender; non-floriferous bracts 2 or 3, 2–2.5 mm long, triangular-subulate, acuminate; rachis 4–14 cm long, angular, sulcate; floral bracts 1.5–1.6 mm long, narrowly elliptic to triangular, subulate, acuminate, spreading. **Flowers** *c.* 3–3.2 mm long, quaquaversal, sepals and petals cinnamon-orange with blackish purple warty spots, lip translucent greenish yellow, with white tip and a reddish brown subapical spot, column green, rostellum white, anther-cap purple. **Pedicel** and **ovary** 1.5 mm long, dark green. **Dorsal sepal** 1.5–1.7 × 0.7–0.8 mm, oblong, obtuse, concave, erect, incurved, 1-nerved. **Lateral sepal** 1.4–1.6 × 0.8 mm, obliquely oblong, obtuse to somewhat acute, 1-nerved. **Petals** 1.4–1.7 × 1–1.3 mm, broadly subovate, obliquely rounded, concave, apex recurved and convex, 1-nerved. **Lip** 2.25–3 mm long, 1.5–1.7 mm deep, immobile, adnate to column, shallowly 3-lobed, laterally compressed, with a large, irregular saccate portion without inner ornaments, separated from apical part by a fleshy, transverse wall; side lobes minute, triangular, tooth-like, erect; mid-lobe triangular, obtuse, apiculate, concave, incurved; spur *c.* 1 mm long, broadly conical. **Column** *c.* 1.4 mm long; foot absent; rostellum projection longer and broader than rest of column, cordate, acuminate, narrowed at base; anther-cap cucullate; stipes linear, clavate, acute; viscidium absent; pollinia 4, appearing as 2 unequal masses. Plate 1A.

HABITAT AND ECOLOGY: Lowland and hill forest on limestone; also recorded from ultramafic substrate. Alt. 300 to 900 m. Flowering observed from May to July.

DISTRIBUTION IN BORNEO: SABAH: Lamag District; Mt. Kinabalu; Batu Urun.

GENERAL DISTRIBUTION: Thailand, Peninsular Malaysia, Java, Bali, Borneo and the Philippines.

NOTES: *Abdominea* is a monospecific genus noted for its large cordate, acuminate rostellum prolongation which is longer and broader than the column. Seidenfaden (1988) comments that the long, linear, sharply acute stipes appears to act like a sharp needle that fastens to the pollinator replacing the need for a sticky viscidium. The insect is forced toward this by the sharp, tooth-like lip apex.

DERIVATION OF NAME: The generic name is derived from the Latin *abdomen*, referring to the rostellum which is shaped rather like the abdomen of an insect. The specific epithet is derived from the Latin *minimus*, very little, and *floralis*, relating to the flower, describing the diminutive size of the flowers.

2. AGROSTOPHYLLUM GLUMACEUM Hook.f.

Agrostophyllum glumaceum *Hook.f.*, Fl. Brit. Ind. 5: 821 (1890). Types: Peninsular Malaysia, Perak, *Scortechini* 1810; Perak, Ulu Bubong, *King's collector* 10876 (syntypes K).

? *Agrostophyllum khasiyanum* sensu Ridl. in J. Linn. Soc., Bot. 31: 286 (1896), *non* Griff.

Epiphyte. Roots 1–1.5 mm in diameter, branching and forming a dense mass, smooth, young roots bearing villose hairs. **Stems** *c.* 8–35 × 1.8 cm, flattened, with 1- to 3 internodes 6–22 cm long, erect or porrect. **Leaves** mostly grouped at base of stem, subtended by a few basal cataphylls, with one at its apex, and sometimes another half-way along it; blade 15–40(–50) × 0.7–2.2 cm, linear, ensiform, acute, becoming progressively shorter towards stem base; sheath 6–13 × 0.8–1.2 cm, uppermost shortest, imbricate, margins hyaline, dark brown. **Inflorescences** terminal, in a globular head, *c.* 3–3.5 cm in diameter, composed of many crowded branches, the individual spikes with *c.* 3 flowers arranged on a zigzag rachis *c.* 1 cm long; floral bracts *c.* 3.4 mm long, triangular, concave acute. **Flowers** *c.* 6 mm in diameter, white, column edged red. **Pedicel** and **ovary** 5–6 mm long. **Dorsal sepal** 5 × 2 mm, ovate, concave, acute, apex somewhat hooded. **Lateral sepals** 5–6 × 2 mm, slightly obliquely ovate-elliptic, acute, carinate on reverse. **Petals** 4 × 1–1.5 mm, linear-ligulate, acute. **Lip** 6 mm long; hypochile concave, saccate, its apex bilobed to form small, triangular, acute side lobes which are joined transversally and pressed against the column; epichile 4 × 3 mm, shallowly concave, ovoid, shortly acutely tipped. **Column** 4 mm long, with short square wings near the apex; foot rudimentary; anther-cap cordate, cucullate; pollinia 8, attached to a solitary viscidium. Plate 1B.

HABITAT AND ECOLOGY: Lowland and hill dipterocarp forest; mixed forest on sandy soil with scattered limestone rocks; riverine forest; submontane mossy forest; often recorded from forest on limestone; recorded epiphytic upon *Brownlowia* spp. and *Saraca* spp. Alt. sea level to 900 m. Flowering observed from February until April and July until November.

DISTRIBUTION IN BORNEO: BRUNEI: Temburong District, Kuala Belalong area. KALIMANTAN: Mt. Ilas Bungaan. SABAH: Mt. Kinabalu; Pun Batu; Tenom District; Tongod District, Ulu Menanam area. SARAWAK: Gunung Mulu National Park; Kapit District, Batu Tiban Hill; Mukah District, Iju Hill.

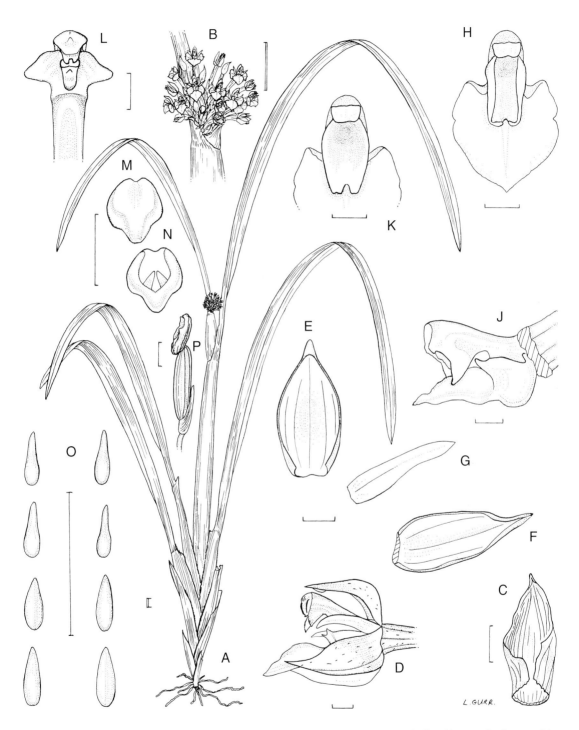

Figure 2. Agrostophyllum glumaceum Hook.f. - A: habit. - B: inflorescence. - C: floral bract. - D: flower, side view. - E: dorsal sepal. - F: lateral sepal. - G: petal. - H: lip and column, anther-cap removed, front view. - J: lip and column, anther-cap removed, side view. - K: lip showing hypochile and lower part of epichile, front view. - L: column with anther-cap, front view. - M: anther-cap, back view. - N: anther-cap, interior view. - O: pollinia. - P: fruit capsule. A & B drawn from *J. & M. S. Clemens* 27481, C–L from *Lamb* SNP 1992 and M–P from *Leche* s.n. by Linda Gurr. Scale: single bar = 1 mm; double bar = 1 cm.

GENERAL DISTRIBUTION: Peninsular Malaysia, Sumatra and Borneo.

NOTES: Agrostophyllums are widespread in the forests of the Far East, but are often overlooked because of their unorchid-like, often rather sedge-like, appearance. About sixteen species have been recorded from Borneo. All have terminal inflorescences of usually white flowers, and in habit superficially resemble the New Guinean genus *Glomera*.

DERIVATION OF NAME: The generic name is derived from the Greek *agrostis*, grass, and *phyllon*, leaf, referring to the grass-like leaves of most species. The specific epithet is derived from the Latin *glumaceus*, resembling the glumes of grass spikelets, in reference to the structure of the inflorescence.

3. ANIA BORNEENSIS (Rolfe) Senghas

Ania borneensis (*Rolfe*) *Senghas* in Schltr., Orch. 1, ed. 3: 863 (1984). Type: Borneo, Sabah, Mt. Kinabalu, Kiau, *Gibbs* 3958 (holotype K, isotype BM).

Ascotainia borneensis Rolfe in Gibbs in J. Linn. Soc., Bot. 42: 154 (1914).
Tainia penangiana J.J. Sm., Orch. Java: 183 (1905), *non* Hook.f.
Tainia malayana J.J. Sm. in Feddes Repert. 31: 76 (1932). Type: New Guinea, Papua (Irian Jaya), Arfak Mountains, Ditschi, 1200 m, 8 June 1928, *Mayr* 158 (holotype L).
Tainia steenisii J.J. Sm. in Blumea 5: 306 (1943). Type: Sumatra, Atjeh, Gajolanden, Gunung Goh Lemboeh, from bivouac Aer Poetih waterfall to bivouac Halfweg, 1400 m, *van Steenis* 8940 (holotype L).
Tainia rolfei P.F. Hunt in Kew Bull. 26 (1): 182 (1971), *nom. nov.*
Ania malayana (J.J. Sm.) Senghas in Schltr., Orch. 1, ed. 3: 863 (1984).

Terrestrial. **Stem** of sterile shoot up to *c*. 5.8 cm long, with *c*. 5 internodes. **Pseudobulb** 2.2–3.6 × 0.5–1.2 cm, consisting of several internodes, conical, erect; scale of subterminal node 11–16.2 cm long, ± persistent, torn open at base. **Leaf-blade** 19–33.4 × 2.1–5.1 cm, elliptic, slightly acuminate, thin-textured; petiole 17–24.7 cm long, with an articulation *c*. half-way. **Inflorescence** arising from base of pseudobulb, 5- to 12-flowered; peduncle 21–54.4 cm long; peduncle scales 5–6, the longest 2.5–5 cm long, acute; rachis 4.5–30 cm long; floral bracts 6–15 mm long, acute, spreading to reflexed. **Flowers** resupinate, pale greenish-yellow with reddish nerves, lip yellow with crimson spots on exterior of spur, column pale yellow spotted crimson adaxially. **Pedicel** and **ovary** 4–17 mm long. **Dorsal sepal** 11.5–20 × 2–4 mm, elliptic to obovate, acute to slightly acuminate. **Lateral sepals** 12–20 × 2.5–4 mm, elliptic, straight to slightly falcate, acute to slightly acuminate. **Petals** 10–20 × 2–3.5 mm, obliquely elliptic to obovate, acute to acuminate. **Lip** distinctly spurred, 10–15 × 6–9 mm, blade entire, obovate, acuminate, slightly undulate distally, glabrous; disc 3- to 5-keeled, median keel elevated plate-like from just below broadest part of lip, decurrent towards tip, highest distally, crest straight, entire, lateral keels elevated plate-like from just below broadest part of lip, decurrent towards tip or terminating rather abruptly, highest distally, crest straight, outer keels (when present) 1–4 mm long. **Column** 7–10 mm long; foot to 1 mm long; stelidia absent or in the form of 2 distal side lobes; apex truncate to slightly semiorbicular, sometimes obscurely toothed; wings seam-like; anther-cap 1–2 × 1–2 mm, abaxially with 2 crests; pollinia 8. Plate 1D.

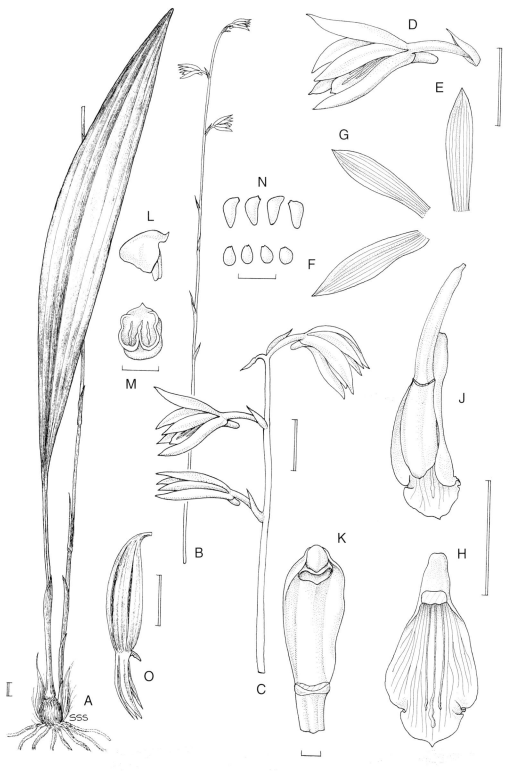

Figure 3. Ania borneensis (Rolfe) Senghas. - A: habit. - B: - inflorescence. - C: - inflorescence apex. - D: flower, side view. - E: dorsal sepal. - F: lateral sepal. - G: petal. - H: lip, front view. - J: pedicel with ovary, lip and column, viewed from above. - K: column, oblique view. - L: anther-cap, side view. - M: anther-cap, interior view. -N: pollinia. - O: fruit capsule. All drawn from *Lamb* AL 321/85 by Susanna Stuart-Smith. Scale: single bar = 1 mm; double bar = 1 cm.

HABITAT AND ECOLOGY: Open secondary forest, under bamboo; open grassy places; sunny areas on steep slopes above rivers. Alt. 270 to 1400 m. Flowering observed in February, June, August and September.

DISTRIBUTION IN BORNEO: SABAH: Mt. Kinabalu; Crocker Range, Sinsuron road.

GENERAL DISTRIBUTION: Sumatra, Java, Maluku (Ambon) and New Guinea.

NOTES: *Ania* is represented by two species in Borneo (Turner, 1992). Both *Ania* Lindl. and *Tainia* Blume are closely related. Turner distinguishes *Ania* by the pseudobulbs which are swollen throughout and usually consist of several internodes. The petiole usually has an articulation more or less half-way. The inflorescence is lateral and the flowers usually have a spurred lip (absent in *A. ponggolensis*). In *Tainia* the pseudobulbs are thin, cylindrical and sometimes swollen towards the base. The petioles lack any articulation. The inflorescence is usually terminal (although lateral in *T. paucifolia*). The lip, at the very most, is only very slightly saccate.

In several collections of *A. borneensis* from Borneo there are distinct lateral lobes on the distal portion of the column.

DERIVATION OF NAME: The generic name is derived from the Greek *ania*, trouble, alluding perhaps to the uncertain taxonomic position of this concept at the time of its establishment. The specific epithet refers to the island of Borneo.

4. ANIA PONGGOLENSIS A. Lamb in H. Turner

Ania ponggolensis *A. Lamb in H. Turner* in Orchid Monogr. 6: 58, fig. 30, pl. 5d (1992). Type: Borneo, Sabah, Batu Ponggol, *c.* 900 m, 29 April 1984, *Lamb* AL 204/84 (holotype K, isotype L).

Terrestrial. **Stem** of sterile shoot *c.* 5 mm long, creeping. **Pseudobulb** 0.9–1.4 × 0.8–1.6 cm, consisting of one internode, gourd-shaped, erect; scale of subterminal node 1.5–3 cm long. **Leaves** liver-coloured when young; blade 9–10.5 × 1.9–3.5 cm, elliptic, acuminate, thin-textured; petiole 2.9–3.7 cm long, articulated *c.* half-way, magenta. **Inflorescence** lateral, *c.* 2-flowered, erect, purple blotched; peduncle 1.3–1.8 cm long; peduncle scales 2, up to 7 mm long, tubular, acute, slightly pilose; rachis *c.* 3.3 cm long, pilose; floral bracts 7–8 mm long, triangular to ovate, acute, pilose, spreading, purple. **Flowers** resupinate, with dark greenish purple pedicel and ovary, sepals translucent greenish yellow, with green nerves, sometimes with purplish spots, petals translucent yellow, sometimes with purple spots, lip yellow, sometimes spotted purple laterally and towards tip, keels greenish yellow, column with a basal patch of purple spots, anther-cap purple. **Pedicel** and **ovary** 14–18 mm long, pilose. **Dorsal sepal** 14.5–16 × 4.5–8 mm, triangular, acute, tip slightly hooded, slightly pilose on reverse. **Lateral sepals** 12–16 × 7–8 mm, triangular, slightly falcate, acute, slightly pilose on reverse. **Petals** 12–14 × 5–6 mm, obliquely elliptic, slightly falcate, obtuse to acuminate, glabrous. **Lip** without a spur; blade 9.5–13 × 5.5–8 mm, elliptic, 3-lobed to entire with a distinct sinus; mid-lobe 3–5 × 3.5–4.5 mm, orbicular, obtuse, margins crenate, undulate, minutely papillose, slightly pilose below; side lobes (when present) *c.* 1 mm long, obliquely triangular, obtuse, margins entire, glabrous; disc with 3 keels, median keel raised plate-like from basal part of lip, decurrent distally, highest half-way, crest gently undulate, entire, glabrous; lateral

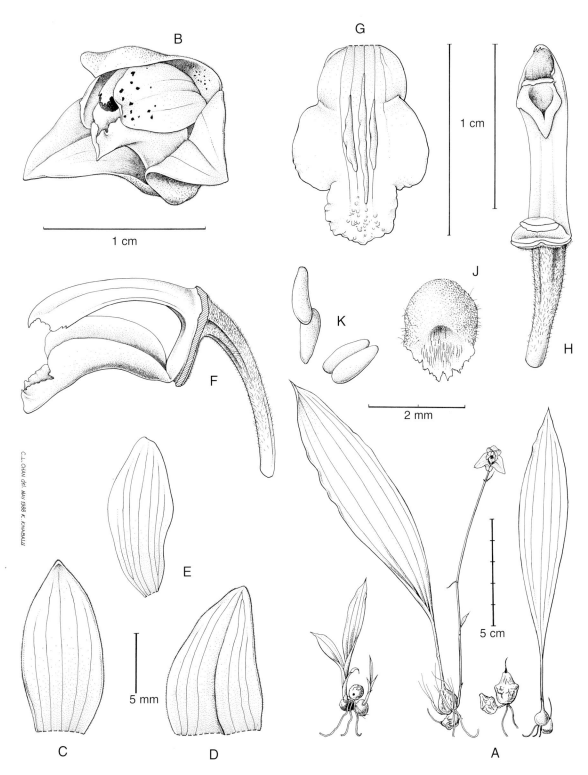

Figure 4. Ania ponggolensis A. Lamb in H. Turner. - A: habits. - B: flower, oblique view. - C: dorsal sepal. - D: lateral sepal. - E: petal. - F: pedicel and ovary, lip and column, side view. - G: lip, front view. - H: pedicel and ovary and column, anther-cap removed, oblique view. - J: anther-cap, back view. - K: pollinia. All drawn from *Lamb* AL 204/84 (holotype) by C.L.Chan.

keels similar, highest proximally, crest slightly undulate. **Column** *c.* 10 mm long; foot *c.* 4 mm long, papillose-pilose; stelidia absent; apex triangular, ± entire; wings seam-like; abaxially with a distinct longitudinal ridge; anther-cap without crests; pollinia 8. Plate 1E.

HABITAT AND ECOLOGY: In shady places among limestone rocks. Alt. *c.* 900 m. Flowering observed in April.

DISTRIBUTION IN BORNEO: SABAH: Pensiangan District, Batu Ponggol.

GENERAL DISTRIBUTION: Endemic to Borneo.

NOTES: Turner (1992) admits that the generic position of this species is not very clear. The absence of a spur indicates an affinity with *Tainia*, while the articulated petiole points to a relationship with *Ania*. The shape of the pseudobulbs and the lateral inflorescence are also characters found in *Ania*. A unique feature of *A. ponggolensis* is the pilose-hairy inflorescence and flowers.

DERIVATION OF NAME: The specific epithet refers to Batu Ponggol, a limestone rock, which is the type locality.

5. APHYLLORCHIS MONTANA Rchb.f.

Aphyllorchis montana *Rchb.f.* in Linnaea 41: 57 (1876). Types: Sri Lanka, Ambagumowa District, *Thwaites* C.P. 3189 (syntype W); *Mrs Walker* s.n. (syntype W).

Apaturia montana auct., *non* Lindl.
? *Aphyllorchis odoardi* Rchb.f., Bot. Centralbl. 28: 345 (1886). Type: Papua New Guinea, *Beccari* s.n. (holotype W).
Aphyllorchis prainii Hook.f., Fl. Brit. Ind. 6: 117 (1890). Type: India, Assam, Naga Hills, Nesama, August 1886, *Prain* 68 (holotype CAL, isotype K).
Aphyllorchis borneensis Schltr. in Bull. Herb. Boissier 2, 6: 299 (1906). Type: Borneo, Kalimantan Timur, Long Sele, *Schlechter* 13520 (holotype B, destroyed).
Aphyllorchis benguetensis Ames, Orchidaceae 2: 49 (1908). Type: Philippines, Luzon, Benguet Province, Baguio, 24 August 1906, *Curran* 5086 (holotype AMES).
Aphyllorchis tanegashimensis Hayata in J. Coll. Sci. Imp. Univ. Tokyo 30, 1: 344 (1911). Type: Taiwan, *Tanaka* 442 (holotype ?TI or TNS).
Aphyllorchis unguiculata Rolfe ex Downie in Kew Bull. 1925: 415 (1925). Type: Thailand, Doi Suthep, *Kerr* 157 (holotype K).
Aphyllorchis purpurea Fukuy. in Bot. Mag. Tokyo 48: 431 (1934). Type: Taiwan, Taihoku Prefecture, Syo-agyoku-san, Urai, August 1933, *Fukuyama* 4121 (holotype herb. Fukuyama).
Aphyllorchis striata auct., *non* Ridl.

Leafless **saprophyte**. **Rhizome** short, mostly erect. **Roots** 4–7 mm in diameter, fleshy. **Stem** up to 150 cm long (including inflorescence), internodes 0.6–16(–19) × 0.5–1 cm, erect, fleshy,

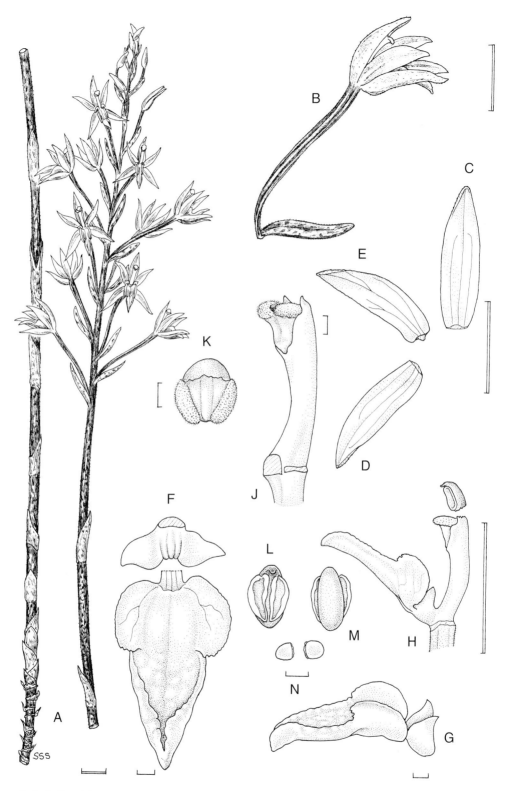

Figure 5. Aphyllorchis montana Rchb.f. A: habit. - B: floral bract and flower, side view. - C: dorsal sepal. - D: lateral sepal. - E: petal. - F: lip, hypochile detatched, front view. - G: lip, side view. - H: lip and column, anther-cap detatched, side view. - J: column, anther-cap removed, oblique view. - K: anther-bed viewed from above. - L: anther-cap, interior view. - M: anther-cap, back view. - N: pollinia. A drawn from *Othman & Yii* S. 48805 and B–N from *Lamb* s.n. by Susanna Stuart-Smith. Scale: single bar = 1 mm; double bar = 1 cm.

cream or yellowish cream, spotted and streaked with purple or violet, basal 85 cm or so with sheathing bracts. **Bracts** 1.5–4 × 0.5–0.7(–1) cm, ovate-elliptic or ovate, acute to acuminate, clasping and sheathing stem, sometimes sparsely papillose-hairy, nerves often raised, particularly on lower bracts, creamy, heavily streaked and spotted purple. **Inflorescence** many flowered (often up to *c*. 30), flowers borne *c*. 3 cm apart at base, 0.5–1.5 cm apart higher up; rachis (15–)20–35 cm long; floral bracts 0.5–3.5 × 0.5–0.07 cm, narrowly ovate-elliptic to oblong-ligulate, acute to acuminate, strongly deflexed, greenish with purple nerves, flecks and spotting. **Flowers** opening widely, 2–3 cm in diameter, whitish cream with various purplish or brownish markings, particularly on reverse of sepals, petals with purple median nerve, lip off white, edged and marked mauve-violet and yellow, column violet, cream at base, pale yellow distally, anther-cap pale yellow to cream. **Pedicel** and **ovary** 1.8–3.5 cm long, narrowly clavate. **Dorsal sepal** 2 × 0.5–0.6 cm narrowly elliptic, apiculate, concave, curving over column. **Lateral sepals** 1.9–2 × 0.4–0.45 cm, narrowly elliptic, apiculate, concave, spreading. **Petals** 1.5–1.7 × 0.4–0.45 cm, narrowly oblong-elliptic, apiculate, flat, spreading. **Lip** *c*. 1.4–1.5 cm long; hypochile 2 mm long, with raised triangular, acute auricles varying in size, usually *c*. 3.5 × 4 mm; epichile *c*. 1.2 × 5 mm, subentire to 3-lobed, side lobes erect, broad, rounded, mid-lobe oblong-elliptic, obtuse, margins thickened and fleshy, somewhat involute. **Column** *c*. 1.3 cm long, curved; anther-cap erect; pollinia 2, powdery. Plate 1C.

HABITAT AND ECOLOGY: Lowland dipterocarp forest on limestone and sandstone; mixed hill dipterocarp forest on sandstone; lower montane forest. Alt. sea level to 1500 m. Flowering observed from January until March and from July until October.

DISTRIBUTION IN BORNEO: KALIMANTAN BARAT/KALIMANTAN TENGAH border: Mt. Raya. KALIMANTAN TIMUR: Kutai, Long Sele. SABAH: Mt. Kinabalu; Tenom District, Mandalom Forest Reserve. SARAWAK: Belaga District, Dema Hill; Bintulu District, Merurong Plateau; Kapit District, Hose Mountains; Kuching District, Mt. Matang; Lawas District, Ba Kelalan; Lawas District, Kota Forest Reserve; Lawas District, Lawas River; Limbang District, Medalom River; Lundu District, Lundu; Marudi District, Mt. Dulit; Marudi District, Mt. Dulit/Long Kapa; Marudi District, Mt. Mulu; Miri District, Lambir Hills.

GENERAL DISTRIBUTION: Sri Lanka, India and the Himalayan region to China, Taiwan and Ryukyu Islands, south to Thailand, Peninsular Malaysia, Sumatra, Borneo, the Philippines and possibly New Guinea.

NOTES: Five species of *Aphyllorchis* have been recorded from Borneo, of which the poorly known *A. kemulensis* J.J. Sm. and *A. spiculaea* Rchb.f. are thought to be endemic. *Aphyllorchis montana* is a widespread species with tall, robust stems which are often found grouped together in large colonies. The size of the plant, flower length and shape of the hypochile side lobes vary a great deal over its extensive range. The equally widespread, but much smaller and delicate *A. pallida* Blume, usually occurs as scattered solitary individuals.

DERIVATION OF NAME: The generic name is derived from the Greek *aphyllos*, leafless, and *orchis* orchid. The specific epithet is derived from the Latin *montanus*, pertaining to or growing on mountains.

6. APPENDICULA BILOBULATA J.J. Wood

Appendicula bilobulata *J.J. Wood* in Contr. Univ. Michigan Herb. 21: 315, fig. 1 (1997). Type: Borneo, Sarawak, route from Ba Kelalan to Mt. Murud, near Camp III, 28 September 1967, *Burtt & Martin* B. 5287 (holotype E, isotypes K, SAR).

Trailing, mat-forming **epiphyte. Roots** filiform, wiry, elongate, simple, very minutely papillose, or hirsute, produced at intervals along stem. **Stem** 12–25 cm long, very slender, branching distally, internodes 2–3 mm long, enclosed in persistent leaf sheaths. **Leaves** 0.7–1 × 0.2–0.3 cm, narrowly oblong-elliptic, minutely obliquely retuse, mucronate, thin-textured; sheaths 2–3 mm long. **Inflorescences** lateral and/or terminal, one flower open successively; peduncle *c.* 5 mm long, enclosed by 2 or 3 tubular, acute to acuminate, non-floriferous bracts; rachis 0.5–*c.* 1.2 cm long, fractiflex; floral bracts 2.5–4 mm long, lanceolate, narrowly acuminate. **Flowers** white, tip of lip pale purple. **Pedicel** and **ovary** 4.8–5 mm long, narrowly clavate, gently curving. **Sepals** 3-nerved. **Dorsal sepal** 3 × 1.6–1.7 mm, ovate, concave, acuminate, cuspidate. **Lateral sepals** 6 × 1.8–2 mm, obliquely oblong, acuminate, cuspidate. **Mentum** 5 mm long, oblong, obtuse. **Petals** 2.4–2.5 × 0.9 mm, narrowly oblong, subacute, 1-nerved. **Lip** hypochile: 5.8–6 mm long, tube-like, with erect sides, margins fleshy and sulcate, especially distally, provided with a smooth, fleshy, U-shaped basal appendage; epichile 3 mm long, 4.5–5 mm wide across lobules, flabellate, bilobulate, lobules each *c.* 3 × 2.5 mm, broadly oblong, rounded, margin minutely irregular. **Column** 0.5–0.6 mm long; foot 5 × *c.* 0.7–0.8 mm; wings 0.4–0.5 mm long, oblong, obtuse, fleshy; apex truncate; rostellum acute, tooth-like; anther-cap 0.8–0.9 × 0.9–1 mm, ovate, cucullate, acute, smooth; pollinia 6, obliquely clavate.

HABITAT AND ECOLOGY: Lower montane forest. Alt. 1740 m. Flowering observed in September.

DISTRIBUTION IN BORNEO: SARAWAK: Mt. Murud.

GENERAL DISTRIBUTION: Endemic to Borneo.

NOTES: The genus *Appendicula*, although unlikely to be familiar to most growers, is sometimes encountered in a few specialist collections and in botanical gardens. About sixty species are distributed from tropical Asia to the Pacific islands, with the majority occurring in Indonesia, the Philippines, and New Guinea. Most inhabit lowland forest or mid-elevation forest in the mountains, particularly favouring shady areas with high humidity. The leaves are arranged in two rows along the entire length of the stem. These are often twisted at the base, so that the blade is lying in one plane and at right angles to the low intensity light source prevalent. The flowers are very small and usually of a whitish or greenish hue, although often flushed with pink or purple. In one Bornean species, *A. torta* Blume, these are subtended by colourful overlapping lilac-pink floral bracts. The lip always has a variably shaped, often appendage-like callus situated on the upper surface just above the base. *Appendicula* is distinguished from the closely related *Podochilus* by having six instead of four pollinia.

Appendicula bilobulata is known only from the type material, which was collected in 1967 by Bill Burtt, of the Royal Botanic Garden, Edinburgh, on Mt. Murud (2,438 m), the highest mountain in the Malaysian state of Sarawak. The general appearance is that of a *Podochilus*, which I thought

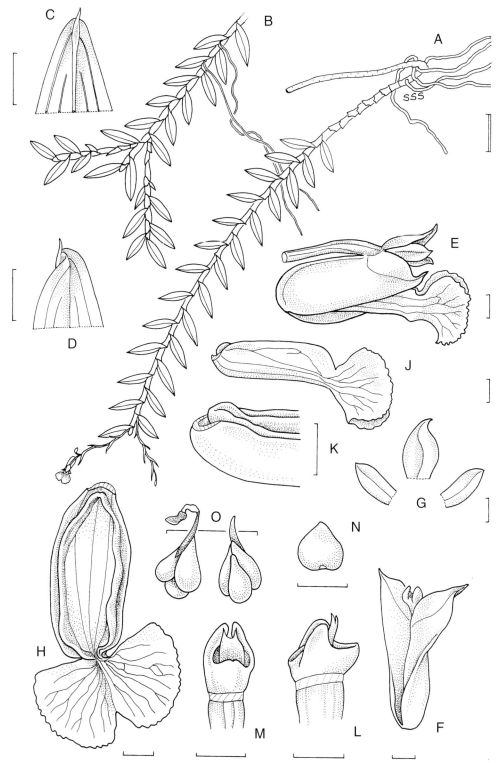

Figure 6. Appendicula bilobulata J.J. Wood. - A & B: habits. - C: leaf apex, back view. - D: leaf apex, front view. - E: flower, side view. - F: lateral sepals and column, oblique view. - G: dorsal sepal and petals. - H: lip, flattened. - J: lip, natural position, side view. - K: base of lip showing callus. - L: pedicel and ovary and column, anther-cap removed, side view. - M: pedicel and ovary and column, anther-cap removed, back view. - N: anther-cap, back view. - O: pollinaria. All drawn from *Burtt & Martin* B.5287 (holotype) by Susanna Stuart-Smith. Scale: single bar = 1 mm; double bar = 1 cm.

it to be on first sight. Examination of the flowers, however, revealed it to be a hitherto undescribed species of *Appendicula* related to *A. fractiflexa* J.J. Wood (see *Orchids of Borneo*, volume 3: figs. 30 & 31), also recently described from Borneo, and *A. undulata* Blume, which occurs in Peninsular Malaysia, Sumatra, Java, and Borneo. *Appendicula bilobulata* is distinguished from both by the distinctive bilobed lip.

DERIVATION OF NAME: The generic name is Latin for a little appendix, referring to the appendiculate callus on the lip. The specific epithet is derived from the Latin *bi*, meaning two, and *lobulus*, a lobule, referring to the bilobed lip epichile.

7. APPENDICULA CALCARATA Ridl.

Appendicula calcarata *Ridl.* in J. Linn. Soc., Bot. 31: 302 (1896). Types: Borneo, Sarawak, *Haviland* 2334 & 3145 (syntypes SING, isosyntypes K).

Podochilus calcaratus (Ridl.) Schltr. in Mém. Herb. Boissier 21: 41 (1900).

Epiphyte. **Roots** elongate, shortly tomentose, forming a dense mass. **Stems** 30–55 cm long, branching, branches 2.5–20 cm long, erect, becoming pendulous, compressed, leafy. **Leaves** distichous, stiffly coriaceous; blade 1.5–2.5(–3.5) × 0.3–0.8(–1) cm, narrowly oblong to narrowly oblong-elliptic, obtusely unequally bilobed, with a mucro in the sinus; sheath 0.8–1.8 cm long. **Inflorescences** terminal, subtended by 1 or 2, 1–1.5 cm long, apiculate leafy sheaths, simple or occasionally with an additional branch, densely *c.* 6- to 13-flowered, flowers concealed by floral bracts; peduncle abbreviated; rachis 2–3 cm long; floral bracts 1–2 cm long, foliaceous, narrowly elliptic, somewhat falcate, obtuse to acute, finely nerved, distichous, yellowish green or cream. **Flowers**: pedicel and ovary greenish white, sepals yellow, edged darker yellow, mentum creamy, petals yellowish white, edged crimson to purple, lip pale yellow or pale orange, light brown distally, keels reddish, column pale yellow, marked dark red, anther-cap purple. **Pedicel** and **ovary** 5–5.5 mm long. **Dorsal sepal** 4.2–5 × 2.5–2.6 mm, ovate-elliptic or ovate, acute, mucronate, sometimes dorsally erose-carinate, concave. **Lateral sepals** 9–10 mm long, free portion 5–5.1 × 2 mm, very unequally triangular-ovate, acute, strongly dorsally erose-carinate to alate. **Mentum** 4–5 mm long, narrowly saccate, obtuse, parallel to ovary. **Petals** 3.8 × 1–1.1 mm, narrowly elliptic, acute. **Lip** *c.* 9 mm long, claw 1 mm long, proximal portion *c.* 1.8 mm wide, apical blade 2 × 3.2 mm, narrow, abruptly dilated into a flabellate, rotundate, erosulate, slightly undulate blade; disc with 2 parallel lamellate keels united at and protruding beyond base of lip, terminating on mid-lobe. **Column** 1–1.1 mm long; foot 4–5 mm long; rostellum *c.* 1 mm long, sharply bifurcate; stigmatic cavity cymbiform; anther-cap 2.8–3 mm long, acuminate. Plate 3A.

HABITAT AND ECOLOGY: Hill forest; lower montane forest. Alt. 700 to 900 m. Flowering observed from July until September.

DISTRIBUTION IN BORNEO: SABAH: Mt. Kinabalu. SARAWAK: Marudi District, Tama Abu Range.

GENERAL DISTRIBUTION: Endemic to Borneo.

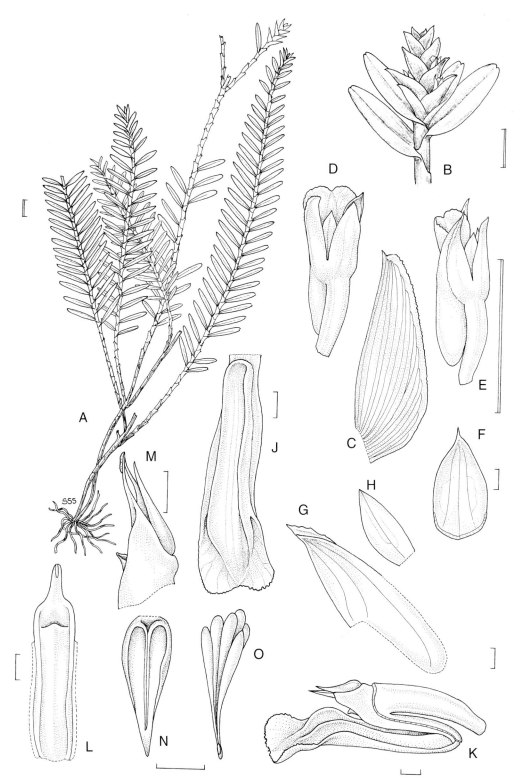

Figure 7. Appendicula calcarata Ridl. - A: habit. - B: upper leaves and inflorescence. - C: floral bract. - D: flower, viewed from above. - E: flower, oblique view. - F: dorsal sepal. - G: lateral sepal. - H: petal. - J: lip, front view. - K: pedicel with ovary, lip and column, side view. - L: column, anther-cap removed, front view. - M: apex of column showing anther-cap and pollinia, side view. - N: anther-cap, interior view. - O: pollinia. A drawn from *Haviland* 3145 (syntype) and B–O from *Bailes & Cribb* 571 by Susanna Stuart-Smith. Scale: single bar = 1 mm; double bar = 1 cm.

NOTE: *Appendicula calcarata* is one of several species having conspicuous imbricate floral bracts.

DERIVATION OF NAME: The specific epithet is derived from the Latin *calcar*, a spur, in reference to the spur-like mentum.

8. APPENDICULA CONGESTA Ridl.

Appendicula congesta *Ridl.* in Stapf in Trans. Linn. Soc. London, Bot. 4: 239 (1894). Type: Borneo, Sabah, Mt. Kinabalu, Tinekuk (Penokok) River, 1200 m, *Haviland* 1302 (holotype SING, isotype K).

Podochilus congestus (Ridl.) Schltr. in Mém. Herb. Boissier 21: 59 (1900).
Chilopogon kinabaluensis Ames & C. Schweinf., Orchidaceae 6: 141, pl. 93 (1920). Type: Borneo, Sabah, Mt. Kinabalu, Marai Parai Spur, *J. Clemens* 230 (holotype AMES, isotypes BM, K).
Appendicula kinabaluensis (Ames & C. Schweinf.) J.J. Sm. in Bull. Jard. Bot. Buitenzorg, ser. 3, 5: 65 (1922).

Epiphyte. **Roots** slender, fibrous, finely tomentose or glabrous, forming a dense mass. **Stems** 7–25 cm long, *c*. 1.5 mm in diameter, internodes 0.3–1 cm long, clump-forming, spreading, porrect or slightly curved, leafy. **Leaves** very variable in size, smallest at base, distichous; blade 0.7–3.2 × 0.3–1.2 cm, narrowly elliptic, oblong-elliptic or ovate-elliptic, apex obtuse, minutely equally or unequally bilobed, lobes obtuse to acute, with a mucro in the sinus, base clasping, rigid, somewhat coriaceous, articulated to sheath; sheath 0.3–1 cm long, infundibuliform, scarious. **Inflorescences** terminal, 1.5–2.8 cm long, 1–1.8 cm in diameter, densely many-flowered, capituliform, simple or with a solitary short branch, subtended by a few 0.4–1 cm long oblong, mucronate, foliaceous bracts; peduncle abbreviated; rachis 1.5–2.8 cm long; floral bracts 4–10 × 3.5–5 mm, ovate to ovate-oblong, truncate, mucronate, crisped, erose, membranous, 3-nerved, translucent pale green, whitish green at base. **Flowers**: pedicel and ovary pale green, sepals translucent white to yellow, mentum white, petals purple, lip translucent white, yellow centrally, callus red, sometimes with pink or purple-red markings elsewhere. **Pedicel** and **ovary** *c*. 1 mm long. **Sepals** and **petals** connivent. **Dorsal sepal** 3.5 × 1.5 mm, ovate-elliptic, acute, mucronate, concave. **Lateral sepals** 3.5 × 2 mm, ovate-elliptic, very oblique, sharply mucronate, forming a saccate mentum. **Petals** 3 × 1 mm, ligulate, obtuse, apiculate. **Lip** 4 × 2.4 mm when flattened, ovate-oblong, broadly rounded, concave-involute, in natural position folded against column and enclosed by mentum; anterior portion provided with a broad, transverse, bilobed, horseshoe-shaped plate which is barbed with long slender hairs on the posterior margin, with a slender papilliform tubercle above the sinus of the plate; disc with a stout median keel running from base of lip to the plate. **Column** *c*. 1.75 mm long, stout, with 4 sharp teeth, central pair longer than lateral; foot broad, membranous; anther-cap narrowly elliptic. Plate 2A & B.

HABITAT AND ECOLOGY: Hill forest; lower montane forest, frequently on ultramafic substrate; oak-laurel forest; often epiphytic low down on tree trunks. Alt. 700 to 2100 m. Flowering observed throughout the year.

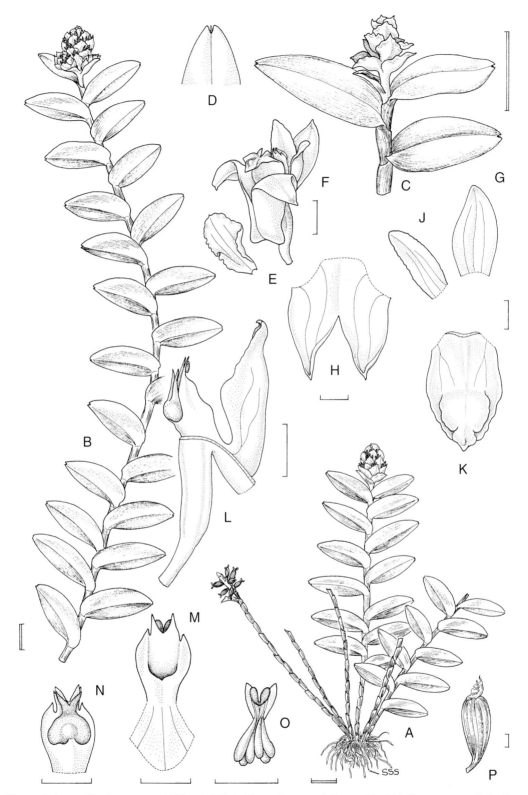

Figure 8. Appendicula congesta Ridl. - A & B: habits. - C: apex of shoot with old inflorescence. - D: leaf apex, adaxial surface, enlarged. - E: floral bract. - F: flower, oblique view. - G: dorsal sepal. - H: synsepalum, adaxial surface. - J: petal. - K: lip, front view. - L: pedicel with ovary, lip and column, side view. - M: column, front view. - N: column apex, back view. - O: pollinarium. - P: fruit. A & E–P drawn from *J. & M. S. Clemens* 50352, B & D from *J. & M.S. Clemens* 27138 and C from *Bailes & Cribb* 823 by Susanna Stuart-Smith. Scale: single bar = 1mm; double bar = 1cm.

DISTRIBUTION IN BORNEO: SABAH: Mt. Kinabalu; Mt. Lumaku.

GENERAL DISTRIBUTION: Endemic to Borneo.

NOTES: Ames and Schweinfurth compared their *Chilopogon kinabaluensis* with the closely related *C. merrillii* (Ames) Ames (= *Appendicula merrillii* Ames) from the Philippines. The broader truncate, mucronate floral bracts and differently coloured flowers with smaller petals and a longer pair of central teeth on the column distinguish *A. congesta* from *A. merrillii*.

DERIVATION OF NAME: The Latin specific epithet refers to the densely flowered terminal inflorescences.

9. APPENDICULA FOLIOSA Ames & C. Schweinf.

Appendicula foliosa *Ames & C. Schweinf.*, Orchidaceae 6: 145 (1920). Type: Borneo, Sabah, Mt. Kinabalu, Kiau, *J. Clemens* 361 (holotype AMES, isotypes BM, K, SING).

Epiphyte or **lithophyte**, rarely **terrestrial**. **Stems** 30–90 cm long, internodes 1–2 cm long, up to *c.* 6 mm in diameter, simple, very rarely with small lateral branches, erect, leafy, entirely concealed by leaf sheaths. **Leaf-blade** (4–)6–9(–10.5) × (1–)2.2–2.5(–3.2) cm, elliptic to oblong-elliptic, obtusely and obscurely unequally bilobed, mid-nerve extended to form an apical mucro, thin-textured, chartaceous, minutely dark-brown punctate-ramentaceous on reverse, main nerves 7–9; sheath 1–2 cm long, striate-sulcate, minutely densely irregularly dark brown punctate-ramentaceous, becoming fibrous with age. **Inflorescences** very numerous, mostly lateral and opposite leaf sheaths, sometimes terminal also, sometimes two-branched, arcuate, horizontal or decurved, few-flowered; peduncle 0.6–2 cm long; non-floriferous bracts 0.3–1.5 cm long, lowermost leafy, ovate-elliptic, obtuse and mucronate to acute, imbricate, sparsely to densely brown punctate-ramentaceous, entirely clothing peduncle; rachis 2–4 cm long; floral bracts 3–6 mm long, ovate, apex obtuse and erose, sulcate, concave, scarious, reflexed, prominently nerved. **Flowers** variously described as greenish white, cream and yellow, green and yellow, light green or light yellow, lip with maroon markings, column yellow with red margin. **Pedicel** and **ovary** 0.6–1.2(–1.5) cm long, cylindrical to narrowly clavate. **Dorsal sepal** 3 × 2 mm, broadly ovate, acute, strongly concave, 3-nerved. **Lateral sepals** 3 × 5 mm, broadly triangular, oblique, acute, 3-nerved. **Mentum** 2–4 mm long, saccate, obtuse. **Petals** 2.6 × 1.3–1.75 mm, oblong or obovate-oblong, obtuse and minutely apiculate, obscurely 3-nerved. **Lip** *c.* 6 mm long, 3 mm wide near base, slightly narrower distally, oblong-pandurate, terminating in a pair of rounded lobes, rounded at base; disc with a thin basal horseshoe-shaped keel, highest at base, with parallel arms extending to anterior portion. **Column** 0.8–1 mm long, broadly winged; foot 2–4 mm long, unciform; rostellum short, triangular, retuse. **Fruit** green, striped purple. Plate 2C & D.

HABITAT AND ECOLOGY: Hill dipterocarp forest; lower montane forest, sometimes on ultramafic substrate; submontane forest on sandstone; riparian forest; oak-laurel forest; mossy boulders. Alt. 800 to 2400 m. Flowering observed in January, from March to May, September, November and December.

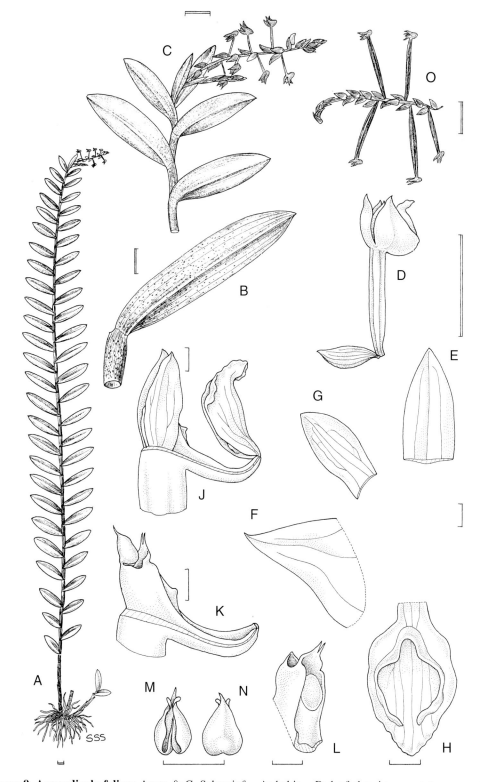

Figure 9. Appendicula foliosa Ames & C. Schweinf. - A: habit. - B: leaf showing punctate - ramentaceous indumentum on sheath and abaxial surface. - C: shoot apex with inflorescence. - D: floral bract and flower, side view. - E: dorsal sepal. - F: lateral sepal. - G: petal. - H: lip, front view. - J: portion of ovary, petals, lip and column, side view. - K: column with anther-cap, side view. - L: column apex, anther-cap removed, oblique view. - M: anther-cap, interior view. - N: anther-cap, back view. - O: infructescense. A & C–N drawn from *Beaman* 9782 and B & O from *Yii* S.44421 by Susanna Stuart-Smith. Scale: single bar = 1 mm; double bar = 1 cm.

DISTRIBUTION IN BORNEO: SABAH: Mt. Kinabalu; Lahad Datu District. SARAWAK: Lundu District, Mt. Gading; Mt. Murud.

GENERAL DISTRIBUTION: Endemic to Borneo.

NOTES: *Appendicula foliosa* is closely related to the widespread *A. pendula* Blume from which it can be distinguished by the generally narrower leaves with minutely dark brown-spotted sheaths, denser inflorescences and an anterior callus on the lip. The size and shape of the leaves is rather variable. For example, *J. Clemens* 63 (AMES) has large leaves up to 12.8 × 3.55 cm, while those of *J. Clemens* 222 (AMES) are much narrower.

Ames & Schweinfurth (1920) commented that "The close but irregular dark brown dots that are universal on the sheathed stems and also occur somewhat on the leaves and even on the bracts and flowers of the type material are probably fungus growth." Close study, however, reveals that this "fungus growth" is actually a covering of numerous minute chaffy outgrowths from the epidermis.

DERIVATION OF NAME: The specific epithet is derived from the Latin *foliaceus*, leafy, and refers to the habit.

10. APPENDICULA PILOSA J.J. Sm.

Appendicula pilosa *J.J. Sm.* in Icon. Bogor. 2: 53, t. 110A (1903). Type: Borneo, Kalimantan Timur, Sungai Tjehan, *Nieuwenhuis* s.n. (holotype BO).

Appendicula niahensis Carr in Gard. Bull. Straits Settlem. 8: 95 (1935). Type: Borneo, Sarawak, Niah, *Synge* S.601 (holotype SING, isotype K).

Epiphyte. **Stems** 25–45 cm long, internodes 0.5–0.8 cm long, 2–3 mm in diameter, simple, mostly pendent, entirely concealed by leaf sheaths. **Leaf-blade** 1.7–3.4 × 0.6–1.35 cm, uppermost smaller, ovate, ovate-elliptic or oblong-elliptic, very shortly conduplicate and nearly equally, obtusely, very shortly bilobed at apex, twisted at base, rather rigid, main nerves 3, glossy greyish green above, brownish below; sheath 0.5–0.8 cm long, tubular, funnel-shaped distally, glossy brownish. **Inflorescences** terminal and lateral from uppermost nodes, short, densely few-flowered (up to *c.* 12); peduncle 0.5–1 cm long; non-floriferous bracts 3–4, 2–3 mm long, sheathing; rachis 1–1.5 cm long; floral bracts 4.7–5.7 × 2.8–4.5 mm, ovate or oblong-obovate, obtuse to subacute, concave, carinate, furfuraceous on exterior, reflexed, apex incurved. **Flowers** white or yellowish green, unscented. **Pedicel** and **ovary** 3–3.7 mm long, straight. **Dorsal sepal** 4.7 × 4.7 mm, oblong, obtuse, cucullate at apex, concave, 3-nerved. **Lateral sepals** 4.3 × 3–3.3 mm, triangular-ovate, obtuse, concave, 3-nerved. **Mentum** *c.* 3.5 mm long, narrowly oblong, rounded. **Petals** 3–4 × 2–3 mm, oblong-obovate, obtuse, concave, punctate, 3-nerved. **Lip** *c.* 4.7 × 2.3 mm (when flattened), pandurate; lower half a broadly elliptic claw provided with a subquadrate callus produced in front to 2 short keels which diverge to margins at base of blade; upper half a shortly ovate-triangular, subacute blade, margins shortly adnate at base to column-foot, concave, strongly recurved above base, provided above base with a fleshy median keel which is sharply attenuate towards apex, margins above bend incurved, apex incurved and minutely apiculate. **Column** 3–3.5 mm long,

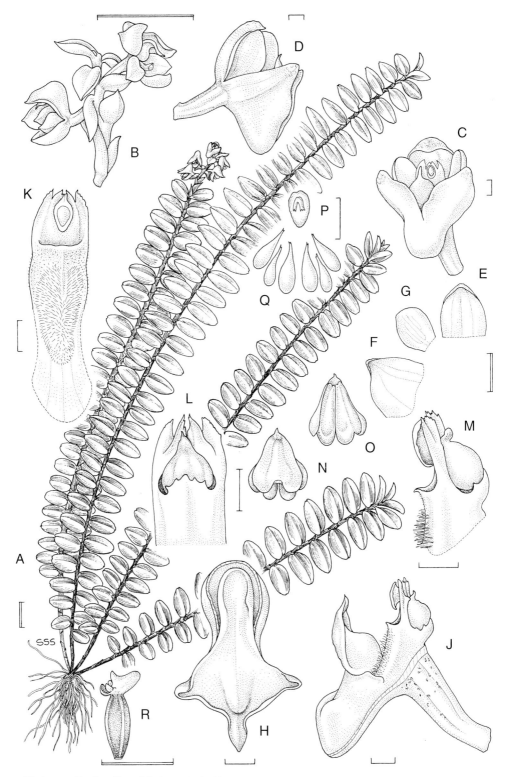

Figure 10. Appendicula pilosa J.J. Sm. - A: habit. - B: inflorescence. - C: flower, front view. - D: flower, side view. - E: dorsal sepal. - F: lateral sepal. - G: petal. - H: lip, front view. - J: pedicel with ovary, lip and column, side view. - K: column, front view. - L: apex of column, front view. - M: apex of column, with anther-cap, side view. - N: anther-cap, back view. - O: anther-cap, interior view. - P: viscidium, front view. - Q: pollinia. - R: fruit. A–Q drawn from *Hansen* 350 and R from *Lugas* 2127 by Susanna Stuart-Smith. Scale: single bar = 1 mm; double bar = 1 cm.

pubescent on inner surface; foot 3–3.7 mm long, margins shortly adnate at base to lip, pubescent on inner surface; rostellum bifid; anther-cap cordate, acute to acuminate.

HABITAT AND ECOLOGY: Lowland forest on steep sandstone slopes and scree; lowland dipterocarp forest; hill forest on limestone and sandstone; epiphytic on trunks near ground. Alt. sea level to 700 m. Flowering observed in February and November.

DISTRIBUTION IN BORNEO: KALIMANTAN TIMUR: Cehan (Tjehan) River. SARAWAK: Marudi District, Gunung Mulu National Park; Miri District, Niah.

GENERAL DISTRIBUTION: Endemic to Borneo.

NOTES: Carr (1935) compared his *A. niahensis* with *A. latibracteata* J.J. Sm., now considered conspecific with the widespread *A. pendula* Blume. He was apparently unaware that Smith had also examined material from Kalimantan and already described it as *A. pilosa*.

DERIVATION OF NAME: The specific epithet is derived from the Latin *pilosus*, with distinct long ascending hairs, and refers to the pubescent inner surface of the column.

11. APPENDICULA ROSTELLATA J.J. Sm.

Appendicula rostellata *J.J. Sm.* in Icones Bogor. 4: 13, t. 305 (1910). Type: Borneo, Kalimantan Tengah, Liangangang (Liang Gagang), *Hallier* s.n. (holotype BO).

Epiphyte. **Stems** up to 35 cm long, internodes 5–6 mm long, uppermost shorter, up to 2 mm in diameter, branching, erect, leafy, entirely concealed by leaf sheaths. **Leaf-blade** 0.8–1.9 × 0.2–0.47 cm, linear-ligulate or linear-oblong, unequally obtusely bilobulate and mucronate, somewhat twisted at base, slightly curved, thin, coriaceous, with many fine nerves, mid-nerve prominent on reverse; sheath 5–6 mm long, tubular, striolate, nervose, becoming fibrous with age. **Inflorescences** terminal, 6- to 7-flowered; peduncle abbreviated; rachis 0.6–1 cm long, fractiflex; floral bracts 6–9 × 0.8 mm, ovate, acute, concave, adpressed, imbricate, semi-pellucid, greenish-white. **Flowers** *c.* 7–7.3 mm long, white, sometimes flushed palest pink, petals with a purple margin, anther-cap purple. **Pedicel** and **ovary** 2–3.5 mm long, cylindrical. **Dorsal sepal** *c.* 4 × 2.3 mm, ovate, obtuse and mucronate, concave, 3-nerved. **Lateral sepals** *c.* 7 × 2.4 mm, obliquely triangular, acute and mucronate, concave, 3-nerved. **Mentum** 3 × 1.7 mm, saccate, obtuse. **Petals** *c.* 3.5 × 1 mm, elliptic, acute, slightly oblique. **Lip** 6 mm long, *c.* 2.5 mm wide across mesochile, *c.* 3.4 mm wide across mid-lobe, spathulate, subtrilobed, clawed, subacute, apiculate, distal portion recurved, transversely ovate, claw linear, concave; disc with 2 ribs, produced below into an oblong, retuse, concave, somewhat undulate appendage. **Column** 2.7–3 mm long (with rostellum), wings triangular, obtuse; foot 3 mm long, concave, incurved; rostellum *c.* 1.6 mm long, linear, narrowly bidentate; anther-cap *c.* 2 mm long, cucullate, ovate, rostrate. Plate 3B.

HABITAT AND ECOLOGY: Hill and lower montane forest on sandstone; recorded as epiphytic on *Parashorea* (Dipterocarpaceae). Alt. 300 to 1500 m. Flowering recorded in June and July.

Figure 11. Appendicula rostellata J.J. Sm. - A: habit. - B: inflorescence and upper leaves. - C: floral bract and flower, side view. - D: flower, front view. - E: dorsal sepal. - F: petal. - G: lip, front view. - H: lip, back view. - J: pedicel with ovary, lip and column, side view. - K: column, front view. - L: ovary and column, from above. - M: anther-cap, back view. - N: pollinarium. - O: viscidium and stipes. A drawn from *Madani* SAN 88849, B–F & H–O from after *J.J.Smith & R.N. Naladipoera del., M. Kromohardjo lith.*, and G from *deVogel & Cribb* 9173 by Linda Gurr. Scale: single bar = 1 mm; double bar = 1 cm.

DISTRIBUTION IN BORNEO: KALIMANTAN TENGAH: Liangangang. SABAH: Lamag District, Namatoi River area; Mt. Kinabalu.

GENERAL DISTRIBUTION: Endemic to Borneo.

NOTES: *Appendicula rostellata* is another of the Bornean species with conspicuous imbricate floral bracts. It is distinguished from the closely related *A. calcarata* Ridl. by its less robust habit, narrower rhizome, smaller leaves, floral bracts shorter than or equalling the flowers and broader lip with a shorter claw and larger basal appendage.

DERIVATION OF NAME: The specific epithet is derived from the Latin *rostellum*, a beak, and refers to the conspicuous elongate rostellum.

12. ASCIDIERIA LONGIFOLIA (Hook.f.) Seidenf.

Ascidieria longifolia *(Hook.f.) Seidenf.* in Nordic J. Bot. 4 (1): 44 (1984). Type: Peninsular Malaysia, Perak, Ulu Batang Padang, *Wray* 1541 (holotype K).

Eria longifolia Hook.f., Fl. Brit. Ind. 5: 790 (1890).
Pinalia longifolia (Hook.f.) Kuntze, Rev. Gen. 2: 679 (1891).
Eria verticillaris Kraenzl. in Bot. Jahrb. Syst. 44, Beibl. 101: 29 (1911). Type: Borneo, Sarawak, *Beccari* 2453 (holotype B, destroyed, isotype FI).
Cymboglossum longifolium (Hook.f.) Brieger, Die Orchideen, ed. 3, 1: 649 (1981).

Erect, tufted, clump-forming **epiphyte** or **lithophyte**. **Stems** (6–)15–25 cm long, slender, swollen at base, internodes few, 2- to 3-leaved, covered with long, pale brown scarious sheaths 7–15 cm long. **Leaves** 15–25 × 0.5–1.2 cm, apical, linear-ligulate, narrowed at base, obliquely acute to acuminate. **Inflorescences** 1 or 2, emerging just below stem apex, erect, densely many-flowered, flowers arranged in whorls of *c*. 10, each whorl borne *c*. 5 mm apart, whorls not apparent in young inflorescences; peduncle and rachis covered in dense white woolly hairs; peduncle 3–6 cm long; rachis 6–12 cm long; floral bracts 1 mm long, ovate, obtuse, hairy at base. **Flowers** white, non-resupinate, sepals densely white woolly-hairy on outer surface. **Ovary** 1–1.5 mm long, sessile, white woolly-hairy. **Dorsal sepal** 3 × 2 mm, ovate-elliptic, obtuse or subacute. **Lateral sepals** 5 × 2.5 mm, obliquely ovate, obtuse, partially connate along the base with underside of lip sac. **Mentum** 3–4 mm long, obtuse. **Petals** 2 × 1 mm, elliptic, obtuse. **Lip** attached to base of column-foot by its upper edge, blade 2.5 × 2.5 mm, ovate, subacute, with 2 lines of hairs, free saccate base 2–2.5 mm long, obtuse. **Column** 1 mm long; foot 1.5 mm long; anther-cap cucullate. Plate 3C.

HABITAT AND ECOLOGY: Lower montane forest; oak-laurel forest; sand forest; mossy rocks; sometimes on ultramafic substrate. Alt. 900 to 1900 m. Flowering observed throughout the year.

DISTRIBUTION IN BORNEO: BRUNEI: Mt. Retak. SABAH: Crocker Range; Kinabalu Park; Sipitang District, Long Pasia area. SARAWAK: Gunung Mulu National Park; Mt. Dulit; Mt. Pueh.

GENERAL DISTRIBUTION: Peninsular Malaysia, Thailand, Sumatra and Borneo.

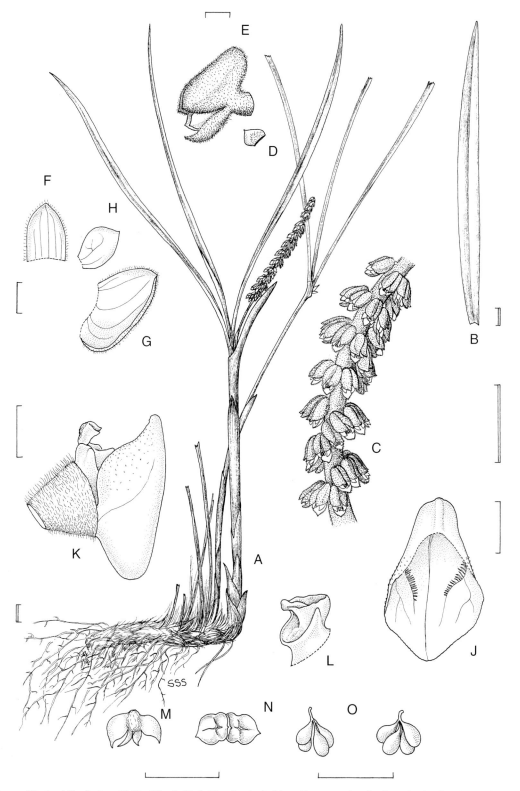

Figure 12. Ascidieria longifolia (Hook.f.) Seidenf. - A: habit. - B: example of a broader leaf. - C: portion of inflorescence. D: floral bract. - E: flower, side view. - F: dorsal sepal. - G: lateral sepal. - H: petal. - J: lip, front view. - K: pedicel with ovary, lip and column, anther-cap attached, side view. - L: column with anther-cap detached, oblique view. - M: anther-cap, back view. - N: anther-cap, interior view. - O: pollinia. A drawn from *Lamb* SAN 89695, B from *Gibot* SAN 68555 and C–O from *Clements* 3231 by Susanna Stuart-Smith. Scale: single bar = 1 mm; double bar = 1 cm.

NOTES: Seidenfaden (1984) created the monospecific genus *Ascidieria* to accommodate *Eria longifolia* Hook.f. on account of the absence of a column-foot and the virtually free lateral sepals which are attached directly, with the lip, to the base of the column. These observations were made after study of flowers from a plant cultivated at Penang Botanic Garden (Kew Spirit Collection no. 31446). Study of material from Borneo shows that some plants have a short column-foot and agree well with the description of Peninsular Malaysian plants provided by Holttum (1964). The flowers of these Bornean plants are somewhat longer and the lateral sepals are partially connate with the underside of the lip sac. In addition, the inflorescence is rather dense and the flowers are not always arranged in distinct whorls. This, of course, may also be true of young inflorescences that have yet to expand fully. It is clear that further study of variation over the entire range is required before the generic status of *Ascidieria* and the specific delimitation of *Eria longifolia* Hook.f. and *E. verticillaris* Kraenzl. can be clarified.

DERIVATION OF NAME: The generic name is derived from the Latin *ascidium*, a pitcher, in reference to the pitcher-shaped, cymbiform lip. The Latin specific epithet refers to the narrow, elongate leaves.

13. BRACHYPEZA INDUSIATA (Rchb.f.) Garay

Brachypeza indusiata (Rchb.f.) Garay in Bot. Mus. Leafl. 23 (4): 164 (1972). Type: "Insul. Sondaic", *Messrs. Linden* (holotype W).

Thrixspermum indusiatum Rchb.f. in Gard. Chron., ser. 2, 25: 585 (May 1886).
Thrixspermum platyphyllum Rchb.f. in Bot. Centralbl. 28: 343 (Nov. 1886). Type: Papua New Guinea, *Beccari* s.n. (holotype W).
Sarcochilus stenoglottis Hook.f., Fl. Brit. Ind. 6: 34 (1890). Types: Peninsular Malaysia, ? Perak, *Scortechini* s.n. (syntype K, not located); Sumatra, *King's Collector* s.n. (syntype K, not located).
Sarcochilus sigmoideus Ridl. in J. Linn. Soc. 31: 298 (1896). Type: Borneo, Kalimantan Barat, Pontianak, *native dealer* s.n. (holotype SING).
Sarcochilus keyensis J.J. Sm., Ic. Bogor. 3: 49, t. 219 (1906). Type: Indonesia, Kai (Kei) Islands, 1893, *Treub* s.n. (holotype BO).
Sarcochilus indusiatus (Rchb.f.) Carr in Gard. Bull. Straits Settlem. 8: 121 (1935).
Pteroceras stenoglottis (Hook.f.) Holttum in Kew Bull. 14: 271 (1960).
Brachypeza stenoglottis (Hook.f.) Garay in Bot. Mus. Leafl. 23 (4): 164 (1972).

Epiphyte. Stems 1–4 cm long, entirely enclosed by persistent leaf sheaths, producing many elongate roots. **Leaves** up to *c.* 8, (7–)15–30 × (2–)3.8–7.2 cm, decurved, ligulate, often widest distally, apex broadly rounded, asymmetrical, emarginate, surface almost flat to gently undulate, coriaceous. **Inflorescences** simple, pendulous, densely many flowered, bearing one to three flowers open in succession; peduncle 7–9 cm long, terete, dull purple, bearing 2–3 remote non-floriferous triangular-ovate, obtuse bracts *c.* 2 mm long; rachis 9–12 cm long, somewhat thicker than peduncle; floral bracts 1 mm long, ovate, acute. **Flowers** 1.3–1.7 cm in diameter, lasting one day only, sepals and petals translucent pale primrose-yellow, with a few faint red or purple streaks or bars at base inside, lip white, pale yellow near base, side lobes sometimes with a few pale red

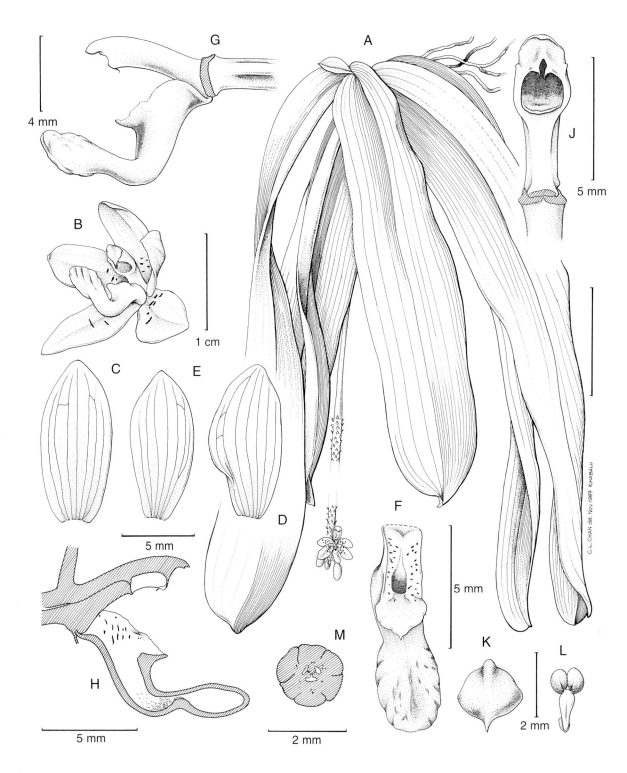

Figure 13. Brachypeza indusiata (Rchb.f.) Garay. - A: habit. - B: flower, oblique view. - C: dorsal sepal. - D: lateral sepal. - E: petal. - F: lip, front view. - G: pedicel with ovary, lip and column, side view. - H: lip and column, longitudinal section. - J: column, anther-cap removed, front view. - K: anther-cap, back view. - L: pollinarium. - M: ovary, transverse section. All drawn from material cultivated at *Tenom Orchid Centre* (TOC 2194) by C.L. Chan.

streaks, column white, pale yellow at base. **Pedicel** and **ovary** 8 mm long, slender. **Sepals** and **petals** opening widely. **Dorsal sepal** 0.7–10 × 4.5 mm, oblong-ovate, obtuse. **Lateral sepals** 0.7–10 × 4.5 mm, obliquely oblong-ovate, obtuse. **Petals** 9 × 4 mm, obovate, subacute. **Lip** *c.* 10 mm long, very narrow, laterally flattened, curved, shortly clawed, with two small, rounded auriculate side lobes beyond the claw; spur *c.* 7 mm long, ellipsoid, held in line with remainder of lip, resembling a terminal lobe, its entrance a perforation between the side lobes. **Column** *c.* 7 mm long, narrowly winged to near its base; foot short; apex acute; rostellum triangular; anther-cap cucullate, shortly beaked; pollinia 2, sulcate. Plate 3D.

HABITAT AND ECOLOGY: Hill forest. Alt. 200 to 300 m. Flowering observed in June.

DISTRIBUTION IN BORNEO: SABAH: Tenom District.

GENERAL DISTRIBUTION: Peninsular Malaysia, Sumatra, Borneo and Kai Islands.

NOTES: *Brachypeza* was established by Garay in 1972 and comprises about seven species distributed from Thailand and Laos through Malaysia and Indonesia, north to the Philippines and east to New Guinea. The elongate column with a short foot, hanging lip, and sulcate pollinia readily differentiate this genus from *Pteroceras*.

Three species have been recorded from Borneo, of which *B. zamboangensis* (Ames) Garay was illustrated in *Orchids of Borneo*, volume 3, fig. 10.

DERIVATION OF NAME: The generic name is derived from the Greek *brachys*, short, and *peza*, foot, in reference to the characteristically short column-foot. The specific epithet is derived from the Latin *indusiatus*, an indusium, literally a woman's under-garment, referring to the spur which, according to Reichenbach, has "a kind of bucket at its apex".

14. BROMHEADIA BORNEENSIS J.J. Sm.
var. BORNEENSIS

Bromheadia borneensis *J.J. Sm.* in Bull. Jard. Bot. Buitenzorg, ser. 2, 25: 18 (1917). Type: Borneo, Kalimantan Timur, Mt. Samonggaris, *Amdjah* 980 (holotype BO, isotypes BO, K, L)

var. **borneensis**

Terrestrial. **Stems** up to 120 cm long, internodes 3–7 cm long, 4–6 mm wide; flowering stems with 4–6 bladeless basal sheaths, central portion with 6–9 leaves, upper portion with 7–11 bladeless sheaths. **Leaf-blade** 7–14 × 2–4 cm, elliptic to obovate, attenuate at base, very shallowly bilobed, with a tiny curved mucro in the sinus; sheath 3–5 cm long, with a 7–12 mm deep, V-shaped incision with a tiny mucro opposite blade. **Inflorescences** terminal, up to 10 cm long, sometimes branched, with 25(–100) flowers opening one or few simultaneously; rachis 5–12 mm long, internodes 2–3 mm long; floral bracts 3–4 mm long, quadrangular to triangular, acute, upper margin abruptly incurved. **Flowers** cream, pale yellow or reddish, lip yellowish inside, side lobes sometimes reddish, mid-lobe yellowish, column yellowish, sometimes spotted red on inner surface, anther-cap white. **Pedicel** and **ovary** 4–5 mm long. **Dorsal sepal** 1.4–2.1 × 0.2–0.5 cm,

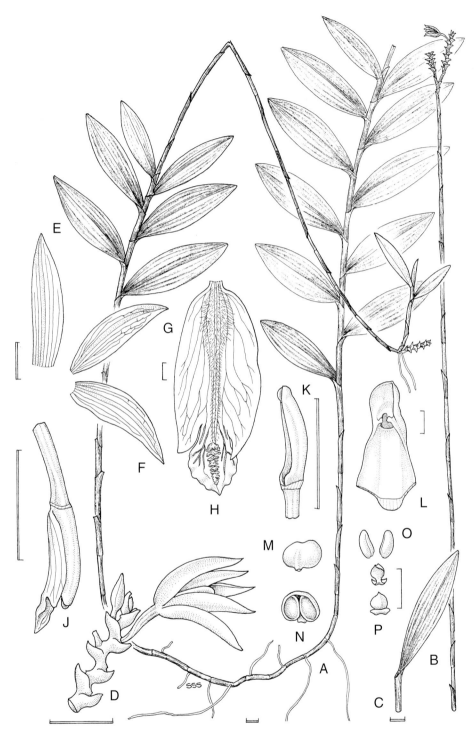

Figure 14. Bromheadia borneensis J.J. Sm. var. **borneensis** – A & B: habit. - C: example of a narrower leaf. - D: inflorescence. - E: dorsal sepal. - F: lateral sepal. - G: petal. - H: lip, flattened, front view. - J: pedicel with ovary, lip and column, side view. - K: upper portion of pedicel with ovary and column, anther-cap detached, side view. - L: column with anther-cap detached, front view. - M: anther-cap, back view. - N: anther-cap, interior view. - O: pollinarium, pollinia detached. - P: viscidium. A & B drawn from *Vermeulen* 816 and C–P from *Lamb* SAN 93455 by Susanna Stuart-Smith. Scale: single bar = 1 mm; double bar = 1 cm.

elliptic to linear-lanceolate, acute, sometimes dorsally carinate. **Lateral sepals** 1.4–1.9 × 0.2–0.5 cm, elliptic, slightly subfalcate, subacute, often dorsally carinate. **Petals** 1.3–1.7 × 0.2–0.4 cm, elliptic to linear-lanceolate, acute. **Lip** 1–1.3 × 0.45–0.7 cm (when flattened), elliptic to obovate in outline; hypochile elliptic to obovate in outline, thin, central area a little swollen and hairy, side lobes 1.7–2.2 mm long, far projecting in front, broad, rounded; epichile 3–4.5 × 2.5–3.5 mm, semi-orbicular to elliptic, acuminate, central area swollen and with irregularly warty bands, margins shallowly crispate and undulate. **Column** 0.9–1 cm long; anther-cap 1 × 0.8 mm, elliptic; pollinia 2.

HABITAT AND ECOLOGY: Swamp forest; lowland dipterocarp forest; hill forest; kerangas forest. Alt. sea level to 500 m. Flowering observed in February, April, June, August and from October to December.

DISTRIBUTION IN BORNEO: BRUNEI: Belait District. KALIMANTAN TENGAH: Locality unknown. KALIMANTAN TIMUR: Mt. Samonggaris. SABAH: Nabawan area; Pensiangan area; Sandakan District, Kapur River; Tawai Hills; Telupid area. SARAWAK: Marudi District, Gunung Mulu National Park; Serian District, Sadong.

GENERAL DISTRIBUTION: Peninsular Malaysia and Borneo.

NOTES: *Bromheadia borneensis* and *B. divaricata* (Fig. 15) belong to section *Bromheadia*, distinguished by dorsiventrally flattened, bilobed leaves articulated with the sheath more or less perpendicular to the stem. Of the seven species of section *Bromheadia* recognised, three are endemic to Borneo. *Bromheadia borneensis* is sometimes confused with the much more widespread *B. finlaysoniana* (Lindl.) Miq. The latter, however, has an almost terete stem and narrower, distinctly bilobed leaf blades with a shallowly U-shaped incision on the blade.

DERIVATION OF NAME: The generic name honours Sir Edward French Bromhead, described as a zealous student of botany. The specific epithet refers to the island of Borneo.

14a. BROMHEADIA BORNEENSIS J.J. Sm.
var. LONGIFLORA

Bromheadia borneensis *J.J. Sm.* var. **longiflora** Scheindelen & de Vogel in Orchid Monographs 8: 88, fig. 31 (1997). Type: Singapore, Murai River, *Ridley* s.n. (holotype BM).

Flowers dull golden yellow, lip side lobes with crimson nerves, speckled crimson on middle, epichile rich yellow with paler margins. **Sepals** 1–2.8 × 0.35–0.5 cm. **Petals** 1–2.5 × 0.35–0.6 cm. **Lip** 1.7–2.3 × 0.5–0.8 cm (when flattened); hypochile long elliptic to narrowly spathulate in outline, side lobes small, 0.7–1.5 mm long, not far projecting in front, triangular, subacute or rounded; epichile 3.5–6 × 3.5–4.5 mm, semi-orbicular to elliptic or obovate, acuminate, central area with a narrowly triangular, swollen patch with low irregular warty bands, margins undulate. **Column** 1.3–1.5 cm long; anther-cap 1.5 × 1 mm.

HABITAT AND ECOLOGY: Lowland dipterocarp forest; peat forest. Alt. lowlands. Flowering observed in September.

DISTRIBUTION IN BORNEO: KALIMANTAN TENGAH: Locality unknown. SARAWAK: Mt. Pueh.

GENERAL DISTRIBUTION: Singapore and Borneo.

NOTE: Although in habit indistinguishable from var. *borneensis*, var. *longiflora* has consistently much longer flowers with much smaller lip side lobes.

DERIVATION OF NAME: The varietal epithet refers to the longer flowers of this variety.

15. BROMHEADIA DIVARICATA Ames & C. Schweinf.

Bromheadia divaricata *Ames & C. Schweinf.* in Orchidaceae 6: 155 (1920). Type: Borneo, Sabah, Mt. Kinabalu, Marai Parai Spur, *J. Clemens* 389 (holotype AMES, isotypes BM, BO, K, NY, SING).

Terrestrial or **epiphytic**. **Stems** up to 90 cm long, sturdy and robust, flattened, internodes to 4 cm long, 0.8–1.2 cm wide; flowering stems with more than 13 leaves. **Leaf-blade** 10.5–12.5 × 2.2–5 cm, elliptic to obovate, amplexicaul, unequally bilobed, coriaceous; sheath 2.5–5 cm long, strongly nerved, with or without an up to 5 mm deep V-shaped incision opposite the blade. **Inflorescences** lateral and terminal, with up to 150 flowers per inflorescence; lateral inflorescences borne on upper part of stem, subsessile, often with one to few subsessile branches; terminal inflorescence a lax, few- to many-branched panicle branched to the first or second degree, peduncle 1–3.5 cm long, main branch up to 15 cm long, terminal branches 1–10 cm long, bearing flower-bearing rachises 1–7 cm long, internodes 1.7–2 mm long; floral bracts 3.5–6 mm long, triangular. **Flowers** cream with purple-red spotting, lip with pronounced purple-red spots or lines, column greenish yellow. **Pedicel** and **ovary** 1.4–1.8 cm long. **Sepals** 2.1–2.7 × 0.6–0.8 cm, elliptic to ovate, or linear-lanceolate, acute, laterals subfalcate. **Petals** 2.1–2.7 × 0.5–0.8 cm, elliptic to ovate, acute, subfalcate. **Lip** 1.4–2.2 × 0.5–1 cm (when flattened), ovate-lanceolate in outline; hypochile elliptic in outline, central area with a somewhat swollen, hairy band in front of which are a few low warts, side lobes 0.5–1 mm long, rounded; epichile 6–7 × 4–5.5 mm, constricted at base, ovate-elliptic to suborbicular, apiculate, fleshy, conduplicate distally, central area thickened and warty, distal margins shallowly undulate, connate. **Column** 1.5–1.8 cm long; anther-cap 2 × 1 mm, obovate. Plate 4A.

HABITAT AND ECOLOGY: Lower and upper montane forest, on ultramafic substrate. Alt. (1350–)2100 to 3000 m. Flowering observed in February, June and November.

DISTRIBUTION IN BORNEO: SABAH: Mt. Kinabalu.

GENERAL DISTRIBUTION: Endemic to Borneo.

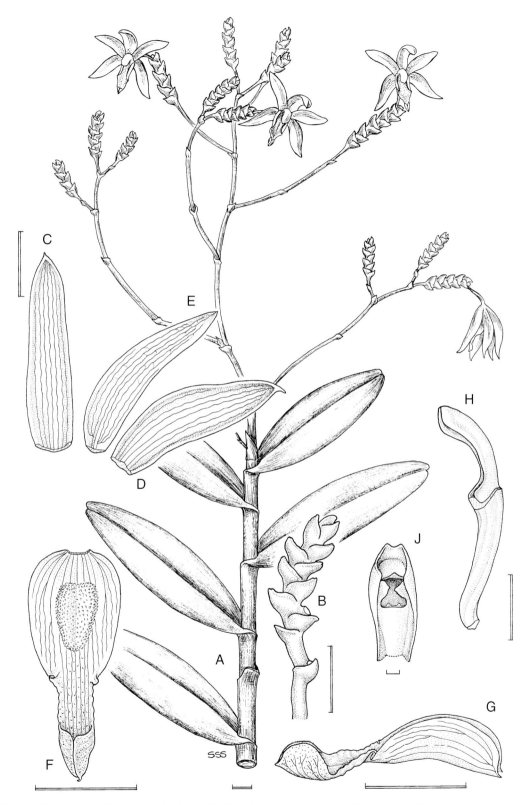

Figure 15. Bromheadia divaricata Ames & C. Schweinf. - A: habit. - B: branch of inflorescence showing imbricate floral bracts. - C: dorsal sepal. - D: lateral sepal. - E: petal. - F: lip, front view. - G: lip, side view. - H: pedicel with ovary and column, side view. - J: column apex, front view. A drawn from *Beaman* 11770 and B–J from *Lamb* AL 355/91 by Susanna Stuart-Smith. Scale: single bar = 1 mm; double bar = 1 cm.

NOTES: *Bromheadia divaricata* is unique in the genus in having a laxly branched paniculate inflorescence producing many raceme-like branches. It is a rare species currently known from only seven collections.

DERIVATION OF NAME: The specific epithet is derived from the Latin *divaricatus*, spreading asunder at a wide angle, and refers to the arrangement of the branches of the inflorescence.

16. BROMHEADIA GRAMINEA Kruizinga & de Vogel

Bromheadia graminea *Kruizinga & de Vogel* in Orchid Monographs 8: 105, fig. 44 (1997). Type: Borneo, Sarawak, *Martin* S.37575 (holotype SAR, isotype SING).

Epiphyte. **Stem** 13–50 cm long, 1.2–3 mm wide, erect, terete at base, with 2 or 3 bladeless sheaths, distal portion little flattened, internodes 1.5–5.2 cm long, terminal internode 1–11 cm long. **Leaf-blade** 7–35 × 0.13–0.26 cm, linear, acute, laterally compressed, straight to slightly curved, rather soft and flexible, 12- to 17-nerved; sheath 1–5 cm long. **Inflorescences** terminal, simple or with up to 5 branches, up to 1.5 cm long, with up to 22 flowers; non floriferous bracts 2–3 at base of raceme, 3–8 × 2.5-3 mm, persistent; floral bracts 1.5–3.4 × 2.6–4 mm, hard, thick, mid-nerve keeled. **Flowers**: sepals yellow with indistinct purple marks down the centre, petals white, lip hypochile cream with fine bright red nerves grading to spots on side lobes, claw of epichile spotted red, callus bright yellow, margins white, column cream with minute reddish violet speckles on ventral surface, anther-cap cream. **Pedicel** and **ovary** 1–1.5 cm long. **Dorsal sepal** 1.8–2.1 × 0.42–0.5 cm, narrowly elliptic, acute, 7-nerved. **Lateral sepals** 1.7–1.9 × 0.44–0.5 cm, narrowly elliptic, acute, 7-nerved. **Petals** 1.6–1.8 × 0.44–0.5 cm, narrowly elliptic, acute, 7-nerved. **Lip** 13.5–15 × 0.85–0.9 cm; hypochile 12–13.2 × 0.85–0.9 cm, elliptic, sparsely hairy, side lobes projecting 2.1–2.3 mm, triangular, acute; epichile 4.2–5 × 5 mm, obovate, slightly emarginate, apiculate, glabrous, callus 4–5 × 2 mm high, broad, swollen, ruminate, brain-shaped, minutely papillose. **Column** 1.3–1.5 cm long, truncate to slightly emarginate; anther-cap 1.2–1.5 × 1.7–2 mm.

HABITAT AND ECOLOGY: Transitional forest between mixed dipterocarp and submontane zones; hill dipterocarp forest; ridge forest. Alt. 700 to 1300 m. Flowering observed in August.

DISTRIBUTION IN BORNEO: KALIMANTAN BARAT: Serawai, Merah River. KALIMANTAN TIMUR: Locality unknown. SARAWAK: Kapit District, Melatai Hills; Kuching District, Mt. Penrissen.

GENERAL DISTRIBUTION: Endemic to Borneo.

NOTES: *Bromheadia graminea* is one of several recently described taxa in section *Aporodes* which is distinguished from section *Bromheadia* by the laterally flattened, acute leaves which are articulated with the sheath more or less parallel with the stem. In a recent revision, Kruizinga *et al.* (1997) recognised 26 taxa in section *Aporodes*, eleven of which were new to science. Of these new taxa, seven, including *B. graminea*, are endemic to Borneo.

DERIVATION OF NAME: The specific epithet is derived from the Latin *gramineus*, grassy, grass-like, in reference to the leaf shape.

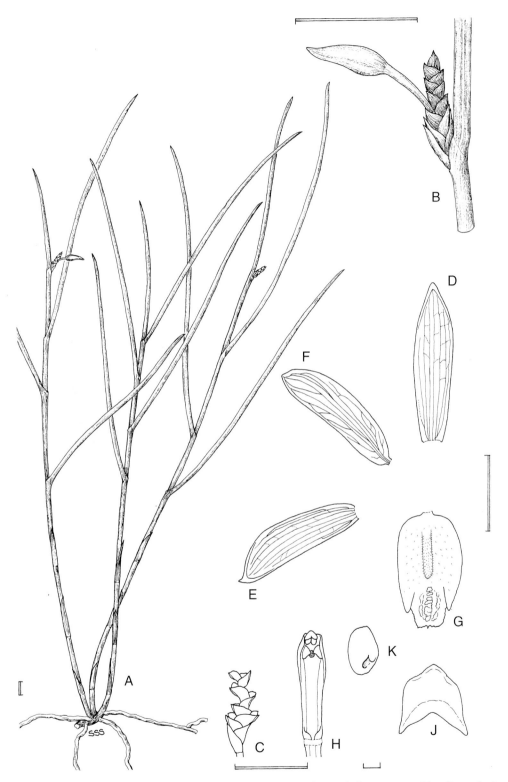

Figure 16. Bromheadia graminea Kruizinga & de Vogel - A: Habit. - B: inflorescence with a flower bud. - C: floral bracts. - D: dorsal sepal. - E: lateral sepal. - F: petal. - G: lip, flattened, front view. - H: column, front view. - J: viscidium. - K: pollinium. A & B drawn from *Laman et al.* TL 794 and C–K from after an illustration of *Leiden cult.* (*deVogel*) 913435 by E.F. de Vogel, by Susanna Stuart-Smith. Scale: single bar = 1 mm; double bar = 1 cm.

17. BULBOPHYLLUM CORIACEUM Ridl.

Bulbophyllum coriaceum *Ridl.* in Stapf in Trans. Linn. Soc. London, Bot., ser. 2, 4: 235 (1894). Type: Borneo, Sabah, Mt. Kinabalu, *Haviland* 1100 (holotype SING, isotype K).

Bulbophyllum kinabaluense Rolfe in Gibbs in J. Linn. Soc., Bot. 42: 148 (1914). Type: Borneo, Sabah, Mt. Kinabalu, Lubang/Paka-paka Cave, *Gibbs* 4252 (holotype BM).
Bulbophyllum venustum Ames & C. Schweinf., Orchidaceae 6: 198 (1920). Type: Borneo, Sabah, Paka-paka Cave, *J. Clemens* 113 (holotype AMES, isotypes BM, K).

Epiphyte or **lithophyte**. **Rhizome** elongate, 2–3 mm in diameter, creeping, tough, terete, flexuous, naked at flowering time but bearing traces of scarious sheaths. **Roots** elongate, *c.* 0.5–0.7 mm in diameter, branching, smooth, forming a dense mass. **Pseudobulbs** 1–2 × 3–5 mm, cylindrical, borne 1.5–10 cm or more apart, yellow. **Leaf-blade** 4.5–11 × 1–3.5 cm, oblong to elliptic, abruptly cuneate towards base, broadly rounded, sometimes shallowly retuse, thick and fleshy, coriaceous; petiole 1–3(–5) cm long, sulcate. **Inflorescences** rather secund, borne beside base of pseudobulb, porrect, arching or sometimes slightly flexuous, laxly many-flowered, flowers borne 4–18 mm apart, distance often varying greatly on a single inflorescence; peduncle 3.5–15 cm long, provided with several 0.5–2 cm long tubular basal sheaths; rachis 4–15 cm long; floral bracts 3–7.5 mm long, narrowly elliptic, concave, long acuminate, spreading. **Flowers** sweetly fragrant, yellow or greenish-yellow, lip deeper yellow to pale orange. **Pedicel** and **ovary** 2–3 mm long, narrowly clavate. **Dorsal sepal** 6–7 × 2.3–2.7 mm, oblanceolate to oblong, complicate, acute, dorsally apically carinate, strongly angulate-sulcate. **Lateral sepals** 6–8 × 3 mm, triangular-elliptic, subfalcate, sulcate, strongly complicate, mucronate, apically strongly carinate. **Mentum** 2 mm long, saccate, obtuse. **Petals** 3 × 1 mm, oblong to linear, slightly falcate, acute, minutely ciliolate and papillose above. **Lip** 3.5–4(–5.2) × 2.4 mm, very shortly clawed, entire, decurved, elliptic-ovate to oblong, acute or obtuse, fleshy, lateral margins slightly decurved, lower sides erect and subconduplicate, disc with a pair of small approximate fleshy calli at base extending longitudinally down lip as fleshy indistinct ridges, 3-nerved. **Column** 1 mm long; foot 2 mm long, stout, arcuate, with a prominent porrect tubercle that terminates a fleshy ridge extending from base; stelidia short, triangular, acuminate; pollinia 4. Plate 4C.

HABITAT AND ECOLOGY: Lower montane forest; upper montane forest; *Leptospermum* scrub between granitic rocks, sometimes on ultramafic substrate. Alt. 1200 to 3500 m. Flowering observed throughout year.

DISTRIBUTION IN BORNEO: SABAH: Mt. Kinabalu.

GENERAL DISTRIBUTION: Endemic to Borneo.

NOTES: *Bulbophyllum coriaceum* belongs to section *Aphanobulbon* and is a relatively common species between 2700 and 3500 metres on Mt. Kinabalu. Rolfe (1914) comments that *B. kinabalunse* Rolfe differs in having more slender racemes of smaller flowers with narrower segments. Ames & Schweinfurth (1920) distinguished *B. venustum* Ames & C. Schweinf. by its remote pseudobulbs, longer and somewhat wider leaves, taller scapes bearing more flowers with

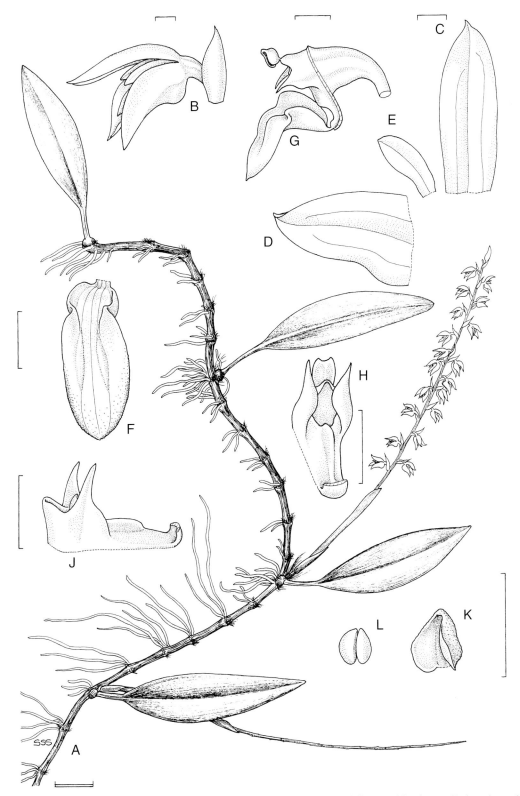

Figure 17. Bulbophyllum coriaceum Ridl. - A: habit. - B: floral bract and flower, side view. - C: dorsal sepal. - D: lateral sepal. - E: petal. - F: lip, front view. - G: pedicel with ovary, lip and column, side view. - H: column, oblique view. - J: column, anther-cap detached, side view. - K: anther-cap, interior view. - L: two of four pollinia. All drawn from *Beaman* 11426 by Susanna Stuart-Smith. Scale: single bar = 1 mm; double bar = 1 cm.

unthickened lip margins. These minor differences, however, appear to fall within the natural variation found between populations on Mt. Kinabalu.

Vermeulen (pers. com.) points out that it is difficult to separate *B. coriaceum* from *B. unguiculatum* Rchb.f., and that both appear to grade smoothly into *B. apodum* Hook.f. via a series of intermediates. Similar plants also occur in Sumatra, Sulawesi and the Philippines. *B. apodum* seems only to differ in that it usually turns red when dried. *Bulbophyllum coriaceum*, which is so abundant on the upper slopes of Mt Kinabalu, may simply represent an alpine form of the complex having a small, sturdy growth-form and relatively large flowers.

DERIVATION OF NAME: The generic name is derived from the Greek *bulbos*, bulb, and *phyllon*, leaf, referring to the leaf-bearing pseudobulbs. The specific epithet is derived from the Latin *coriaceus*, leathery, referring to the texture of the thick, fleshy leaves.

18. BULBOPHYLLUM LEMNISCATOIDES Rolfe

Bulbophyllum lemniscatoides *Rolfe* in Gard. Chron., ser. 3, 8: 672 (1890). Type: Java, locality unknown, cult. *van Lansberge* (holotype K).

Bulbophyllum comosum auct., *non* Collett & Hemsl.

Epiphyte. Pseudobulbs 1.5–3 × 2 cm, ovoid, brownish-green, shiny, borne close together, bifoliate. **Leaves** (5–)8–12 × (0.5–)1–1.5 cm, linear-lanceolate, subacute, almost sessile, borne during the wettest season. **Inflorescences** borne from base of leafless pseudobulbs, densely many flowered (up to *c.* 30), flowers borne 1–2 mm apart; peduncle *c.* 13–30 cm long, terete, with a waxy surface, often dilated distally; non-floriferous bracts 2–3, 5–8 mm long, remote, clasping, acute; rachis 3–5 cm long, pendulous; floral bracts *c.* 2–3 mm long, narrowly elliptic, acuminate, pale brown. **Flowers** *c.* 3 mm in diameter, faintly scented, sepals very dark violet or purple with white hairs on reverse, greenish near base, appendages whitish banded pale mauve, lip crimson or purple, with almost black margins, column white, anther-cap yellow. **Pedicel** and **ovary** 2–2.5 mm long, hirsute. **Sepals** not opening widely, hirsute on reverse. **Dorsal sepal** 2 × 1.1–1.2 mm, ovate, obtuse, concave, versatile appendage 8–9 mm long, borne just below apex, filiform, minutely papillose. **Lateral sepals** 2.4–2.5 × 2 mm, obliquely triangular-ovate, obtuse, concave, appendage as dorsal sepal. **Petals** 1.9–2 × 0.5–0.6 mm, linear to narrowly triangular, acuminate, finely toothed, 1-nerved. **Lip** 2 × 1 mm, entire, oblong to elliptic-oblong, obtuse, fleshy, concave along centre. **Column** 1 mm long; foot 1.1–1.2 mm long; stelidia stout, acute; anther-cap minutely papillose. Plate 4B.

HABITAT AND ECOLOGY: Unknown.

DISTRIBUTION IN BORNEO: KALIMANTAN TIMUR: Sangkulirang area. SABAH: Crocker Range.

GENERAL DISTRIBUTION: Peninsular Malaysia, Thailand, Vietnam, Sumatra, Java and Borneo.

Figure 18. Bulbophyllum lemniscatoides Rolfe - A & B: habits. - C: floral bract. - D: flower, oblique view. -
E: flower, side view. - F: dorsal sepal. - G: lateral sepal. - H: petal. - J: lip, front view. - K: apex of sepaline
appendage with close-up of a portion. - L: pedicel with ovary, petals, lip and column, side view. - M: column
with anther-cap, oblique view. - N: column apex, anther-cap detached, front view. - O: anther-cap, front and
back views. - P: four pollinia above, one pair below. A–B drawn from *Comber* 1081 & 1226A (Javan material)
and C–P from *deVogel & Cribb* 992 by Susanna Stuart-Smith. Scale: single bar = 1 mm; double bar = 1 cm.

NOTES: *Bulbophyllum lemniscatoides* is the sole representative of section *Pleiophyllus* in Borneo. Most species are found on mainland Asia, especially in regions experiencing a seasonal climate. Seidenfaden (1979), for example, lists eleven species from Thailand alone. The flowers appear during the dry season by which time the leaves have fallen. It is curious, therefore, that *B. lemniscatoides* should have been found growing in non-seasonal Borneo.

Bulbophyllum lemniscatoides is unusual among the majority of Asiatic *Bulbophyllum* in having two-leaved pseudobulbs, a character otherwise widespread in species from Africa and Madagascar. The long sepaline appendages (palea) are distinctive and move with the slightest air current. The flowers recall those of *B. lemniscatum* C.S.P. Parish & Hook.f., a species from Myanmar and Thailand, hence the name. In *B. lemniscatum*, however, the appendages consist of six to ten lamellae of thin layers of rectangular cells radiating from a capillary axis.

DERIVATION OF NAME: The specific epithet refers to its resemblance to *B. lemniscatum*.

19. CALANTHE OTUHANICA C.L. Chan & T.J. Barkman

Calanthe otuhanica *C.L. Chan & T.J. Barkman* in Sandakania 9: 29, figs. 1–3 (1997). Type: Borneo, Sabah, Mt. Kinabalu, Kinabalu Lipson, 2550–2800 m, 15 July 1994, *Barkman* TJB 27 (holotype SAN, isotypes K, Kinabalu Park Herbarium, SAN, UNIMAS).

Terrestrial or **lithophyte**. **Roots** elongate, 2–3 mm in diameter, finely hairy. **Pseudobulbs** 1.4–1.6 cm long, with 3–4 nodes, ovoid, concealed by leaf-bases. **Leaf-blade** 14.5–43 × 2.9–5.9 cm, narrowly elliptic, acuminate, rather coriaceous, plicate, glabrous, suberect to arcuate; petiole 4.5–18.5 cm long, slender at base. **Inflorescences** terminal, erect, racemose, laxly 14- to 20-flowered, in total 83–110 cm long; peduncle 34–57 cm long, shortly pubescent; non-floriferous bracts 3–4, 2.9–3.1 mm long, lanceolate, acuminate; rachis shortly pubescent; floral bracts 11.5–18 mm long, narrowly oblong-linear, acuminate, persistent, dark green. **Flowers** strongly sweetly scented, sepals dark glossy green, petals creamy white, flushed green, lip white, spur lime-green, column and anther-cap white. **Pedicel** and **ovary** 2 cm long, finely pubescent. **Dorsal sepal** 7 × 15 mm, elliptic, obtuse, erect. **Lateral sepals** 7 × 15 mm, elliptic-ovate, acute, spreading. **Petals** 5 × 13 mm, elliptic, obtuse. **Lip** 11 × 11 mm, 3-lobed; side lobes 4.5 × 6.5 mm, linear-elliptic, rounded, erose, spreading; mid-lobe 10 × 7 mm, widening from a cuneate base into 2 rounded, erosulate lobules; disc with a complex basal callus, with abundant warty ornamentation; spur *c.* 9 mm long, clavate, retuse, weakly pubescent, parallel with ovary. **Column** *c.* 6 mm long, fleshy, shortly clavate; foot absent; anther-cap ovate, cucullate; pollinia 8, clavate, mealy. Plate 4D.

HABITAT AND ECOLOGY: Upper montane *Dacrydium/Leptospermum* scrub on old landslide areas on ultramafic substrate in full sun. Alt. 2500 to 2900 m. Flowering observed in January and July.

DISTRIBUTION IN BORNEO: SABAH: Mt. Kinabalu.

GENERAL DISTRIBUTION: Endemic to Borneo.

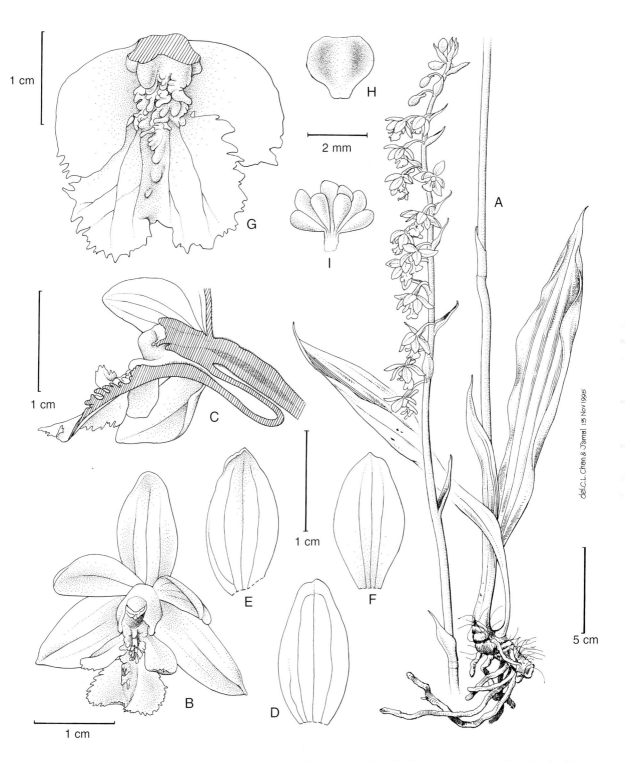

Figure 19. Calanthe otuhanica C.L. Chan & T.J. Barkman - A: habit. - B: flower, front view. - C: pedicel with ovary, lateral sepal, petal and column, longitudinal section. - D: dorsal sepal. - E: lateral sepal. - F: petal. - G: lip, flattened. - H: anther-cap, back view. - J: pollinia. All drawn from *Barkman* TJB 27 (holotype) by C.L. Chan.

NOTES: Few other species of *Calanthe* occur at comparable elevations on Mt. Kinabalu, although *C. transiens* J.J. Sm. can be found growing nearby. *Calanthe otuhanica* grows in association with *Coelogyne papillosa* Ridl., *C. rupicola* Carr, *Pholidota sigmatochilus* J.J. Sm. and the pitcher plant *Nepenthes villosa* Hook.f. Associated scrub trees are *Dacrydium gibbsiae* Stapf (Podocarpaceae) and *Leptospermum recurvum* Hook.f. (Myrtaceae).

DERIVATION OF NAME: The generic name is derived from the Greek *kalos*, beautiful, and *anthe*, flower, in praise of the beautiful flowers of many species. The specific epithet is derived from the local Dusun word *otuhan*, meaning a landslide, referring to the unusual habitat of this species.

20. CHEIROSTYLIS SPATHULATA J.J. Sm.

Cheirostylis spathulata J.J. Sm. in Bull. Jard. Bot. Buitenzorg, ser. 3, 9: 32, t. 5, II (1927). Type: Java, Surabaya, Mt. Kembangan near Gresik, 100 m, 31 Aug. 1923, *Dorgelo* 2072 (holotype L).

Terrestrial. Rhizome up to *c.* 5 cm long, fleshy. **Leaf-blade** 2–6 × 1–2.3 cm, obliquely narrowly elliptic, cuneate below, acute, thin-textured, reddish, sometimes flushed gold; petiole 1–2.5 cm long, sheathing at base. **Inflorescences** erect, 2- to 4(-7)-flowered; peduncle 8–12(–20) cm long, densely glandular pubescent; non-floriferous bracts 2 or 3, 1–1.8 cm long, tubular, acuminate, sparsely pilose; rachis 2–6 cm long, densely glandular-pubescent; floral bracts 0.8–1 cm long, ovate-elliptic, acuminate, concave, glandular-pubescent. **Flowers** with greenish sepals, petals white, lip white with 2 red spots on claw. **Pedicel** and **ovary** 5–8 mm long, densely glandular-pubescent. **Sepals** connate to form a tube, free at apex, pilose, 1-nerved. **Dorsal sepal** 9.5 mm long, free portion *c.* 5 × 3 mm, oblong-triangular, truncate to obtuse, concave. **Lateral sepals** 8.5 mm long, free portion *c.* 4.6 × 3 mm, obliquely oblong-triangular, obtuse. **Petals** *c.* 9.5 × 2.3 mm, obliquely narrowly elliptic from a narrow base, obtuse, glabrous, 1-nerved, connivent with sepals. **Lip** *c.* 1.35 cm long, porrect, spathulate, minutely papillose, 3-nerved, narrowly clawed; epichile blade *c.* 7.5 × 8 mm, deeply bilobed, cuneate-obovate, lobules *c.* 5.5 × 4 mm, obliquely obovate, rounded, irregularly toothed; claw bearing a 5- to 6-lobed, pectinate gland along each lateral nerve. **Column** *c.* 4.7 mm long; stelidia 2, porrect; stigmas 2, lateral under rostellum; anther-cap cordate, acuminate; pollinia 2, sectile. Plate 4E.

HABITAT AND ECOLOGY: Lowland forest, on limestone and sandstone. Alt. below 100 m? Flowering observed in July.

DISTRIBUTION IN BORNEO: SABAH: Mt. Sidungol. Tenom District, Kallang Waterfall.

GENERAL DISTRIBUTION: Thailand, Vietnam, Java and Borneo.

NOTES: The most characteristic feature of this genus, belonging to subtribe *Goodyerinae*, are the sepals which are united along half of their length to form a swollen tube. Bifid or pectinate, papillous glands are found in the lip hypochile claw, one each side of the sac along the lateral nerve.

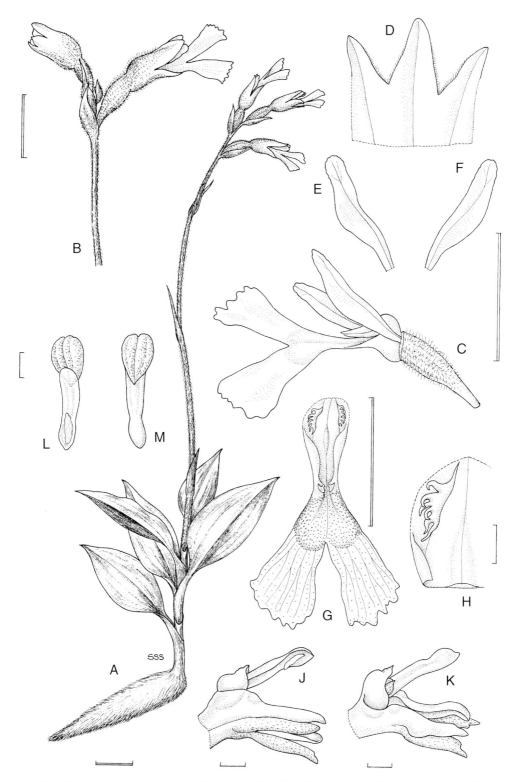

Figure 20. Cheirostylis spathulata J.J. Sm. - A: habit. - B: inflorescence. - C: flower, sepaline tube removed, side view. - D: sepaline tube, spread open. - E & F: petals. - G: lip, front view. - H: close-up of portion of hypochile showing pectinate gland. - J: column, anther raised to expose rostellum, side view. - K: column, anther raised to expose rostellum, oblique view. - L: pollinarium, back view. - M: pollinarium, front view. All drawn from *Keith* 9331 by Susanna Stuart-Smith. Scale: single bar = 1 mm; double bar = 1 cm.

Between 20 and 25 species are mainly centred on South East Asia, with others occurring in East Africa, north to China and Japan, and east to New Guinea and Vanuatu. A second species, *C. montana* Blume, is recorded from Borneo. This has much smaller flowers with three-lobed hypochile glands.

DERIVATION OF NAME: The generic name is derived from the Greek *cheir*, hand, and *stylis*, style, in reference to the anther bed (clinandrium) on the column which is said to bear a fanciful resemblance to a hand. The specific epithet is Latin for spatula-shaped and refers to the overall shape of the lip.

21. CLEISOSTOMA SUAVEOLENS Blume

Cleisostoma suaveolens *Blume*, Bijdr.: 363 (1825). Type: Java, Kambangan Island, *Blume* s.n. (holotype BO).

Cleisostoma longifolium Teijsm. & Binn. in Nat. Tijdschr. Ned. Ind. 5: 494 (1853). Type: Java, Salak, cult. *Bogor Botanic Gardens* (holotype BO).
Sarcanthus suaveolens (Blume) Rchb.f. in Bonplandia 5: 40 (1857).
Sarcanthus robustus O'Brien in Gard. Chron., ser. 3, 55: 21 (1914), *non* Schltr. Type: Borneo, cult. *Rothschild* (holotype not located).
Cleisostoma borneense J.J. Wood in Jermy, Studies Fl. G. Mulu Nat. Park: 17 (1984).

Epiphyte. **Stems** up to 40 cm or more long, *c.* 1 cm in diameter, tough, horizontal or pendulous, producing numerous elongate roots *c.* 3 mm in diameter. **Leaf-blade** 15–26 × 3–3.8 cm, ligulate, unequally obtusely bilobed with an acute mucro in the sinus, thick, rigid and fleshy; sheath 2–4.5 cm long, transversely rugulose. **Inflorescences** horizontal or stiffly pendulous, sometimes branching, branches often decurved, many flowered, flowers borne 3–7 mm apart; peduncle 4–22 cm long, *c.* 0.4 cm in diameter; non-floriferous bracts 2–3, remote, 6 mm long, ovate, obtuse, adpressed to peduncle; rachis 3–15 cm long, 0.4–0.5 cm in diameter, angular, branches 4–6 cm long; floral bracts 2–3 mm long, triangular-ovate, acute, concave. **Flowers** *c.* 1.5 cm in diameter, facing in all directions, dorsal sepal and petals dark red, lateral sepals dark red in upper half, yellow, flushed pale red in lower half, lip and spur dark rose-pink, column white, anther-cap white flushed yellow. **Pedicel** and **ovary** 6–7 mm long, sparsely ramentaceous. **Sepals** fleshy. **Dorsal sepal** 8 × 4.5 mm, oblong, obtuse, concave. **Lateral sepals** 8–8.1 × 5 mm, slightly obliquely oblong, shallowly concave, apex rounded and thickened inside. **Petals** 6 × 2.8–3 mm, narrowly elliptic, obtuse to acute, flat, fleshy, margins hyaline. **Lip** 1.1 cm long (including spur), 5 mm wide across base of mid-lobe, held horizontally, 3-lobed, fleshy; side lobes *c.* 2–2.1 mm long, broadly triangular with narrow, subulate, acuminate apex; mid-lobe *c.* 5 mm long, oblong, slightly auriculate at base, apex abruptly incurved, erose, apiculate; back wall callus narrowly oblong; spur *c.* 5 mm long, obtuse. **Column** 4–4.5 mm long, obtuse; foot 1.8–2 mm long; anther-cap 3 × *c.* 2.2 mm, cucullate, with acuminate projection; pollinia 4, appearing as 2 unequal masses. Plate 5A.

HABITAT AND ECOLOGY: Hill forest on limestone. Alt. 200 m. Flowering observed in May.

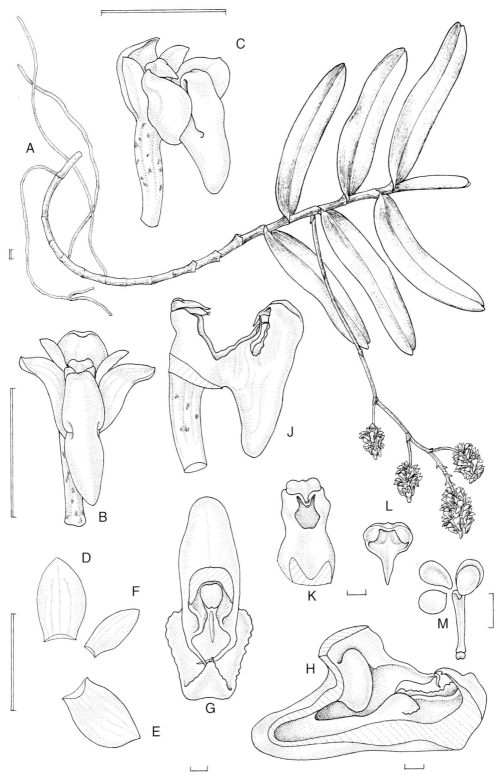

Figure 21. Cleisostoma suaveolens Blume - A: habit. - B: flower, front view. - C: flower, oblique view. - D: dorsal sepal. - E: lateral sepal. - F: petal. - G: lip, front view. - H: lip, longitudinal section. - J: pedicel with ovary, lip and column, side view. - K: column, front view. - L: anther-cap, interior view. - M: pollinarium. A drawn from *Lamb* 7 & B–M from material cultivated at *Tenom Orchid Centre* (TOC *805*) in *Wood s.n.* by Susanna Stuart-Smith. Scale: Single bar = 1mm; double bar = 1cm.

DISTRIBUTION IN BORNEO: SARAWAK: Gunung Mulu National Park; Bungo Range. SABAH: Tenom District, Padas Gorge; widespread.

GENERAL DISTRIBUTION: Sumatra, Java, Bali and Borneo.

NOTES: This is one of at least 22 species of *Cleisostoma* recorded from Borneo. *Cleisostoma discolor* Lindl. was figured in *Orchids of Borneo*, volume 1: fig. 17.

DERIVATION OF NAME: The generic name is derived from the Greek *kleis(t)os*, closed, and *stoma*, mouth, referring to the mouth of the inflated spur which is blocked by two calli. The Latin specific epithet *suaveolens*, means fragrant and refers to the sweetly scented flowers.

22. CORDIGLOTTIS FULGENS (Ridl.) Garay

Cordiglottis fulgens (*Ridl.*) *Garay* in Bot. Mus. Leafl. 23, 4: 176 (1972). Type: Locality uncertain, possibly Sumatra, cult. *Rauch*, Singapore (holotype SING).

Dendrocolla fulgens Ridl. in J. Linn. Soc., Bot. 32: 383 (1896).
Cheirorchis fulgens (Ridl.) Carr in Gard. Bull. Straits Settlem. 7: 46 (1932).

Epiphyte. Stem up to *c.* 10 cm long, slender, entirely enclosed by leaf sheaths. **Leaves** 4–5, erect to pendent; blade up to 30 cm long, terete, shallowly sulcate above, acute, erect to pendent, sometimes flecked or stained purplish; sheath 0.5–1 cm long. **Inflorescences** borne from base of leaf sheath opposite blade, with up to *c.* 8 flowers opening one at a time in succession; peduncle 3.5–7.5 cm long, slender, terete; non-floriferous bracts 2–3, remote; rachis 4–20(?) cm long, fleshy, 2–3 mm thick; floral bracts 0.3– *c.* 0.8 mm long, ovate, acute, closely spaced, quaquaversal, recurved. **Flowers** lasting one day, sepals and petals dark red, paler at margins, lip bright orange-yellow with reddish-purple stripes on side lobes and purple spots on claw, column orange-red, anther-cap yellow. **Pedicel** and **ovary** 0.5–1 cm long, slender. **Dorsal sepal** 0.8–1 × 0.3 cm, oblong-elliptic, subacute, 5-nerved. **Lateral sepals** 0.7 × 0.3–0.4 cm, oblong to ovate-elliptic, subacute or acute, 5-nerved. **Petals** 0.7 × 0.3–0.4 cm, spathulate, obtuse, 3-nerved. **Lip** 0.5–0.6 cm long, *c.* 0.6–0.7 cm wide when flattened, reniform, rounded, with a 1.5–1.6 mm long claw; side lobes obtuse, suberect to erect, remainder of blade gently decurved, glabrous; disc with a rounded, densely pubescent callus situated at junction of claw and side lobes, hairs very slender and narrowly clavate distally. **Column** *c.* 3.5 mm long; foot 2.3–2.5 mm long, gently curved; stigmatic cavity cordate, flat, margins slightly elevate; anther-cap ovate, cucullate; stipes broad, spathulate; pollinia 4, appearing as 2 unequal masses. **Fruit** up to *c.* 3.8 cm long slender. Plate 5B.

HABITAT AND ECOLOGY: Kerangas forest on podsolic sandstone soils. Alt. 400 to 500 m. Flowering observed in January.

DISTRIBUTION IN BORNEO: SABAH: Nabawan area.

GENERAL DISTRIBUTION: ? Sumatra and Borneo.

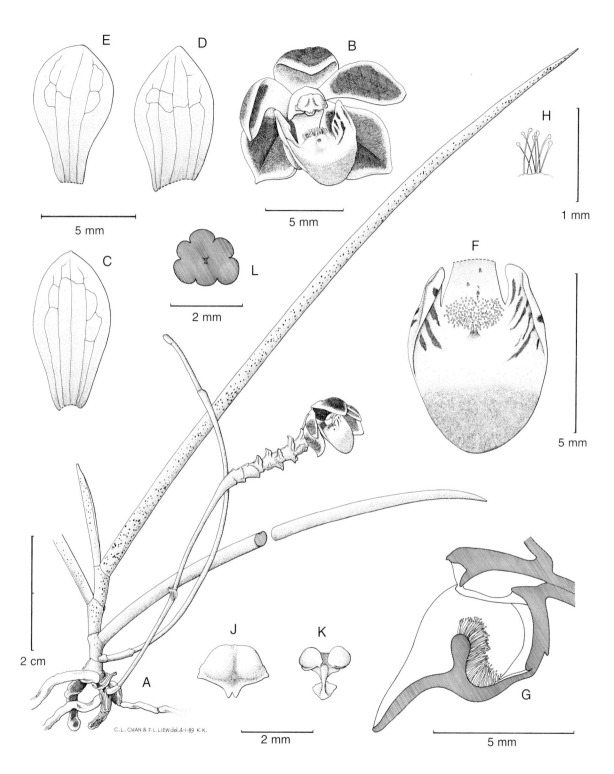

Figure 22. Cordiglottis fulgens (Ridl.) Garay - A: habit. - B: flower, oblique view. - C: dorsal sepal. - D: lateral sepal. - E: petal. - F: lip, front view. - G: lip and column, longitudinal section. - H: clavate hairs on lip. - J: anther-cap, back view. - K: pollinarium. - L: ovary, transverse section. All drawn from a plant cultivated at *Tenom Orchid Centre* by C.L. Chan and Liew Fui Ling.

NOTES: Carr (1932) established the genus *Cheirorchis* to which he referred five taxa including *C. fulgens* Ridl. Garay (1972) found Carr's *Cheirorchis* to be congeneric with *Cordiglottis*, however, which was established by J.J. Smith in 1922 and based upon the Sumatran *C. westenenkii* J.J. Sm. (see fig. 23). Garay added *C. multicolor* (Ridl.) Garay from Sarawak, originally placed by Ridley in *Dendrocolla* Blume, bringing the total of species to seven, five of which occur in Borneo.

The genus is clearly allied to *Thrixspermum*, but is distinguished by the clawed lip which is articulately attached to the prominent column-foot, longer column and terete or laterally compressed leaves.

Of the other species so far recorded from Borneo, *C. multicolor* and *C. westenenkii* also have terete leaves. The leaves of the remainder, viz. *C. major* (Carr) Garay and *C. pulverulenta* (Carr) Garay, are laterally flattened.

DERIVATION OF NAME: The generic name is derived from the Latin *cordis*, heart, and the Greek *glottis*, mouthpiece of a flute, in reference to the heart-shaped (cordate) lip of the type species. The Latin specific epithet *fulgens* means shining or brightly coloured and refers to the attractively coloured flowers of this species.

23. CORDIGLOTTIS WESTENENKII J.J. Sm.

Cordiglottis westenenkii *J.J. Sm.* in Bull. Jard. Bot. Buitenzorg, ser. 3, 5: 95 (1922). Types: Sumatra, Benkoelen, Lebong, Rimbo Pengadang, *Westenenk,* June 1916, *Ajoeb* 223 (syntype L); East coast, Besitan River, Bukit Kubu (Koeboe), *van Alderwerelt van Rosenburgh* (syntype L, sketch).

Epiphyte. Stem up to *c.* 2.5 cm long, internodes 0.4–0.7 cm long, terete. **Leaves** few, 3.5–17 × 0.15–0.18 cm, terete, slender, obtuse to acute, esulcate, curving; sheaths 0.175 cm in diameter, tubular, smooth. **Inflorescences** borne from base of leaf sheath, spreading, densely few- to many-flowered, opening one at a time in succession; peduncle 1–2 cm long, filiform, swollen below apex, glabrous; non-floriferous bracts 1 or 2, *c.* 1 mm long, ovate, acute, clasping, porrect; rachis 0.3–1.5 cm long, thickened, *c.* 1.6 mm wide; floral bracts 1.5–2 mm long, quaquaversal, triangular, concave, dorsally carinate, fleshy, spreading. **Flowers** *c.* 1.1 cm across, described as having red sepals and petals with a yellow midline, the sepals bordered yellow, lip yellow with brown spotting. **Pedicel** and **ovary** 5 mm long. **Dorsal sepal** 6–7 × 2–2.2 mm, oblong, narrowed at base, shortly acute, concave, recurved distally, 5-nerved. **Lateral sepals** 4.6–6.5 × 3–3.5 mm, obliquely oblong-triangular, obtuse, concave, recurved distally, 5-nerved. **Petals** 5–7 × 1.2–1.7 mm, narrowly subspathulate, slightly oblique, obtuse, recurved at middle, 3-nerved. **Lip** 5–7 mm long, 5.1–8.2 mm wide, claw 1–1.7 × 2–2.7 mm, cuneate-quadrangular, concave, blade *c.* 5.4 mm long, broadly subreniform-triangular, obtuse, basal lobes rounded, concave, main nerves 3, disc with a small, papillose, central transverse callus. **Column** 3–3.5 mm long; foot 2–2.5 mm long; anther-cap *c.* 2 mm wide, cucullate.

HABITAT AND ECOLOGY: Kerangas forest on ridges above rivers. Alt. 900 m. Flowering observed in October.

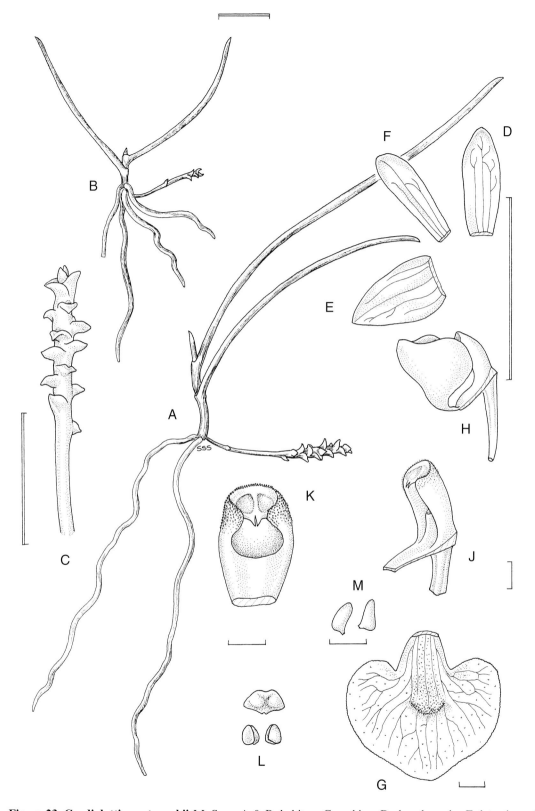

Figure 23. Cordiglottis westenenkii J.J. Sm. - A & B: habits. - C: rachis. - D: dorsal sepal. - E: lateral sepal. - F: petal. - G: lip, flattened, front view. - H: pedicel with ovary, lip and column, side view. - J: column, oblique view. - K: column apex, front view. - L: anther-cap and pollinia. - M: two pollinia. A–K & M drawn from *deVogel & Cribb* 9197 and L after *J.J. Smith* by Susanna Stuart-Smith. Scale: single bar = 1 mm; double bar = 1 cm.

DISTRIBUTION IN BORNEO: KALIMANTAN TIMUR: Apo Kayan, Mt. Sungai Pendan.

GENERAL DISTRIBUTION: Sumatra and Borneo.

NOTE: *Cordiglottis westenenkii* is the type of the genus and appears to be allied to *C. fulgens* from which it differs in having a much smaller, papillose callus.

DERIVATION OF NAME: The specific epithet honours L.C. Westenenk who collected part of the type material.

24. CORYBAS SERPENTINUS J. Dransf.

Corybas serpentinus *J. Dransf.* in Kew Bull. 41 (3): 604, fig. 11, plate 5B (1986). Type: Borneo, Sabah, Lahad Datu District, Bukit Silam, *J. & S. Dransfield* 5847 (holotype K).

Terrestrial. Tuber *c.* 5 mm in diameter, spherical, villous. **Stem** 10–15 mm long, subangular, with a basal sheath up to 3 × 1 mm. **Leaf** to 13 mm from origin to apex, 16 mm from basal auricles to apex, 14 mm wide at widest point, sessile, cordate, acuminate, ± flat or gently undulate at margins, auricles shallow, blade dark green with conspicuous shining white nerves, crystal bundles abundant, visible as pale spots. **Bract** to 5 × 1 mm, erect or reflexed, pale green. **Flower** sessile, to 15 mm high; dorsal sepal translucent whitish outside, marked dark crimson within, except for a narrow band at tip where pale and translucent, lateral sepals whitish, petals whitish, lip with a rich crimson throat patch ringed white, distal to this darker crimson, margin whitish, spurs white. **Dorsal sepal** to 20 mm long, 3 mm wide at base, 11 mm wide at apex, gradually widening and abruptly widened at the very apex, tip truncate with a central, triangular point to *c.* 1 mm long, 7-nerved. **Lateral sepals** *c.* 25 × 1.5 mm, thread-like, acute, ± erect. **Petals** 18–20 × 1.5 mm, thread-like, acute, ± horizontal. **Lip** erect in basal half, bent through ± 180° at middle, margins in basal half forming a tube with sepal claw, expanded in upper half, orbicular, ± 14 mm wide when flattened out; throat V-shaped, or slightly bulging but with a central furrow, surface smooth, margin irregularly very shallowly toothed; spurs 2, *c.* 1.5 × 1 mm. **Column** short, *c.* 2.5 mm high; stigma *c.* 0.75 mm wide; pollinia 2. Plate 5C & D.

HABITAT AND ECOLOGY: Lower montane forest, growing on thin soil and in moss carpets on ultramafic serpentine rock on ridge tops. Alt. *c.* 750 m. Flowering observed in November.

DISTRIBUTION IN BORNEO: SABAH: Lahad Datu District, Mount Silam; Mt. Tawai, Telupid.

GENERAL DISTRIBUTION: Endemic to Borneo.

NOTES: *Corybas serpentinus* is closely allied to *C. pictus* (Blume) Rchb.f. (see *Orchids of Borneo*, volume 1: fig. 20) but differs in the coloration of the lip and the two low swellings on the lip at the mouth of the throat, which is thus shallowly V-shaped rather than flat or concave.

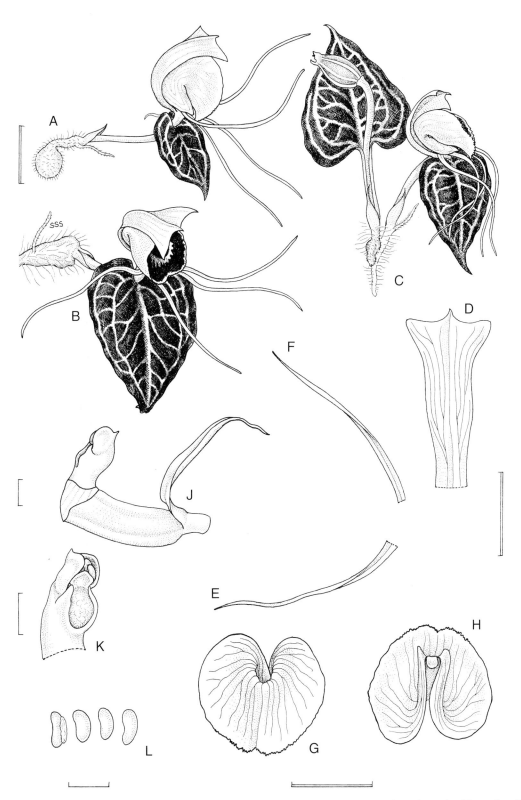

Figure 24. Corybas serpentinus J. Dransf. - A–C: habits. - D: dorsal sepal. - E: lateral sepal. - F: petal. - G: lip, front view. - H: lip, back view. - J: floral bract, ovary and column, side view. - K: column, oblique view. - L: pollinia. All drawn from *Dransfield* 5847 (holotype) by Susanna Stuart-Smith. Scale: single bar = 1 mm; double = 1 cm.

DERIVATION OF NAME: The generic name is derived from the Greek *korybas*, drunken man or, priest of Cybele, which describes the dorsal sepal which simulates a veiled drooping head, alluding to a priest's head-dress or to the nodding of a drunken man. The specific epithet refers to the serpentine rock substrate upon which the plant grows.

25. CYMBIDIUM KINABALUENSE K.M. Wong & C.L. Chan

Cymbidium kinabaluense *K.M. Wong & C.L. Chan* in Sandakania 2: 86 (1993). Type: Borneo, Sabah, Mt. Kinabalu, 1700 m, 20 January 1993, *Chan & Nais* s.n. (holotype SAN, isotype SNP).

Epiphyte. Pseudobulbs up to 5 × 2.5 cm, hidden by leaf bases and cataphylls. **Cataphylls** 3, persistent as fibrous sheaths. **Leaves** 9–12, up to 102 × 1.3–1.8 cm, linear to ligulate, acute, sheathing base articulated 9–10 cm from pseudobulb, with a slightly uneven to erose margin. **Inflorescences** borne from base of pseudobulb, pendulous, 78–83 cm long; peduncle bearing 9–10 imbricate, acute sheaths up to 17 cm long; rachis 31–38 cm long, with 21–23 flowers; floral bracts 3–4 mm long, ovate, acute. **Flowers** *c.* 3.5 cm across, pendulous to horizontal, pedicel and ovary dark olive-green, slightly flushed maroon brown near apex, slightly glossy, sepals and petals pale dull yellowish green, adaxially spotted and blotched rather uniformly throughout with maroon brown, dorsal sepal abaxially with maroon brown spots and streaks, especially in the lower 2/3, glossy at base, lateral sepals abaxially with maroon brown spots and streaks almost restricted to the longitudinal half nearer the dorsal sepal, petals abaxially with pale maroon brown spots, lip darker yellowish green with maroon brown bars, mid-lobe with pale maroon brown stripes, callus ridges yellowish green, column yellowish green flushed maroon brown at apex, sides and lower surface of base with a few maroon brown spots, anther-cap cream, whitish at base, stipes translucent cream, pollinia dark yellow. **Pedicel** and **ovary** 2.5–3.2 cm long, narrowly clavate. **Dorsal sepal** 3.1–3.3 × 1 cm, narrowly lanceolate-obovate, acute, concave, porrect, forming a hood over column. **Lateral sepals** 3.1–3.3 × 1 cm, narrowly lanceolate-obovate, falcate, acute, concave, spreading. **Petals** 2.9–3 × 0.4–0.5 cm, ligulate, acute, curved. **Lip** 3-lobed, adnate to base of column for 4–5 mm, forming a short sac; side lobes 6 mm wide, broadly triangular-falcate, erect, clasping column, glabrous, front margin S-shaped and minutely, irregularly dentate; mid-lobe 7.5–9 × 2–2.5 mm, ligulate, acute, glabrous, strongly recurved, disc with 2 subparallel ridges, not apically fused, glabrous. **Column** 1.8–2 cm long, S-shaped in profile, glabrous; rostellum reflexed; anther-cap ovate; pollinia 4, in 2 unequal pairs; viscidium narrow, rectangular, lacking hair-like processes from lower corners. Plate 5E & F.

HABITAT AND ECOLOGY: Epiphytic in shade amongst mosses near ground level in lower montane forest. Alt. 1700 m. Flowering observed in January.

DISTRIBUTION IN BORNEO: SABAH: Mt. Kinabalu.

GENERAL DISTRIBUTION: Endemic to Borneo.

NOTES: *Cymbidium kinabaluense* belongs to section *Cyperorchis* and is closely related to *C.*

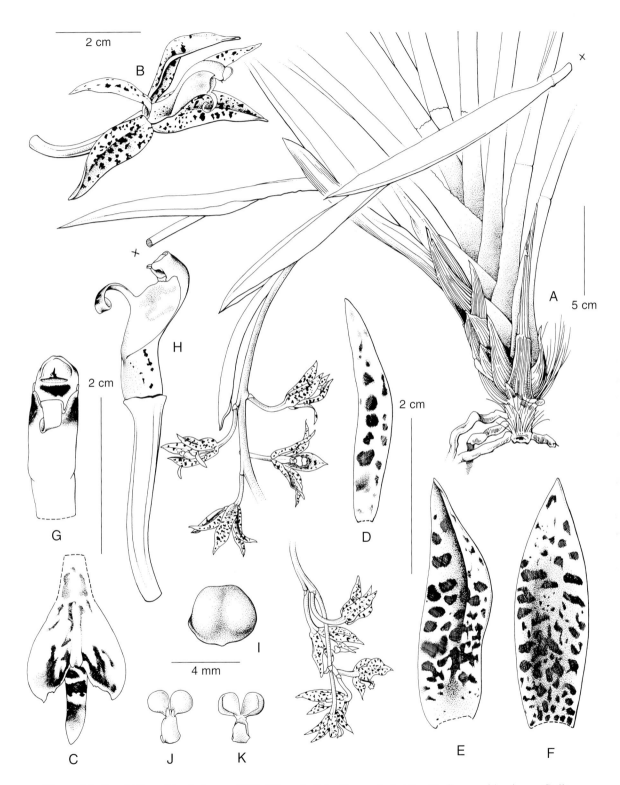

Figure 25. Cymbidium kinabaluense K.M. Wong & C.L. Chan - A: habit. - B: flower, side view. - C: lip, flattened, front view. - D: petal. E: lateral sepal. - F: dorsal sepal. - G: column, back view. - H: pedicel with ovary, lip and column, side view. - I: anther-cap, back view. - J: pollinarium, front view. - K: pollinarium, back view. All drawn from *Chan & Nais* s.n. (holotype) by Yong Ket Hyun after C.L. Chan and Liew Fui Ling.

sigmoideum J.J. Sm., a species from Java and Sumatra. It is distinguished from *C. sigmoideum* by its much longer inflorescence with many more larger flowers having a much shorter column.

DERIVATION OF NAME: Named after Mt. Kinabalu, the type locality.

26. CYSTORCHIS MACROPHYSA Schltr.

Cystorchis macrophysa *Schltr.* in Feddes Repert. 9: 429 (1911). Type: Borneo, Sarawak, Mt. Matang, Sept. 1866, *Beccari* 2636 (holotype B, destroyed, isotype FI).

Physurus glandulosus Lindl. in J. Linn. Soc., Bot. 1: 180 (1857). Type: Borneo, locality unknown, *Lobb* s.n. (holotype K-LINDL), *syn. nov.*
Erythrodes glandulosa (Lindl.) Ames in Merr., Bibl. Enum. Born. Pl.: 140 (1921), *syn. nov.*

Terrestrial. **Rhizome** elongate, decumbent, producing flexuous, villose roots at the nodes. **Stem** 6–13(–17) cm long, puberulus with glandular hairs, bearing 1 or 2 remote acuminate sheaths 0.8–1 cm long, pale brownish. **Leaves** mostly 5–8, distributed on lower and central portion of stem, spreading, lowermost often rather small; blade (1.2–)1.5–4.5 × 0.7–2 cm, obliquely elliptic, acute or shortly acuminate, rounded to cuneate at base, glabrous, purplish green or reddish purple, sometimes with darker patches; petiole 0.5–1.1 cm long, sheathing. **Inflorescence** densely few- to *c.* dozen-flowered, erect; rachis 2–3(–12) cm long; floral bracts 6–9 mm long, narrowly elliptic, acuminate, glandular-papillose, erect to spreading, bright red or brownish. **Flowers** erect to spreading, sepals and petals brownish red near base, yellowish distally, lip red near base, grading to white, then yellow at apex, spur cream to white. **Ovary** sessile, 4–6 mm long, cylindrical, glandular-papillose. **Sepals** glandular-papillose on exterior. **Dorsal sepal** 4–5.1 × 2 mm, narrowly elliptic, obtuse. **Lateral sepals** 7 mm long, 3 mm wide across swollen base, 2 mm wide above, oblong-ligulate, suborbicular and convex below, obtuse. **Petals** connivent with dorsal sepal, 4–5 × 1.2 mm, obliquely ligulate, broadest at base, narrowest at centre, obtuse, minutely papillose on distal exterior. **Lip** spurred; hypochile swollen at base into twin lateral sacs, 2 × 2 mm, sacs each containing a sessile verruculose gland; epichile 5 × 1–1.1 mm, ligulate, margins fleshy, involute, forming a tube, partially minutely papillose, expanded distally into a subreniform blade 0.5–0.9 × 0.9–1 mm; spur 1–2 mm long, conical, laterally compressed, obtuse to subacute. **Column** *c.* 1.7–1.8 mm long; pollinia 2.

HABITAT AND ECOLOGY: Lowland rainforest; mixed lowland dipterocarp forest on sandstone; alluvial forest; often found in damp areas near streams. Alt. sea level to 600 m. Flowering observed in February, March and June.

DISTRIBUTION IN BORNEO: SARAWAK: Kuching area; Mt. Matang; Gunung Mulu National Park.

GENERAL DISTRIBUTION: Endemic to Borneo.

NOTES: The genus *Cystorchis* Blume comprises around 20 species distributed from Thailand and Peninsular Malaysia through Borneo and the Indonesian Archipelago to the Philippines, New

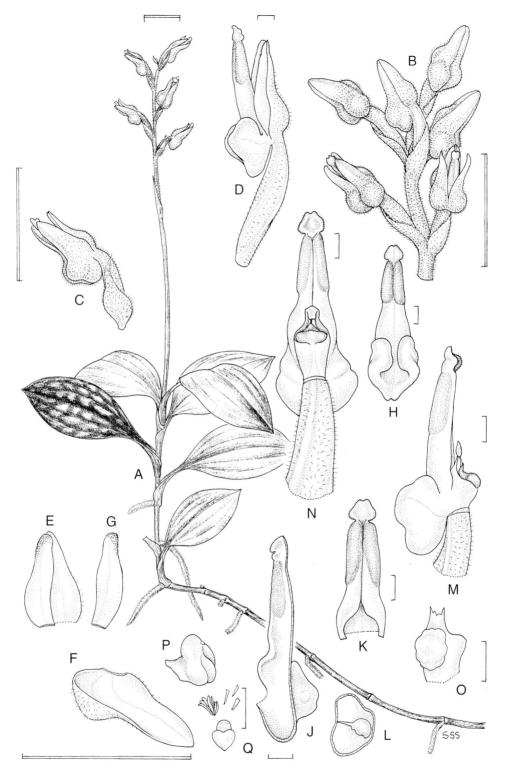

Figure 26. Cystorchis macrophysa Schltr. - A: habit. - B: inflorescence. - C: flower, side view. - D: pedicel with ovary, dorsal sepal, petals, lip and column, lateral sepals removed, side view. - E: dorsal sepal with connivent petal. - F: lateral sepal. - G: petal. - H: lip, front view. - J: lip, longitudinal section. - K: lip epichile, front view. - L: longitudinal section of hypochile sac showing gland. - M: portion of ovary, lip and column, side view. - N: portion of ovary, lip and column, viewed from above. - O: apex of column, oblique view. - P: anther-cap, side view. - Q: pollinarium (pollinia detached). All drawn from *Hansen* 341 by Susanna Stuart-Smith. Scale: single bar = 1 mm; double bar = 1 cm.

Guinea and Vanuatu. Of these, only four are saprophytic, viz. *C. aphylla* Ridl. (see *Orchids of Borneo*, volume 1: fig. 24), widespread in Thailand, Malaysia, Indonesia and the Philippines, *C. peliocaulos* Schltr. from Papua New Guinea, *C. saprophytica* J.J. Sm. from Sabah and Sarawak (see Fig. 25), and *C. salmoneus* J.J. Wood from Sabah (see Fig. 27).

The structure of the lip in *Cystorchis* is distinctive, the hypochile (basal portion) being either spurred or saccate and containing two or more stalkless globular glands. The rather fleshy, often wrinkled, epichile (apical portion) has an involute margin, forming a tube in some species, and is sometimes expanded at the tip. The closely related genus *Vrydagzynea* Blume, also recorded from Borneo (see Fig. 100), is autophytic and distinguished by the spur of the lip which contains two distinctly stalked glands.

Examination of the type material of *Physurus glandulosus*, deposited in the John Lindley Herbarium at Kew, shows it to be conspecific.

DERIVATION OF NAME: The generic name is derived from the Greek *kystis*, a bladder, and *orchis*, orchid, in reference to the inflated, bladder-like saccate base of the lip hypochile. The specific epithet is derived from the Greek *macro*, large, and *physo*, bladdery, in reference to the bladder-like hypochile sacs.

27. CYSTORCHIS SALMONEUS J.J. Wood

Cystorchis salmoneus *J.J. Wood* in Lindleyana 5, 2: 81, fig. 1 (1990). Type: Borneo, Sabah, Sipitang District, Ulu Long Pa Sia, 8 km north west of Long Pa Sia, 23 Oct. 1985, *Wood 625* (holotype K).

Glabrous, leafless **saprophyte**. **Rhizome** 6–7 mm thick, fleshy, shortly branched, bearing numerous imbricate scales. **Stem** 8 cm long, 0.8–0.9 cm thick, fleshy. **Sheaths** 1–1.5 × 0.8 cm, numerous, bract-like, ovate-elliptic, acute, imbricate, salmon-pink. **Inflorescence** densely many-flowered; rachis 2 cm long; floral bracts 2–2.5 cm long, 0.6–0.8 cm broad at base, narrowly elliptic, acuminate, salmon-pink. **Flowers** white. **Pedicel** and **ovary** 8 mm long. **Dorsal sepal** and **petals** connivent to form a hood. **Dorsal sepal** 9 × 2–3 mm, narrowly elliptic, obtuse to subacute. **Lateral sepals** 10 × 3 mm, obliquely narrowly oblong-elliptic, obtuse to subacute, entirely enclosing base of lip. **Petals** 8 × 2 mm, slightly obliquely narrowly-elliptic, acute. **Lip** 9 mm long; hypochile 3 mm long, margins erect, swollen at base into didymous lateral sacs each containing a sessile globular gland; epichile 6 × 2.5 mm, margins involute, fleshy and rugulose, apex obtuse, not expanded. **Column** 2.5 mm long; anther-cap 3 mm long, ovate, acuminate; pollinia 2, sectile. Plate 6A.

HABITAT AND ECOLOGY: Lower montane forest of *Agathis borneensis, Castanopsis, Lithocarpus,* etc., with a field layer of small rattans and gingers. Alt. 1200 to 1300 m. Flowering observed commencing in October.

DISTRIBUTION IN BORNEO: SABAH: Sipitang District, Long Pa Sia area.

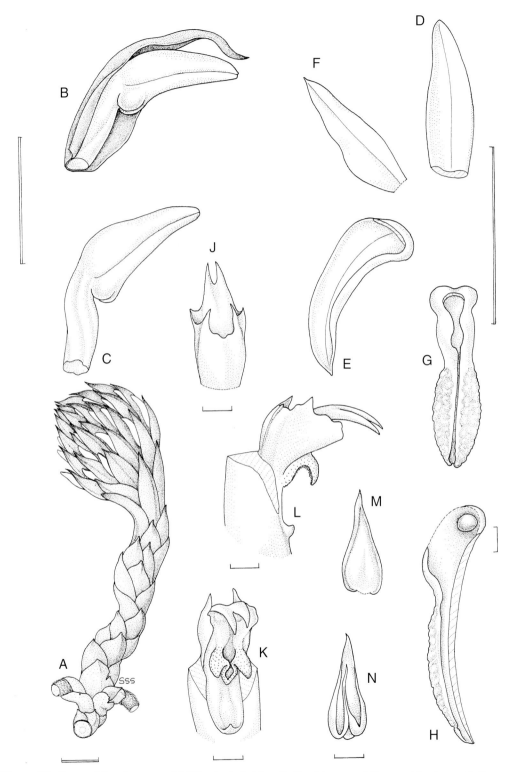

Figure 27. Cystorchis salmoneus J.J. Wood - A: habit. - B: floral bract and flower, side view. - C: flower, side view. - D: dorsal sepal. - E: lateral sepal. - F: petal. - G: lip, front view. - H: lip, longitudinal section. - J: column with anther-cap removed showing bifid rostellum, back view. - K: column with anther-cap removed, front view. - L: portion of ovary and column, side view. -M: anther-cap, back view. - N: anther-cap, interior view. All drawn from *Wood* 625 (holotype) by Susanna Stuart-Smith. Scale: single bar = 1 mm; double bar = 1 cm.

GENERAL DISTRIBUTION: Endemic to Borneo.

NOTES: *Cystorchis salmoneus* is known only from the type collection (*Wood* 625), a search of the type locality to locate further specimens having proved unsuccessful. *Wood* 625 is an immature plant with a partially expanded inflorescence and the measurements provided above should be regarded as only preliminary until specimens of fully mature plants are collected.

From the material at hand it is clear that *C. salmoneus* is quite distinct from the three other saprophytic species so far described. It differs from *C. aphylla* Ridl., (see *Orchids of Borneo*, volume 1: fig. 24), *C. peliocaulos* Schltr. and *C. saprophytica* J.J. Sm. by its much thicker, fleshier stems, salmon-pink floral bracts and pure white flowers. In *C. aphylla* and *C. saprophytica* the floral bracts are much shorter and the apex of the lip epichile is expanded to some degree. The floral bracts are longer and the lip is not expanded in *C. salmoneus*. The New Guinean *C. peliocaulos* Schltr. may also be distinguished from *C. salmoneus* by its undulate petals and lip epichile and the shorter, acute anther-cap.

28. CYSTORCHIS SAPROPHYTICA J.J. Sm.

Cystorchis saprophytica *J.J. Sm.* in Mitt. Inst. Allg. Bot. Hamburg 7: 23, t. 3, fig 13 (1927). Type: Borneo, Kalimantan, Mulu Hill, 1 December 1924, *Winkler* 465 (holotype HBG).

Leafless **saprophyte**. **Rhizome** 1.2–3.25 cm long, 3.5–5 mm in diameter, cylindrical, fleshy, with short irregular branches, villose. **Stem** *c.* 6–12 cm long, internodes 0.5–1.8(–2) cm long, glabrous, sometimes sparsely hirsute distally, whitish or pinkish white. **Sheaths** 4–5, 0.8–1.2 cm long, ovate, acuminate, adpressed, glabrous, whitish pink to pink with dark pink nerves. **Inflorescence** 1- to 3-flowered; rachis 1.3–1.7 cm long, pubescent, whitish or pinkish; floral bracts 8–11 mm long, triangular-ovate, acute to acuminate, concave, ciliolate, 3-nerved, adpressed, pink. **Flowers** 1–1.3 cm long, *c.* 5 mm across, pedicel and ovary orange, pink or brownish, sepals and petals white, or salmon-pink with a whitish base and translucent cream apex, lip red at base, central section translucent golden-yellow, apical lobe white or creamy yellow, reverse reddish distally grading to pink and off white. **Ovary** sessile, 8–13 mm long, fusiform, 3-ridged, glabrous. **Sepals** parallel proximally, distally divergent. **Dorsal sepal** 5–8 × 2.2–2.3 mm, subovate-lanceolate, obtuse, concave distally, apex densely papillose, 1-nerved. **Lateral sepals** 8–9(-10) × *c.* 2.6–2.7 mm, obliquely linear-lanceolate, obtuse, concave distally, basal portion concave and enclosing base of lip, margins incurved, 1-nerved. **Petals** 7 × 0.14 mm, adherent to dorsal sepal, obliquely linear, falcate, obtuse, 1-nerved. **Lip** 10–11 mm long, spurred, longitudinally sulcate below, 3-nerved; hypochile *c.* 3.6 × 2.8 mm, oblong, concave, partly adpressed to column, margins incurved above; mesochile *c.* 3.2 × 2–2.75 mm, oblong, margins incurved, contiguous, fleshy, rugose, pulvinate, papillose; epichile *c.* 2 × 2.4–3 mm, orbicular-obovate, truncate, often 5-angled, constricted and cuneate at base, usually folded in natural position; spur *c.* 2.2–3 mm long, composed of 2 lateral sacs each containing a small, sessile, globose gland. **Column** *c.* 4 mm long; stigma suborbicular; rostellum porrect, elongate; anther-cap ovate, acute; pollinia 2. Plate 6B.

HABITAT AND ECOLOGY: Leaf litter in lower montane ridge forest with *Agathis, Castanopsis, Lithocarpus,* etc., on sandstone; hill forest. Alt. *c.* 700 to 1400 m. Flowering observed in February, March, October and December.

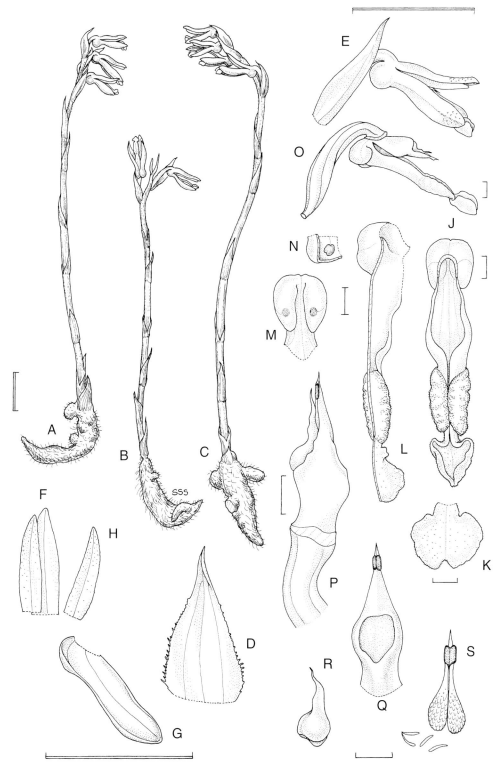

Figure 28. Cystorchis saprophytica J.J. Sm. - A–C: habits. - D: floral bract. - E: floral bract and flower, side view. - F: dorsal sepal with connivent petal. - G: lateral sepal. - H: petal. - J: lip, front view. - K: epichile opened out, front view. - L: lip, longitudinal section. - M: close-up of hypochile sac from below showing glands. - N: portion of hypochile sac cut away to show gland. - O: pedicel with ovary, lip and column, side view. - P: ovary and column, side view. - Q: column, front view. - R: anther-cap, side view. - S: pollinarium. All drawn from *Wood* 795 by Susanna Stuart-Smith. Scale: single bar = 1 mm; double bar = 1 cm.

DISTRIBUTION IN BORNEO: KALIMANTAN BARAT: Tilung Hill. KALIMANTAN (province uncertain): Bidang Menabai, Mehipit Hill. SABAH: Maliau Basin, Mt. Lotung; Sipitang District. SARAWAK: Limbang District, Buli Hill; Marudi District, Sekelun Hill, Pa' Lungan.

GENERAL DISTRIBUTION: Endemic to Borneo.

NOTE: *Cystorchis saprophytica* is easily distinguished from the other saprophytic species of *Cystorchis* by the expanded, orbicular-obovate, truncate lip epichile which resembles the genus *Zeuxine*.

DERIVATION OF NAME: The specific epithet records the saprophytic life-form of this species.

29. DENDROBIUM AURANTIFLAMMEUM J.J. Wood

Dendrobium aurantiflammeum *J.J. Wood* in Orchid Rev. 106 (1224): 340, figs 179 & 181 (1998). Type: Borneo, Sabah, Crocker Range, top of Kimanis road, *c.* 1400 m, August 1979, *Lamb* SAN 89699 (holotype K, isotype SAN).

Dendrobium cinnabarinum Rchb.f. var. *angustitepalum* auct., *non* Carr, 1935.

Clump-forming **epiphyte** with as many as 30 stems per plant. **Roots** numerous, elongate, branching. **Stems** up to 1 metre in length, erect at first, becoming pendent, tough, wiry, producing keikis, consisting of a narrow 0.8–4 cm long basal internode above which are 2–3 swollen, distinctly sulcate (deeply sulcate in dried material) pseudobulbous internodes each (1.5–)3.5–6 cm long, 0.5–1 cm in diameter, beyond which the normal, unswollen, branching stem continues, internodes 0.5–5 × 0.2–0.4 cm, at first enclosed in old pale brown leaf sheaths which soon disintegrate, olive-green to straw-yellow, often becoming dark purplish brown. **Leaf-blade** 6–10 × 0.3–0.6 cm, narrowly linear, narrowly obtuse and minutely mucronate, margins described as sometimes curled-up, apex minutely papillose, main nerves 3, with many fine secondary nerves, dark green; sheaths 0.5–5 cm long. **Inflorescences** 3–10 borne alternately at nodes of leafless distal portion of younger branches, flowers emerging successively from a tuft of scarious, pale brown 3 mm long bracts which soon become fibrous, one flower open at a time per inflorescence, up to 10 flowers open at once per stem and up to *c.* 100 flowers per plant. **Flowers** non-resupinate (always ?), unscented, pedicel and ovary pale pink, purple or orange, mentum vivid dark orange to scarlet, sepals and petals dark orange, nerves often scarlet, less often entirely crimson-scarlet and pale lilac at base, lip yellowish orange, pale orange or apricot, distal portion of mid-lobe dark orange to scarlet, keels scarlet, column yellow, anther-cap white; Sarawak plants described as having crimson-scarlet sepals and petals, pale lilac at base, lip deep crimson with white frilled edge, tip of column yellow. **Pedicel** and **ovary** (0.6–)1–1.1 cm long, narrowly clavate, very shallowly sigmoid. **Sepals** and **petals** not widely spreading. **Dorsal sepal** (2.35–)4.1–4.2(–5) × (0.45–)0.58–0.62 cm, narrowly oblong-ligulate, acute, main nerves 5, parallel. **Lateral sepals** (2.7–)4.6–4.8(–5) × (0.5–)0.6–0.75 cm, obliquely oblong-ligulate, subfalcate, acute, main nerves 5, parallel. **Mentum** (5.5–)6–8 mm long, obtuse. **Petals** (2.28–)4–4.1 × (0.47–)0.6–0.7 cm, narrowly elliptic to oblong-elliptic, acute, main nerves 3–5, divergent. **Lip** (2.2–)3.6–4 cm long, (6–)7–8 mm wide across side lobes; side lobes (6–)8–10 mm long, *c.* 2–3 mm high, erect, obliquely triangular, obtuse; mid-lobe (1.4–)2.5 cm long, 2.8–3 mm wide at widest point, linear-lanceolate, acute, margin strongly undulate, but less so and becoming entire distally; main nerves 3, and several secondary ones, 3 main nerves prominent and developed as smooth narrow, raised ridges or keels running between side lobes, becoming progressively obscure and terminating on basal portion of mid-lobe. **Column** (2.5–)3 mm long; foot (5.5–)6–8 × 2 mm; stelidia triangular, falcate, acuminate, margin rather uneven; anther connective triangular, acute; anther-cap *c.* 2 × 1.3–1.4 mm, rectangular, cucullate; pollinia 4. Plate 6D.

HABITAT AND ECOLOGY: Hill dipterocarp forest; lower montane mossy forest, often on ridges; 'sand forest'; mixed *Gymnostoma* and heath forest. Alt. 900 to 1400 m. Flowering observed in May, August, September and December.

DISTRIBUTION IN BORNEO: BRUNEI: Temburong District, Belalong Hill. SABAH: Crocker Range, Kimanis road; Sinsuron road, Mt. Alab; Mt. Kinabalu. SARAWAK: Belaga District, Mt. Dulit, Ulu Koyan.

GENERAL DISTRIBUTION: Endemic to Borneo.

NOTES: *Dendrobium aurantiflammeum* is one of the most attractive members of section *Crumenata* (formerly *Rhopalanthe*) native to Borneo, where it is most frequently encountered in the Crocker Range in Sabah. It is, for example, a welcome splash of colour among the vivid greens of the mossy lower montane forest which covers the higher ridges along the road that connects Kimanis on the west coast with Keningau in the interior.

The Sabahan collections of *D. aurantiflammeum* received at Kew during the 1980's were referred to *D. cinnabarinum* Rchb.f. var. *angustitepalum* Carr and have been determined thus in recent publications. Carr (1935) disagreed with Kraenzlin's (1910) views that *D. sanguineum* Rolfe was conspecific with *D. cinnabarinum*, saying that "while agreeing that the two plants cannot be kept separate as distinct species, there are at the same time such marked differences as to render it advantageous to regard the present plant as a distinct variety based upon the very narrow sepals and petals." *Dendrobium sanguineum* was described from a cultivated plant sent to Rolfe by the famous London nurserymen Messrs. Hugh Low & Co. It was said to have originated from the island of Labuan which is situated off the west coast of Sabah. The provenance seems doubtful since Labuan is low-lying and has no suitable montane habitat anywhere, and the plant was therefore most probably collected from a mountainous area somewhere on neighbouring Borneo. Carr considered a collection from Sarawak (*Richards* S.481) to be referable, albeit at varietal level, to Rolfe's *D. sanguineum* and established *D. cinnabarinum* var. *angustitepalum* which, according to the rules of botanical nomenclature, has to be based upon the same type as *D. sanguineum*. The single flower preserved from the original importation by Low from Labuan upon which *D. sanguineum* was based, contained in a packet on the type sheet at Kew, has, unfortunately, suffered damage sometime during the intervening years.

Rolfe, when establishing *D. sanguineum*, was evidently unaware that his specific epithet had already been used by Swartz in 1799 for a plant now referable to the Jamaican orchid *Broughtonia sanguinea* (Sw.) R. Br. *Dendrobium sanguineum* Rolfe is therefore a later homonym of *D. sanguineum* Sw. and a new name, *D. holttumianum*, was proposed for it by Hawkes & Heller in 1957.

The original description of *D. sanguineum*, prepared by Rolfe, clearly states that the sepals and petals are "crimson in front, passing towards the base to whitish, spotted and marbled with crimson; wholly crimson behind." This corresponds with certain colour forms of *D. cinnabarinum*

79

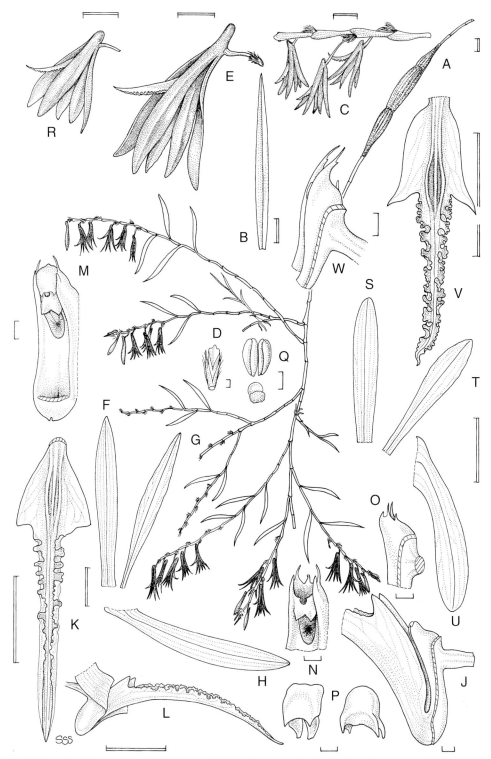

Figure 29. Dendrobium aurantiflammeum J.J. Wood - A: habit. - B: leaf. - C: portion of inflorescence. - D: floral bract. - E: flower, side view. - F: dorsal sepal. - G: petal. - H: lateral sepal. - J: pedicel with ovary, lower portion of lip, mentum and column, side view. - K: lip spread out, front view. - L: mentum and lip, side view. - M: column, oblique view. - N: apex of column, oblique view. - O: apex of column, side view. - P: anther-cap, back and side views. - Q: pollinarium (pollinia detached). - R: flower, side view. - S: dorsal sepal. - T: petal. - U: lateral sepal. - V: lip spread out, front view. - W: column, side view. A–C drawn from *Lamb* SAN 89699 (holotype), D–Q from *Dransfield* 5528 and R–W from *Richards* S. 481 by Susanna Stuart-Smith. Scale: single bar = 1 mm; double bar =1 cm.

(see Wood & Cribb, 1994, plate 7C; *Orchids of Borneo*, volume 3, page 241). In addition, the petals are described as spathulate-obovate, 11 lines (*c.* 9 mm) wide and the mid-lobe of the lip as "oblong, obtuse, undulate." Although the flowers of Carr's Sarawak plant and those of the type of *D. sanguineum* are similar, the petal width and lip mid-lobe morphology clearly are not. The petal width in the Sarawak plant is only 4.7 mm, while the mid-lobe of the lip tapers to a distinctly acute apex. The petals of the Sabah plants measure around 6 or 7 mm in width and the mid-lobe of the lip also tapers to an acute apex. The general lip and column morphology of both is identical, except that the lip side lobes are slightly narrower in the Sarawak collection. The only appreciable difference between the Sabah and Sarawak plants is in flower size and colour. The ground colour of the Sabah plants is predominantly orange, while the length of the sepals and petals is generally between 4.1 and 4.8 cm. The Richards collection from Sarawak is crimson- to scarlet-flowered and the sepals and petals measure only 2.28 to 2.7 cm in length. The single flower from Richards S.481, preserved in alcohol at Kew, may be immature.

Lamb, in a field sketch, describes a specimen from Mt. Alab in Sabah as having bright scarlet-red sepals and petals. All other Sabahan collections hitherto examined have bright orange flowers, albeit sometimes with darker red nerves. Flower colour is variable in the closely related *D. cinnabarinum* (see Fig. 31, Plate 7A–D) and similar variation no doubt occurs in some populations of *D. aurantiflammeum*.

DERIVATION OF NAME: The specific epithet is derived from the Latin *aurantiacus*, orange, and *flammeus*, flame-coloured, which aptly describes the flower colour.

30. DENDROBIUM BIFARIUM Lindl.

Dendrobium bifarium *Lindl.*, Gen. Sp. Orch. Pl.: 81 (1830). Type: Peninsular Malaysia, Penang, *Wallich* 2002 (holotype K-WALL).

Dendrobium excisum Lindl. in Bot. Reg. 27: 77, misc. 165 (1841). Type: Singapore, *Loddiges* 331 (holotype K-LINDL).
Callista bifaria (Lindl.) O. Kuntze, Rev. Gen. Pl. 2: 654 (1891).

Epiphyte. **Stems** *c.* 10–30(–40) cm long, internodes 0.5–1.27 cm long, forming an erect clump, often swollen at the very base, then thin before becoming thicker distally, entirely enclosed by leaf sheaths. **Leaves** distichous; blade 1–3.2 × 0.5–0.8 cm, narrowly elliptic or oblong-elliptic, unequally obtusely bilobed, largest leaves towards middle and distal part of stem; sheath 0.5–1.27 cm long. **Inflorescences** borne at nodes opposite leaf blades, few-flowered, one flower open at a time; peduncle and rachis *c.* 1 mm long, enclosed by 2–3, 2–4 mm long, pale brownish grey, acuminate, persistent bracts. **Flowers** *c.* 1.3 cm in diameter, rigid and fleshy, sweetly scented, white to yellowish green, lip greenish yellow, often with a central brownish ochre streak, and faint red lines at base, whole flower fading to orange. **Pedicel** and **ovary** 0.9–1.3 cm long, narrowly clavate. **Sepals** and **petals** somewhat recurved distally. **Dorsal sepal** 4.9–6.9 × 1.7–3.6 mm, narrowly oblong, acute. **Lateral sepals** 1.1–1.3 × 3.8–4 mm, obliquely triangular-ovate, acute, apiculate. **Mentum** 3–6 mm long, cylindrical, obtuse, held parallel to ovary. **Petals** 6–7 × 2.2–2.5 mm, narrowly linear-elliptic, acute, sometimes minutely erose. **Lip** 1.2–1.5 cm long, 5–5.1 mm wide across rudimentary side lobes, 6.5–7 mm wide across mid-lobe, spathulate in outline, very

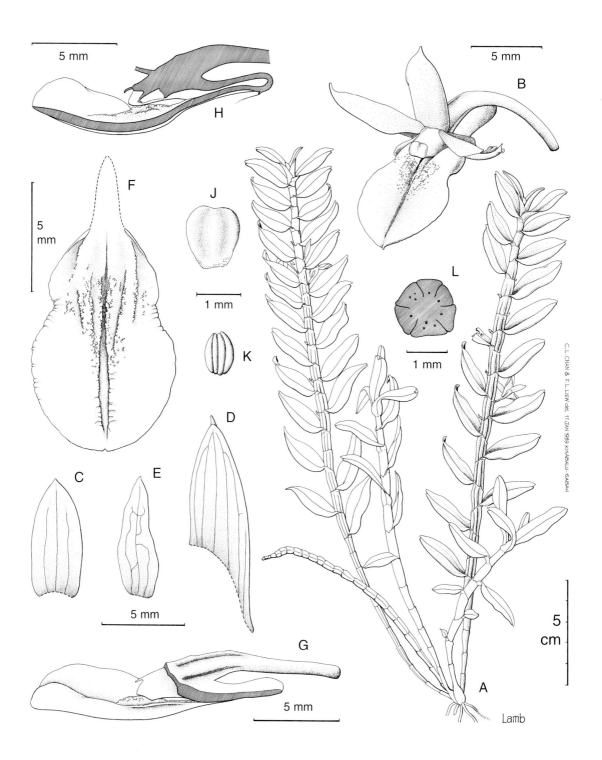

Figure 30. Dendrobium bifarium Lindl. - A: habit. - B: flower, oblique view. - C: dorsal sepal. - D: lateral sepal. - E: petal. - F: lip, front view. - G: pedicel with ovary, lip and column, side view. - H: pedicel with ovary, lip and column, longitudinal section. - J: anther-cap, back view. - K: pollinia. - L: ovary, transverse section. All drawn from *Vermeulen & Surat* s.n., cultivated at Tenom Orchid Centre by C.L. Chan and Liew Fui Ling

shallowly v-shaped at centre, claw gradually broadened within mentum and cuneate, then abruptly broadened into the lamina, with a rounded bulge on each side (rudimentary side lobes), apical portion almost circular, retuse, sometimes itself incised midway along margin; disc with 3 vague warty ridges and a large patch of irregular warts along most of lamina, outer 2 ridges usually more prominent. **Column** 2 mm long; anther-cap 1.5 × 1.1 mm, oblong, minutely papillose. Plate 6C.

HABITAT AND ECOLOGY: Lower montane forest; oak-laurel forest; Fagaceae, *Dacrydium*, *Leptospermum* forest on sandstone, ultramafic and dioritic substrates. Alt. 600 to 1800 m. Flowering observed throughout the year.

DISTRIBUTION IN BORNEO: KALIMANTAN SELATAN: Banjarmasin. SABAH: Mt. Kinabalu; Mt. Napotong; Mt. Trus Madi; Meliau River, Labuk; Pun Batu area. SARAWAK: Bau District, Buso. Kapit District, Batu Tiban. Kuching District, Bako; Mt. Matang. Marudi District, Mt. Dulit. Sri Aman District, Triso Protected Forest.

GENERAL DISTRIBUTION: Peninsular Malaysia, Singapore, Thailand, Sumatra, Borneo and Maluku (Ambon).

NOTES: *Dendrobium bifarium* belongs to section *Distichophyllum* and is obviously closely allied to *D. connatum* (Blume) Lindl., from Sumatra, Java and Borneo, with which it is easily confused. Comber (2001) distinguishes *D. connatum* by the lip having a linear claw and a lamina with a small fold or cleft along the margin each side, and an entire apex. The leaves of *D. connatum* are also generally slightly shorter and broader. The degree and alignment of the papillosity on the lip varies among many Bornean collections of this complex that I have seen. I am not totally convinced that such minor characters are sufficiently important or constant enough to maintain two species. The plant figured here was collected near Pun Batu in Sabah and appears, if two separate entities are to be recognised, to fall within the concept of *D. bifarium*.

DERIVATION OF NAME: The specific epithet is derived from the Latin *bifarius*, in two rows, in reference to the arrangement of the leaves on the stem.

31. DENDROBIUM CINNABARINUM Rchb.f.

Dendrobium cinnabarinum *Rchb.f.* in Gard. Chron. 14: 166 (1880). Type: Borneo, locality unknown, cult. *Messrs. J. Veitch & Sons* (holotype W; specimen likely to have come either from the same plant or the same importation as the holotype, ?isotype K).

Dendrobium sanguineum Rolfe in Gard. Chron. 18: 292 (1895), *non* Sw. Type: Borneo, Labuan, cult. *Messrs. H. Low & Co.* (holotype K).
D. holttumianum A.D. Hawkes & A.H. Heller in Lloydia 20 (2): 121 (1957).
D. cinnabarinum Rchb.f. var. *angustitepalum* Carr in Gard. Bull. Straits Settlem. 8: 103 (1935). Type: Borneo, Labuan, cult. *Messrs. H. Low & Co.* (holotype K).

Epiphyte, sometimes **terrestrial** in thick moss. **Habit** identical to *D. aurantiflammeum*. **Stems** up to 1 metre or more in length, lower 2-3 internodes swollen, each (1.5–)2–5 cm long, 0.4–0.8(–1) cm in diameter, unswollen internodes 0.2–0.3(–0.4) cm in diameter, similarly coloured to *D. aurantiflammeum*. **Leaf-blade** (3–)9–14 × (0.2–)0.6–0.8 cm, narrowly linear, obtuse to minutely unequally bilobed, main nerves 3, with many fine secondary nerves, dark green; sheaths

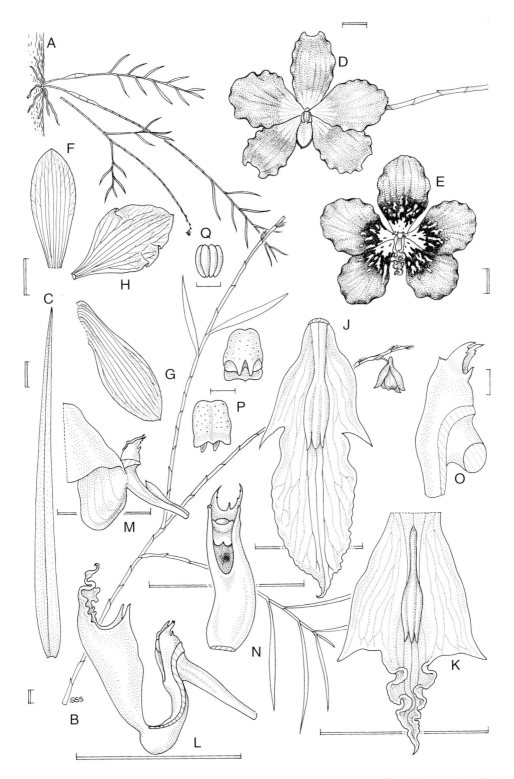

Figure 31. Dendrobium cinnabarinum Rchb.f. - A: habit (without scale). - B: flowering branch. - C: leaf. - D: flower, unspotted form, front view. - E: flower, spotted form. - F: dorsal sepal. - G: lateral sepal. - H: petal. - J: lip, form with large mid-lobe, front view. - K: lip, form with smaller, undulate mid-lobe, front view. - L: pedicel with ovary, lip and column, side view. - M: floral bract, pedicel with ovary, mentum and column, side view. - N: column, oblique view. O: base of pedicel with ovary and column, side view. P: anther-cap, front and back views. - Q: pollinia. A–C drawn from *Wood* 743, D, F–J & M–Q from *Wood* 812, E from a photograph by *R.S. Beaman*, and K–L from *Leche* 131 by Susanna Stuart-Smith. Scale: single bar = 1 mm; double bar = 1 cm.

(1.5–)3–5 cm long. **Inflorescences** as *D. aurantiflammeum*. **Flowers** resupinate or non-resupinate, wide-opening, thin-textured, fragile, unscented, colouring variously described as: sepals and petals glistening coral-red, white at base, mentum dark purple-nerved, lip very pale apricot, column white; sepals and petals deep pink, petals spotted deep crimson near base, lip pale orange with very deep crimson tip, basal ridge bright orange, column white; sepals and petals salmon-pink, glistening on outside, base of lateral sepals shiny deep purple on exterior, interior of sepals and petals paler, briefly lined with purple at base, side lobes of lip custard-yellow, mid-lobe blackish, callus yellow, with four reddish lamellae between callus and apex of mid-lobe, column white; sepals and petals rose-red fading to cream and finally white at centre, mentum striped dark purple, side lobes of lip white to cream, mid-lobe cream with a dark purple tip and three yellow keels; sepals and petals coral-red, white at base, mentum with dark purple nerves, lip very pale apricot, edged dark purple distally, callus apricot, column white; sepals and petals red to pink, base white with dark purple blotches, mentum white with dark purple bands, lip orange, darker orange at base, side lobes white, purple at base, mid-lobe orange with purple blotches, column white; sepals and petals deep pink, petals spotted deep crimson near base, lip pale orange with very deep crimson tip, basal ridge bright orange, column white. **Pedicel** and **ovary** 0.8–1.5 cm long, narrow, slightly curved. **Sepals** and **petals** spreading. **Dorsal sepal** (2.25–)3.3–4.1 × (0.45–)1.5–2.5 cm, oblong-obovate, spathulate, obtuse, main nerves 10–12. **Lateral sepals** (2.2–)3–4.1 × (0.55–)1.6–2.5 cm, obliquely oblong-elliptic, oblong-obovate, sometimes broadly elliptic, often distinctly clawed, obtuse, main nerves 10–12. **Mentum** 6–7 × 6–7 mm, obtuse. **Petals** (2.28–)3–4 × (0.47–)2–2.5 cm, oblong-spathulate to suborbicular, shortly clawed, obtuse, main nerves 6. **Lip** 2.2–2.5 cm long, 3 mm wide at base, 6–7 mm wide across side lobes, glabrous, rarely papillose at base; side lobes 0.8–1 cm long, 2–3 mm high, erect, oblong-triangular, narrowly caudate, shortly acute or uncinate; mid-lobe 1.4–1.5 × (0.28–)0.7–0.8 cm, oblong-ligulate to lanceolate, obtuse, acute or narrowly acuminate, strongly undulate-crisped, particularly distally, often appearing distorted, sometimes only slightly undulate; median nerve sometimes elevate; base of lip sometimes with 1 or 2 short keels on each side; disc between side lobes with a raised, oblong to oblong-elliptic, shallowly 3-ridged callus borne above base of lip and terminating on base of mid-lobe, distal portion thickest, apex of callus with 3 rounded teeth. **Column** 3(–4) mm long; foot 5–6 × 2 mm, sometimes keeled inside, a conical tubercle below apex sometimes present; stelidia narrowly triangular, falcate, acuminate, almost caudate, often recurved, sometimes denticulate; anther connective triangular, acute; rostellum bilobed; anther-cap *c.* 2 × 1.3–1.4 mm, rectangular, cucullate, shallowly tridentate, apex shallowly retuse. Plate 7A–D.

HABITAT AND ECOLOGY: Lower montane mossy ridge forest, with *Rhododendron* spp.; lower montane *Agathis/Gymnostoma/Podocarpus* forest on peaty soils over sandstone; mixed *Gymnostoma* heath forest near streams; margins of upland kerangas forest; kerangas forest with *Agathis*, *Gymnostoma* and *Podocarpus*, grading into lower montane forest rich in *Cyathea*, *Nepenthes* and numerous orchids ; terrestrial in peaty accumulations between limestone pinnacles; terrestrial in low, scrubby vegetation on humus covering limestone pinnacles; terrestrial in deep moss field layer in mossy forest; epiphytic on trees overhanging streams; often found in somewhat open areas. Alt. 900 to 1900 m. Flowering observed in February, from April to June, September and October.

DISTRIBUTION IN BORNEO: SABAH: Kinabatangan District, Maliau Basin; Sipitang District, Mt. Lumaku, Maligan area. SARAWAK: Mt. Dulit; Gunung Mulu National Park.

GENERAL DISTRIBUTION: Endemic to Borneo.

NOTES: It is always exciting to find this species whose large coral-red flowers glow among the subdued greens of the mossy forest. Although individual flowers open for a short period only, a large plant can look very decorative when several of the graceful bamboo-like stems are covered in blooms.

A wide range of variation in flower size, shape and colour exists. The sepals are typically oblong-obovate, but in *Hartley* S.518, collected on Mt. Dulit in Sarawak, the lateral sepals are broadly elliptic and distinctly clawed, while the petals are suborbicular in outline. The degree of undulation of the lip mid-lobe varies considerably from only slightly to markedly so and appearing distorted. The triangular claw of the mid-lobe varies in length and the shape of the blade may be anything from oblong to narrowly elliptic. The width and length of the acuminate apex of the side lobes varies also.

The intensity and distribution of colouring of the flower also varies a great deal. One of the most attractive variants was photographed in the Maliau Basin in Sabah. Here the sepals and petals are blotched with deep purple-red and could almost be mistaken for a species of *Hibiscus*.

DERIVATION OF NAME: The specific epithet is Latin for vermilion-red which refers to one of the many hues of colour found in the flowers.

32. DENDROBIUM DEAREI Rchb.f.

Dendrobium dearei *Rchb.f.* in Gard. Chron., new ser., 18: 361 (1882). Types: Philippines, *Colonel Deare* s.n. (syntype W); cult. *H. Low & Co.* (syntype W); *Veitch* s.n. (syntype W).

Dendrobium ovipostoriferum auct., *non* J.J. Sm.: Wood & Cribb, Checklist Orchids Borneo: Plate 9A (1994).

Epiphyte. **Stems** clustered, up to 1.5 m long, 1–1.5 cm in diameter, internodes 1.5–3 cm long, leafy throughout, enclosed by leaf sheaths. **Leaves** (5–)6.5–18 × 1.5–3.8 cm, oblong-elliptic to narrowly elliptic, very shortly obtusely or subacutely unequally bilobed, glabrous, thinly coriaceous, main nerves 7–8, pale green. **Inflorescences** emerging from base of leaf sheaths opposite blade, pendulous, glabrous, subdensely 10- to 26- flowered; peduncle 2–5 cm long; non-floriferous bracts 2–3 at base, and 1 in middle, 1–3 mm long, ovate, acute; rachis (2–) 20 cm long; floral bracts 2–4 mm long, triangular-ovate, acute to acuminate. **Flowers** with a rather unpleasant musty odour, up to 5.5 cm across, white, lower part of lip including side lobes and base of mid-lobe yellowish green to yellowish, apex of mentum flushed greenish, column flushed greenish, anther-cap green. **Pedicel** and **ovary** 4.5–5.5 cm long, narrowly clavate, strongly triquetrous, wings somewhat undulate, gently curved, white. **Sepals** and **petals** spreading, somewhat recurved, sepals stiff-textured. **Dorsal sepal** 2.5–3.5 × 0.7–1cm, triangular-ovate, acuminate, dorsally strongly carinate, keel confluent with pedicel and ovary and undulate towards base, 6-nerved. **Lateral sepals** (4–) 4.5–4.6 cm long, free portion 2.5–2.6 cm long, *c.* 1.3–1.4 cm wide at base, obliquely triangular–ovate, acuminate, dorsally strongly carinate, keel confluent with pedicel and ovary and undulate towards base, 6- to 7-nerved. **Mentum** 1.5–2 cm long, (1–) 4–5 mm wide at base, narrowly conical, infundibuliform, obtuse, almost straight, parallel and in line with lip. **Petals** 2.8–3.9 × 2–2.3 cm, oblique, broadly elliptic or oblong-elliptic, obtuse, apiculate, somewhat

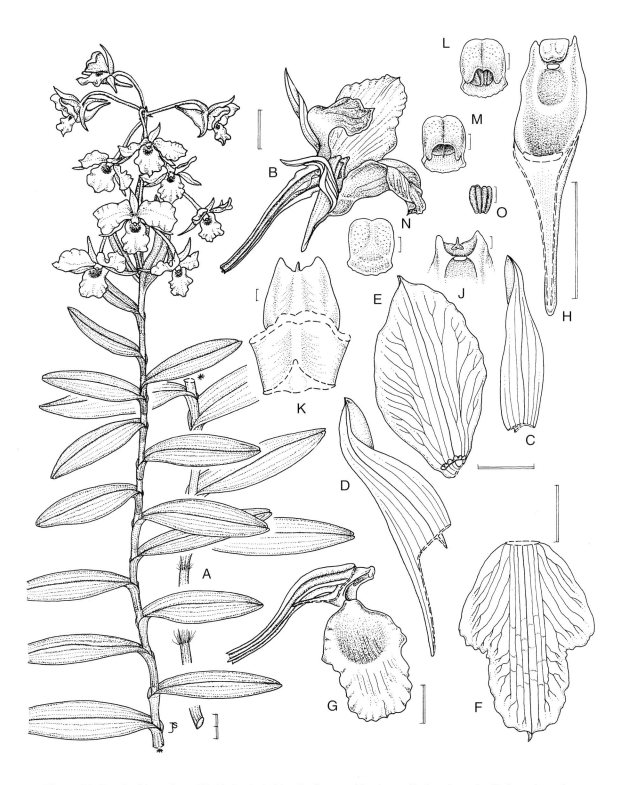

Figure 32. Dendrobium dearei Rchb.f. - A: habit.- B: flower, side view. - C: dorsal sepal. - D: lateral sepal. - E: petal. - F: lip, front view. - G: pedicel with ovary, lip and column, side view, lip spread out. - H: column, front view. - J: apex of column, front view. - K: apex of column, back view. - L: anther-cap with pollinia, front view. - M: anther-cap, pollinia removed, front view. - N: anther-cap, back view. - O: pollinia. All drawn from *Lamb* AL1375/91 by Judi Stone. Scale: single bar = 1 mm; double bar = 1 cm.

undulate, papillose at base, thin-textured, main nerves 5. **Lip** shallowly 3-lobed, main nerves 5–6, 3–3.9 cm long, 1.5–2.2 cm wide across side lobes, claw 1.5–2 cm long, hairy; side lobes 1.7–2.2 × 0.4 cm long, obliquely rounded, sometimes slightly retuse, erect; mid-lobe 1.3–1.6 × 1.3–1.6 cm, oblong, rectangular to square, truncate, retuse to emarginate, apiculate, margin irregular, often dentate; disc smooth, keels absent. **Column** 5 × 5–6 mm; foot 5–7 mm long, 7 mm wide at entrance to mentum, wings obtusely rounded to triangular, subacute, margin papillose; anther-cap 4 × 3 mm, oblong, cucullate, apex truncate, minutely ciliate, ventral surface sulcate. Plate 7E.

HABITAT AND ECOLOGY: Coastal forest. Alt. near sea level. Flowering recorded from May until January.

DISTRIBUTION IN BORNEO: SABAH: Dent Peninsula, Tambisan area, offshore islands off Sandakan.

GENERAL DISTRIBUTION: Borneo and the Philippines.

NOTES: The plant figured here won best species First Prize and Best Malaysian Species at the 1991 Sandakan Orchid Show. It was referred to *D. ovipostoriferum* J.J. Sm. until quite recently when, after a closer study, it was shown to be a showy form of *D. dearei*, a related species originally described from the Philippines. The provenance of the Bornean material is thought to be from along the coast of the Dent Peninsula, south east of Sandakan. This would seem logical since this part of Sabah is not too distant from know localities in the southern Philippines.

Dendrobium dearei is one of the members of section *Formosae* lacking the characteristic covering of blackish hairs on the leaves and inflorescences. The Philippine islands are the home of some of the most beautiful species in this section, the showiest of which is probably the flamboyant *D. sanderae* Rolfe. *Dendrobium dearei*, although belonging to the same alliance, appears more closely related to *D. schuetzei*, another Philippine species described by Rolfe. This can be distinguished from *D. dearei* by its much broader petals and lip, and shorter, broader mentum.

O'Byrne (2000) suspects that the Bornean form of *D. dearei* may be a polyploid. It differs from the typical Philippine plant in having much taller stems up to at least 110 cm long, which are leafy throughout their entire length. The leaves may attain 18 cm in length. The pendulous inflorescences are typically 15 to 25 cm long and bear up to 26 flowers that are described as smelling of "wet socks"! Each flower is about 5 cm across and has an almost square to rectangular lip mid-lobe, often with an irregularly toothed margin.

Dendrobium ovipostoriferum is a closely allied plant described from Kalimantan by J.J. Smith. It differs from *D. dearei* in having much smaller stems up to only 22 cm long, shorter (up to 4 cm), black-hairy leaves and few-flowered, or possibly solitary, inflorescences. The size, shape, and length to width ratio of the floral segments also differs. The flowers have a gold spot at the centre of the lip which has a crenulate margin to the mid-lobe. The basal claw bears five longitudinal ribs, which are lacking in *D. dearei*. The spur-shaped mentum is aligned parallel to but not in a straight line with the lip as in *D. dearei*. The little known *D. takahashii* Carr, also from Kalimantan, may prove to be conspecific with *D. ovipostoriferum*.

DERIVATION OF NAME: The specific epithet honours a Colonel Deare from whom H.G. Reichenbach first received a petal and a couple of dried whole flowers.

33. DENDROBIUM DERRYI Ridl.

Dendrobium derryi *Ridl.*, Mat. Fl. Mal. Penins. 1: 52 (1907). Type: Peninsular Malaysia, Larut Hills, *Derry* s.n. (holotype SING).

Dendrobium cinereum J.J. Sm. in Bull. Jard. Bot. Buitenzorg, ser. 3, 2: 78 (1920). Types: Borneo, Kalimantan, locality unknown, *Nieuwenhuis expedition 1897*, cult. Bogor no. 1386 (syntype L); Sungai Merase, Nieuwenhuis expedition 1899, *Sakeran* s.n., cult. Bogor no. 1821 (syntype L).

Dendrobium groeneveldtii J.J. Sm. in Bull. Jard. Bot. Buitenzorg, ser, 3, 2: 79 (1920). Types: Sumatra, Ophir District, Taloe, *Groeneveldt*, cult. under no. 393 (syntype L); Agam, Bukit Batu Banting, *Groeneveldt,* cult. under no. 1783 (syntype L).

Epiphyte. **Stem** 40–80(–100) cm long, internodes 1.8–3.5 cm long, *c.* 0.5–0.8 cm in diameter, subterete, longitudinally sulcate, glossy, dingy green, pendulous. **Leaves** shiny ash-grey or pearl-grey above, dark violet below, sheaths dirty green, suffused violet, with greenish violet nerves; blade 7.5–10(–12) × 2–3.5 cm, oblong to narrowly elliptic, apex suboblique, acute, thinly textured, mid-nerve prominent on abaxial surface; sheath 3–3.5 cm long, tubular, apex truncate. **Inflorescences** mostly borne from the nodes of leafless stems, 1- to 2(–3)-flowered, sessile; non-floriferous bracts 3, 2–3 mm long, ovate, apiculate; rachis 3 mm long, fractiflex, olive-green or dirty violet; floral bracts 2–3 mm long, triangular, acute, pale green. **Flowers** 2.7–3 cm across, scented of *Dianthus caryophyllus* L. (carnation, clove pink) according to J.J. Smith, hardly discernible according to Lamb (annotation on unpublished sketch), pedicel and ovary pink or greenish mauve, sepals and petals ivory coloured, palest greenish white or palest greenish yellow, lip white, sometimes pale greenish lemon at centre, column creamy white to yellowish, with orange apical auricles, anther-cap white. **Pedicel** and **ovary** 1.5 cm long. **Dorsal sepal** 1.1–1.5 × 0.55–0.85 cm, oblong, obtuse, erect, 7-nerved. **Lateral sepals**: posterior margin 1.2–1.5 cm long, anterior margin 1.8–2.5 cm long, 0.6–0.8 cm wide, obliquely triangular-oblong, obtuse, minutely apiculate, 7-nerved. **Mentum** 1.2–1.4 cm long, subcylindrical, obtuse, slightly laterally compressed. **Petals** 1.1–1.2 × 0.45–0.6 cm, oblong or spathulate, obtuse, minutely erose, reflexed. **Lip** 2.3–2.9 cm (total length), blade usually *c.* 1.5 × 0.8–1.2 cm, spathulate, broadly rounded, bilobed, with a minute mucro in the sinus, recurved, strongly convex, margin minutely erose, undulate, somewhat revolute; claw *c.* 1–1.4 × 0.2–0.35 cm, linear, canaliculate; disc of blade with 2 longitudinal low, raised central keels separated by a groove; disc of claw with 2 tall, thin, parallel lamellae extending from below middle, becoming divergent and uniting with basal margins of blade. **Column** 3.5–3.7 × 2.9–3.5 mm, auricles obliquely quadrangular, obtuse; foot 1.2–1.4 cm long, canaliculate; anther-cap *c.* 2 × 2–2.1 mm, cucullate, quadrangular. Plate 8A–C.

HABITAT AND ECOLOGY: Lower montane forest; riparian forest. Alt. 1400 to 1500 m. Flowering observed from October to December.

DISTRIBUTION IN BORNEO: KALIMANTAN TIMUR: Apokayan. SABAH: Crocker Range. SARAWAK: Kapit District, Hose Mountains. Kuching District, Mt. Serapi. Limbang District, Mt. Pagon. Marudi District, Bario/Pa Berang.

GENERAL DISTRIBUTION: Peninsular Malaysia, Sumatra and Borneo.

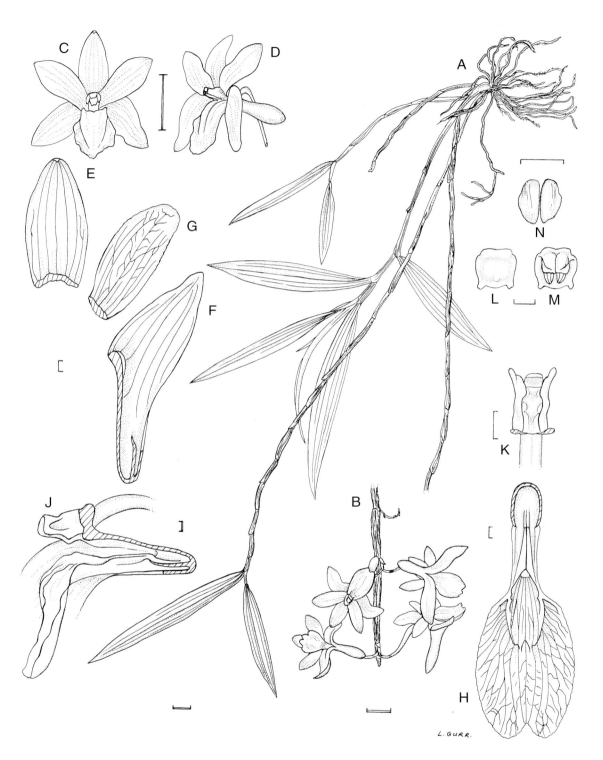

Figure 33. Dendrobium derryi Ridl. - A: habit. - B: inflorescence. - C: flower, front view. - D: flower, oblique view. - E: dorsal sepal. F: lateral sepal. - G: petal. - H: lip, flattened, front view. - J: pedicel with ovary, lip and column, partial longitudinal section. - K: column, front view. - L: anther-cap, back view. - M: anther-cap, interior view showing pollinia. - N: pollinia. A drawn from *Beaman* 7929, B from *Lamb s.n.*, and C–N from *deVogel & Cribb* 561 by Linda Gurr. Scale: single bar = 1 mm; double bar = 1 cm.

NOTES: *Dendrobium derryi* Ridl. belongs to section *Calcarifera* which is represented by at least 21 species in Borneo. Ridley (1907) comments that the flowers of Peninsular Malaysian material of this species are variable in size. Subsequently, Smith (1920) described *D. groeneveldtii* from Sumatra distinguishing it from *D. derryi* by its ovate sepals, much shorter mentum, erose petals and two-lobed lip. He also described *D. cinereum* from Borneo in the same publication, alongside *D. groeneveldtii*, but compared it with *D. hymenopterum* Hook.f. and *D. calicopis* Ridl. Recent investigations by Schuiteman at Leiden (pers. comm.) have shown both to fall within the variation of *D. derryi*. The individual flowers are very attractive and it is a pity that they are not produced in greater abundance.

DERIVATION OF NAME: The specific epithet refers to a Mr Derry who collected the type material.

34. DENDROBIUM HALLIERI J.J. Sm.

Dendrobium hallieri *J.J. Sm.* in Bull. Jard. Bot. Buitenzorg, ser. 2, 8: 40 (1912). Type: Borneo, Kalimantan Barat, Kelam Hill, 1894, *Hallier* 2313 (holotype BO, isotype K).

Epiphyte. **Stems** 30–45 cm long, 0.5–1.1 cm in diameter, rigid, flexuose, somewhat fractiflex, particularly distally, sulcate, erect, internodes 3.5–4 cm long, leafy. **Leaves** 4–9(–12.5) × 1.6–3.5 cm, oblong-elliptic, oblong-ovate or oblong, shortly contracted and sulcate at base, apex obliquely semi-obtuse, median nerve prominent on reverse, sulcate above, surface brown-hirsute above and below, upper surface of mature leaves becoming glabrous, sheaths brown-hirsute, becoming glabrous. **Inflorescences** erect, abbreviated, emerging on upper portion of stem from base of leaf sheaths opposite the blades, 2- to 7-flowered; peduncle and rachis 0.8–2.6 cm long, flexuose, glabrous; floral bracts (1.5–)1.7–1.8 × 0.5–0.8 cm, oblong-ovate, lowermost ovate, acute to subulate-acuminate, concave, 9-nerved, nigrohirsute on reverse. **Flowers** up to *c.* 2.5 cm across, stiff and waxy, pale primrose-yellow to greenish white, lip creamy white with rows of orange or scarlet warts on mid-lobe and red streaks between keels in throat, column yellow, foot orange-yellow, stigma sometimes flushed red, anther-cap yellow and white, pedicel and ovary yellowish, pale green below, sweetly scented. **Pedicel** and **ovary** 4–5.2 cm long, slender, narrowly clavate, glabrous. **Sepals** recurved. **Dorsal sepal** 1.6–1.7 × 0.75–0.9 cm, oblong, apex narrowly obtuse, recurved, 5- to 7-nerved. **Lateral sepals** posterior margin 1.7–2.8 cm long, anterior margin (2.8–)3.1–3.2 cm long, 1.3–1.8 cm wide at base, *c.* 0.2 cm wide at apex, obliquely triangular-ovate, revolute, apex acute to acuminate and dorsally carinate, 7-nerved. **Mentum** (2.3–)2.8–2.9 cm long, narrowly conical, ovipositor-shaped, narrowly obtuse, descending, slightly recurved. **Petals** 1.8–2 cm long, 1.1–1.4 cm wide at base, obliquely ovate, somewhat contracted distally, ± obtuse, undulate, shortly decurrent with column-foot, projecting forward, main nerves 5. **Lip** shallowly 3-lobed, *c.* 2.3–2.6 cm long, 1.6 cm wide measuring across side lobes, mid-lobe *c.* 0.6–0.7 × 1.1–1.2 cm; claw 5–9 mm long, narrowly linear; upper and lower surface verruculose, except around central area; main nerves 5, verruculose; side lobes broadly obliquely rounded; mid-lobe broadly semi-orbicular, rounded, undulate; disc with a complex arrangement of low, indistinct keels along median line, 2 keels start at tip of claw and become broader, uniting in middle to form a broad, low ridge, 2 low narrow keels emerge from this ridge and extend forwards to base of mid-lobe. **Column** 5 × 5 mm; foot *c.* 2.9 cm long, proximal portion 1.6 cm long, concave, cymbiform;

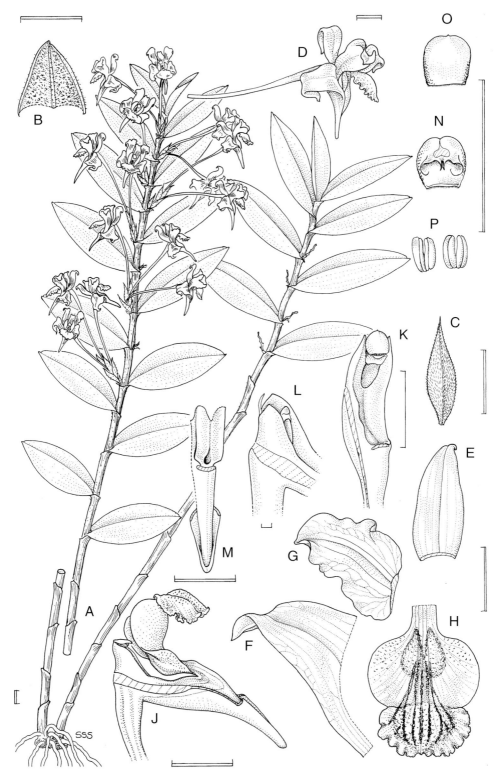

Figure 34. Dendrobium hallieri J.J. Sm. - A: habit. - B: leaf apex, abaxial surface. - C: floral bract. - D: flower, side view. - E: dorsal sepal. - F: lateral sepal. - G: petal. - H: lip. - J: pedicel with ovary, lip and column, side view. -K: column with anther-cap, oblique view. - L: apex of column with anther-cap removed, oblique view. - M: column with portion of mentum, front view. - N: anther-cap, interior view. - O: anther-cap, back view. - P: pollinia. All drawn from *Alston* 13205 by Susanna Stuart-Smith. Scale: single bar = 1 mm; double bar =1 cm.

stigmatic cavity quadrangular; stelidia broadly triangular, acute; anther-cap 3.1 × 3.2 mm, ovate, cucullate, minutely papillose, apex truncate, minutely ciliate.

HABITAT AND ECOLOGY: Not recorded. Alt. near sea level. Flowering observed in January and from July to August.

DISTRIBUTION IN BORNEO: KALIMANTAN BARAT: Kelam Hill. KALIMANTAN TENGAH: Sampit area.

GENERAL DISTRIBUTION: Endemic to Borneo.

NOTES: *Dendrobium hallieri* is a poorly known species which appears to have been recollected only once, from the Sampit River area in 1954, since it was discovered in West Kalimantan in 1912. Recently, however, a specimen was spotted among collections in a Javan orchid nursery by nurseryman Yusof Alsagoff (O'Byrne, 1998). Further collections have since been made in Kalimantan by Alsagoff. O'Byrne (2001: 64) provides an excellent colour photograph of this species.

One of the few yellow-flowered members of section *Formosae* (syns. *Nigrohirsutae*, *Oxygenianthe*), it is closely related to *D. erythropogon* Rchb.f. and *D. lowii* Lindl. (see *Orchids of Borneo*, volume 3: fig. 26), from which it is easily distinguished by its shallowly-lobed, verruculose lip lacking hairs on the mid-lobe.

DERIVATION OF NAME: The specific epithet honours Johann Gottfried Hallier (1868–1932), a German botanist who accompanied the Dutch explorer and ethnologist A.W. Nieuwenhuis on several expeditions in former Dutch Borneo, now Indonesian Kalimantan. Usually cited in the literature as H. Hallier, the initial "H" standing for "Hans", he was known to be a promoter of the German race superiority theory.

35. DENDROBIUM HENDERSONII D.A. Hawkes & A.H. Heller

Dendrobium hendersonii *A.D. Hawkes & A.H. Heller* in Lloydia 20: 120 (1957). Type: Sumatra, Riau Province, Lalah River, Inderagiri, *Schlechter* 13297 (holotype B, destroyed).

Dendrobium fugax Schltr. in Bull. Herb. Boissier, ser. 2, 6: 455 (1906), *non* Rchb.f.
Dendrobium ridleyanum Kerr in Bull. Misc. Inf. Kew 1927: 218 (1927), *non* Schltr. Type: Thailand, Betong, Pattani, *Kerr* 0102 (holotype K).
Dendrobium rudolphii A.D. Hawkes & A.H. Heller in Lloydia 20: 123 (1957).
Aporum hendersonii (A.D. Hawkes & A.H. Heller) Rauschert in Feddes Repert. 94: 439 (1983).

Epiphyte. **Stems** up to 60 cm long, clustered, porrect to pendulous, lowermost 1 or 2 internodes narrow, the next 2 or 3 internodes swollen, quadrangular, 3–8 × 0.8–2 cm, remaining internodes somewhat flattened, 2.5–4 × 0.3–0.6 cm, leafy. **Leaves** sessile; blade (3.5–)5.5–10.8 × 1–1.8(–2.8) cm, oblong-ligulate to narrowly elliptic, unequally obtusely bilobed; sheath 2.5–4 cm long. **Inflorescences** 1- to 2-flowered, borne from tufts of chaffy bracts on leafless stems, subsessile; peduncle and rachis 2–5 mm long, concealed by floral bracts; floral bracts 5–6 mm long, becoming fibrous, persistent. **Flowers** *c.* 1 cm in diameter, white, flushed pink on reverse of sepals and petals, lip pale pink, mid-lobe white, ridges pale yellow. **Pedicel** and **ovary** 0.7–1.3 cm

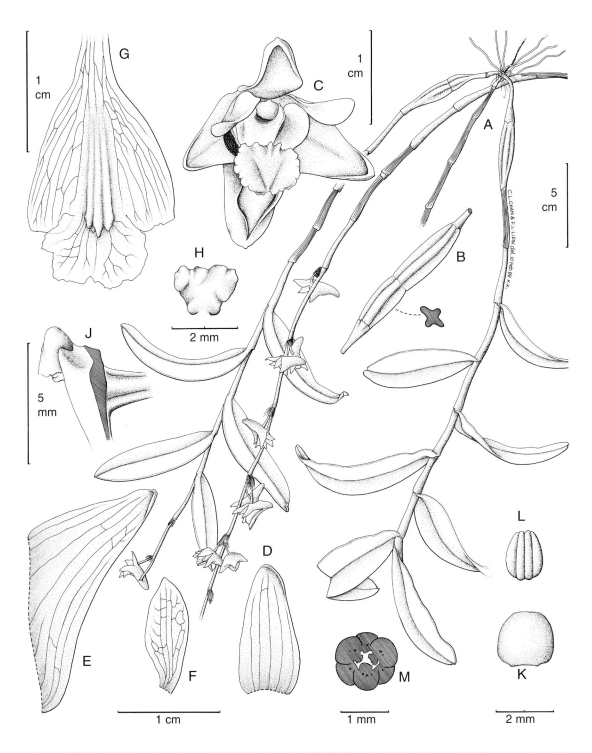

Figure 35. Dendrobium hendersonii A.D. Hawkes & A.H. Heller - A: habit. -B: pseudobulbous portion of stem showing a transverse section. - C: flower, oblique view. - D: dorsal sepal. - E: lateral sepal. - F: petal. - G: lip, front view. - H: irregular gland at base of column-foot. - J: apex of column and anther-cap, side view. - K: anther-cap, back view. - L: pollinia. - M: ovary, transverse section. All drawn from material cultivated at *Tenom Orchid Centre* by C.L. Chan and Liew Fiu Ling.

long, narrowly clavate. **Dorsal sepal** 1–1.2 × 0.35–0.55 cm, ovate-oblong, obtuse. **Lateral sepals** 2 cm long, 1.6 cm wide across base, 0.3 cm wide below apex, broadly triangular-ovate, somewhat falcate, obtuse. **Mentum** 1.2–1.5 cm long, broadly conical, obtuse. **Petals** 1 × 0.25–0.4 cm, oblong-elliptic, acute. **Lip** 3-lobed, 1.5–2 cm long, 1.3–1.4 cm wide across side lobes; side lobes 1–1.4 × 0.46–0.48 cm, erect, rounded; mid-lobe 0.5–0.55 × 0.7–1 cm, subquadrate, margin uneven to irregularly toothed; disc with three often obscure raised, fleshy ridges, highest and terminating as an irregular callosity at base of mid-lobe. **Column** 3 mm long; foot 1.3–1.4 cm long, with an irregular gland just above the base; anther-cap 2 × 2 mm, oblong, cucullate. Plate 8D.

HABITAT AND ECOLOGY: Riverine and hill forest. Alt. 200 to 300 m. Flowering observed in March and June.

DISTRIBUTION IN BORNEO: KALIMANTAN SELATAN: Banjarmasin. SABAH: Tenom District, above Kallang Waterfall; Padas Gorge. SARAWAK: Bau District, Bau, Krokong; Bidi; Siburan. Belaga District, Belaga. Kuching District, Kuching. Limbang District, Buyo Hill.

GENERAL DISTRIBUTION: Peninsular Malaysia, Thailand, Vietnam, Sumatra and Borneo.

NOTES: *Dendrobium hendersonii* belongs to section *Crumenata* and resembles a pale pink-flowered pigeon orchid (*D. crumenatum* Sw.). It was first described as *D. fugax* by Schlechter whose epithet, as pointed out by Hawkes & Heller, is a later homonym, having already been used in 1871 by Reichenbach for a plant now referred to *Flickingeria*.

DERIVATION OF NAME: The specific epithet honours Murray Ross Henderson (1899–1982), a former Director of Singapore Botanic Gardens.

36. DENDROBIUM KURASHIGEI Yukawa

Dendrobium kurashigei *Yukawa* in Lindleyana 13 (1): 28, fig. 2 (1998). Type: Borneo, Sabah, Mt. Kinabalu, Mesilau, 1745 m, 2 Sept. 1996, *Kurashige & Yukawa* 31 (holotype TNS).

Clump-forming **epiphyte**. **Roots** elongate, filiform, undulate, branching sparingly, smooth. **Stems** *c.* 8–30 cm long, *c.* 1.2 mm wide, erect to pendent, wiry, producing keikis, consisting of a narrow 0.5–1 cm long basal internode above which are 1 or 2 swollen, shiny straw-brown 4-ridged, fusiform pseudobulbous internodes each 1–2.5(–5) × 0.5–0.72 cm, beyond which the normal, unswollen, laterally compressed, branching stem continues, internodes *c.* 0.9–2.5(–3) cm long, at first enclosed in leaf sheaths, yellowish brown or straw-coloured when exposed. **Leaf-blade** 2–7 × 0.1–0.18(–0.2) cm, dorsiventral, linear, sometimes almost acicular, shallowly v-shaped, minutely acutely unequally bilobed, rigid; sheaths *c.* 0.9–2.5(–3) cm long. **Inflorescences** several borne alternately at nodes along distal portion of stem, flowers emerging successively one at a time from a tuft of scarious, pale brown or greyish brown, 1.5–2.5 mm long, glumaceous bracts. **Flowers** 6–8 mm across, sepals and petals cream or yellowish, lip cream, side lobes edged with or entirely dull purple, mid-lobe and underside of lip yellow, column yellowish, stained dull purple, anther-cap white. **Pedicel** and **ovary** 3.5–4 mm long, narrowly clavate. **Sepals** and **petals** spreading. **Dorsal sepal** 3.9–4 × 2–2.2 mm, triangular-ovate, subacute, incurved, 3-nerved. **Lateral sepals** *c.* 4.2–4.5

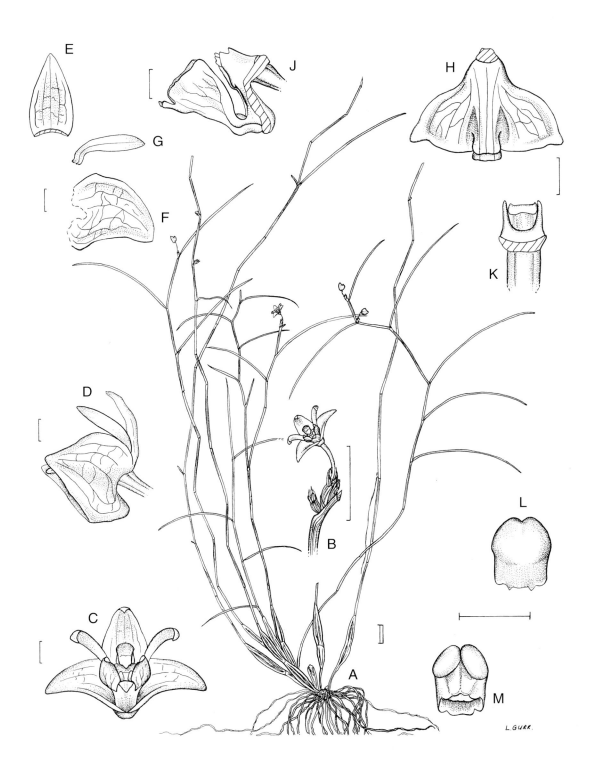

Figure 36. Dendrobium kurashigei Yukawa - A: habit. - B: inflorescence. - C: flower, front view. - D: flower, side view. - E: dorsal sepal. - F: lateral sepal. - G: petal. - H: lip, flattened, front view. - J: pedicel with ovary, lip and column, side view. - K: column, front view. - L: anther-cap, back view. - M: anther-cap, front view. A drawn from miscellaneous material, B, E–G & J from *Wood 648 & 832,* C–D, H and K–M from *Wood 832* by Linda Gurr. Scale: single bar = 1 mm; double bar = 1 cm.

mm long, 3–3.5 mm wide at base, obliquely triangular-ovate, obtuse, incurved, 5-nerved. **Mentum** 1.5–2 mm long, inflated, obtuse. **Petals** 3.8–3.9 × 0.6–0.8 mm, linear to linear-oblanceolate, acute, incurved, 1-nerved. **Lip** 3-lobed, 4 mm long, 3.5–4 mm wide across side lobes, cuneate-triangular in outline; side lobes 2.8 × 1.5 mm, obliquely triangular, rounded, erect and enveloping column; mid-lobe 1.2 × 0.8 mm, oblong-elliptic, obtuse, often truncate, slightly deflexed; disc minutely papillose, with a fleshy transverse, apically truncate callus situated at base of mid-lobe, sometimes with 2 low ridges just below callus. **Column** 1–1.5 mm long; foot 1.5–1.8 mm long, with a protuberant basal gland; stelidia short, triangular; anther-cap 1 × 0.9 mm, oblong, cucullate, glabrous. Plate 8E.

HABITAT AND ECOLOGY: Lower montane forest, often along ridges, with *Agathis*, small rattans, etc.; upper montane forest, with *Drimys*, rattans, etc.; oak-laurel forest. Alt. 1300 to 2100 m. Flowering observed throughout the year.

DISTRIBUTION IN BORNEO: KALIMANTAN TIMUR: Apo Kayan, Mt. Sungai Pendan. SABAH: Mt. Kinabalu; Crocker Range, Mt. Alab; Sipitang District, Ulu Long Pa Sia.

GENERAL DISTRIBUTION: Endemic to Borneo.

NOTES: This tiny-flowered member of section *Crumenata* is closely related to *D. lamelluliferum* J.J. Sm. but can be readily distinguished by its very narrow, linear leaves, narrower lateral sepals, shorter mentum, entire petals, oblong-elliptic lip mid-lobe and truncate callus.

DERIVATION OF NAME: The specific epithet honours Yuji Kurashige, of Akagi Nature Park in Japan, who collected the type.

37. DENDROBIUM LAMRIANUM C.L. Chan

Dendrobium lamrianum *C.L. Chan* in Sandakania 5: 69, figs. 1–3 (1994). Type: Borneo, Sabah, Mt. Kinabalu, Silau Silau Trail, 1700 m, 12 December 1993, *Sumbin & Gunsalam* SNP 5134 (holotype SAN, isotypes K, SNP).

Epiphyte. **Roots** 2.5–5 mm in diameter, white. **Stems** up to 123.5 cm long, internodes 1.5–5 cm long, 4–7 mm in diameter, entirely covered by persistent leaf sheaths, except along lower portion of older stems. **Leaves** 12–15, distichous, sheaths and blade covered by dark brown hairs when young; blade 6.2–15 × 1.1–2 cm, ligulate, acutely unequally bilobed, articulate; sheaths 1.5–5 cm long. **Inflorescences** 1-flowered, emerging through leaf sheaths on upper part of stem; peduncle 6 mm long, covered by 3 adpressed sterile bracts; floral bract 5 mm long, ovate, pale brown. **Flowers** non-resupinate, up to 6.4 cm across, fleshy, very strongly scented, lasting for about one week, sepals orange brown, with dark brown nerves and with a flush of green at centre and towards apex, turning dark orange with age, petals brown, greenish at centre, lip white with olive-green nerves and dark green keels, apex of mid-lobe pale green, column creamy white, pale green from base to foot, anther-cap cream. **Pedicel** and **ovary** up to 3.2 cm long. **Dorsal sepal** 1.6 × 2.3 cm, ovate-elliptic, acute, porrect, slightly reflexed at apex, concave at base. **Lateral sepals** 3.1 × 4.5 cm, obliquely oblong-triangular, acute and concave near apex, spreading. **Mentum** 2.5

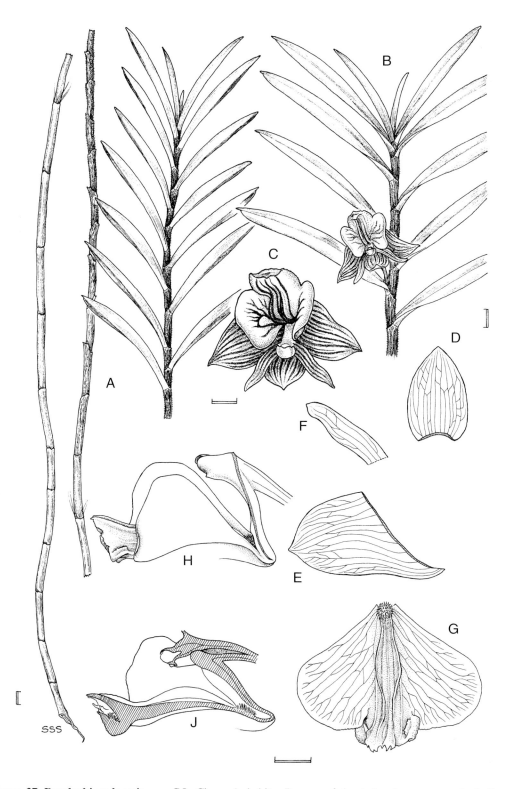

Figure 37. Dendrobium lamrianum C.L. Chan - A: habit. - B: apex of shoot showing non-resupinate flower. - C: flower, natural position, front view. - D: dorsal sepal. - E: lateral sepal. - F: petal. - G: lip, flattened, front view. - H: pedicel with ovary, lip and column, side view. - J: pedicel with ovary, lip and column, longitudinal section, side view. A drawn from *Sumbin & Gunsalam* SNP 5134 (isotype), B from *Lamb & Phillipps* AL 161/83 and a photograph by C.L. Chan, and C–J adapted from a drawing by C.L. Chan and Liew Fui Ling of *Sumbin & Gunsalam* SNP 5234 by Susanna Stuart-Smith. Scale: single bar = 1 mm; double bar = 1 cm.

cm long, conical, edges thickened near base. **Petals** 2.7 × 0.6 cm, oblong, obtuse, spreading, slightly reflexed. **Lip** 3-lobed, 4.8 × 5.7 cm (when flattened), pubescent at base; side lobes erect, obliquely obovate to rounded, minutely erose in front; mid-lobe subquadrate, deflexed, excavated in front, very fleshy, papillose-verrucose, with 2–4 small appendages at the sides; disc with 2 erect outer keels with the central, much weaker keel turned upwards and extending beyond apex of mid-lobe. **Column** 1.1 cm long, fleshy, laterally obtuse at apex, with 2 small teeth on lower margin; foot 2.6 cm long, slightly sigmoid; anther-cap 6 × 6 mm, cucullate, minutely papillose. Plate 9A & B.

HABITAT AND ECOLOGY: Lower montane forest. Alt. 1700 to 1800 m. Flowering observed from October until December.

DISTRIBUTION IN BORNEO: SABAH: Mt. Kinabalu.

GENERAL DISTRIBUTION: Endemic to Borneo.

NOTES: This is the largest flowered representative among the numerous species of section *Distichophyllum* from Borneo. Three other large flowered species of this section are also featured in this series, viz. *D. olivaceum, D. piranha* and *D. sandsii* (see *Orchids of Borneo*, volume 1: figs. 27, 28 and 29 respectively).

DERIVATION OF NAME: Named for Datuk Lamri Ali, Director of Sabah Parks.

38. DENDROBIUM LAWIENSE J.J. Sm.

Dendrobium lawiense *J.J. Sm.* in Bull. Jard. Bot. Buitenzorg, ser. 2, 3: 60 (1912). Types: Borneo, Sarawak, Limbang District, Batu Lawi, May 1911, *Moulton* 8 & 9 (syntypes BO).

Dendrobium cinnabarinum Rchb.f. var. *lamelliferum* Carr in Gard. Bull. Straits Settlem. 8: 103 (1935), *nom. nud.*

Epiphyte. **Stems** 25–60 cm or more long, erect at first, becoming pendent, slender, wiry, branching, producing keikis, internodes (1.2–)2–3.5 cm long, lower 2–3 internodes swollen, 4–10 cm long. **Leaf-blade** (1.3–)3–5.2(–6.5) × (0.1–)0.2–0.3 cm, narrowly linear, obtusely or acutely minutely unequally bilobed, main nerves 3, with many fine secondary nerves; sheath 2–3.5 cm long. **Inflorescences** borne alternately at nodes of leafless distal portion of younger branches, flowers emerging one at a time successively from a tuft of scarious, pale brown 3 mm long bracts which soon disintegrate to become fibrous. **Flowers** thin-textured, fragile, brilliant scarlet. **Pedicel** and **ovary** 0.8–1.1 cm long, slender. **Sepals** and **petals** spreading. **Dorsal sepal** 1.5–2 × 0.46–0.8 cm, oblong to obovate, obtuse, main nerves 5. **Lateral sepals** 1.7–2.9 × 0.6–1.35 cm, obliquely oblong or oblong-obovate, obtuse, main nerves (7–)8–10. **Mentum** 5–6 mm long, obtuse. **Petals** 1.6–2.1 × 0.6–0.9 cm, oblong-spathulate to suborbicular, shortly clawed, obtuse, main nerves 5–6. **Lip** 2.15–2.4 cm long, 7–8 mm wide across flattened side lobes; side lobes 1.2 cm long, *c.* 3.2 mm high, erect, oblong-triangular, subacute, erose distally; mid-lobe 1.1–1.2 × 0.45–0.6 cm, oblong to narrowly elliptic, subacute, slightly undulate at base; lower half of lip bearing a dense covering of

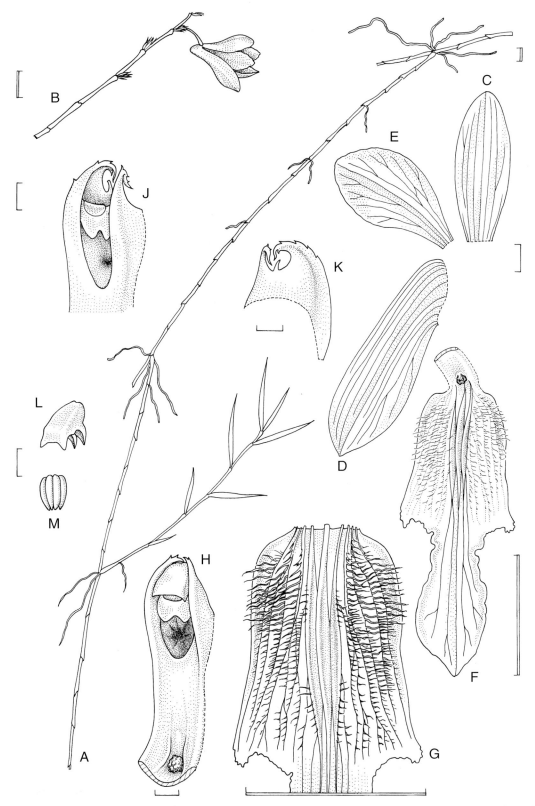

Figure 38. Dendrobium lawiense J.J. Sm. - A: habit. - B: flowering shoot. - C: dorsal sepal. - D: lateral sepal. - E: petal. - F: lip, front view. - G: lip, basal portion. - H: column, oblique view. - J: apex of column, oblique view. - K: apex of column, back view. - L: anther-cap, side view. - M: pollinia. All drawn from *Synge S. 473* by Susanna Stuart-Smith. Scale: single bar = 1 mm; double bar = 1 cm.

long ciliate hairs and finely laciniate lamellae; disc with 3 thin, sharp ridges extending from base of lip and terminating a little way below distal portion of lower half of lip. **Column** 1.6–3 × 2.5 mm; foot 5–6.5 × 2.5 mm, with a subapical subglobose, warty gland; stelidia narrowly triangular, falcate-cuneiform, acuminate, denticulate; anther-cap *c.* 1.8–2 × 1.4 mm, bucket-shaped, tridentate.

HABITAT AND ECOLOGY: moderately open spots in lower montane mossy ridge forest. Alt. *c.* 1300 m. Flowering observed in September.

DISTRIBUTION IN BORNEO: SARAWAK: Limbang District, Mt. Batu Lawi; Marudi District, Mt. Dulit.

GENERAL DISTRIBUTION: Endemic to Borneo.

NOTES: *Dendrobium lawiense* is distinguished from the closely related *D. cinnabarinum* Rchb.f. by the lower half of the lip bearing a dense covering of long ciliate hairs and finely laciniate lamellae. The disc has three thin, sharp ridges which extend from the base of the lip and terminate a little way below the distal portion of its lower half.

Carr (1935), while determining orchids collected by members of the 1932 Oxford University Expedition to Sarawak, considered *Synge* S.473 to be distinct from *D. cinnabarinum*, albeit at varietal level, and proposed the name var. *lamelliferum*. He omitted a Latin diagnosis, however, which became mandatory when describing a new taxon as from 1 January 1935 and his epithet therefore becomes a *nomen nudum* (naked name) and consequently invalid. Examination of *Synge* S.473 confirms it as belonging to *D. lawiense*.

DERIVATION OF NAME: The specific epithet refers to the type locality, Mt. Batu Lawi, a distinctive pillar-shaped sandstone rock located in the Kelabit Highlands.

39. DENDROBIUM METACHILINUM Rchb.f.

Dendrobium metachilinum *Rchb.f.* in Bonplandia 3: 222 (1855). Type: Peninsular Malaysia, Malacca, or Singapore, *Cuming* 2057 (holotype W, isotype K).

Dendrobium rorulentum Teijsm & Binn., Cat. Hort. Bogor.: 44 (1866), *nom. nud.*
Callista metachilina (Rchb.f.) Kuntze, Rev. Gen. Pl. 2: 655 (1891).

Epiphyte. Stems *c.* 28–46 cm long, internodes 1–2.5 cm long, 0.3–0.5 cm in diameter, entirely enclosed by leaf sheaths, pendulous. **Leaf-blade** 2–7 × 0.6–1.2 cm, linear to narrowly oblong, unequally obtusely or acutely bilobed, stiffly spreading; sheath 1–2.5 cm long. **Inflorescences** 2- to 4-flowered, flowers borne 1–2 mm apart; peduncle 1–2 mm long; rachis 0.5–1 cm long; floral bracts 1 mm long, ovate, acute, concave. **Flowers** stiffly fleshy, greenish or pale dull orange, with brownish nerves on sepals and petals, lip brownish, keels with minute white papillae, greenish distally. **Pedicel** and **ovary** 1.5–1.8 cm long, narrowly clavate, sigmoid. **Dorsal**

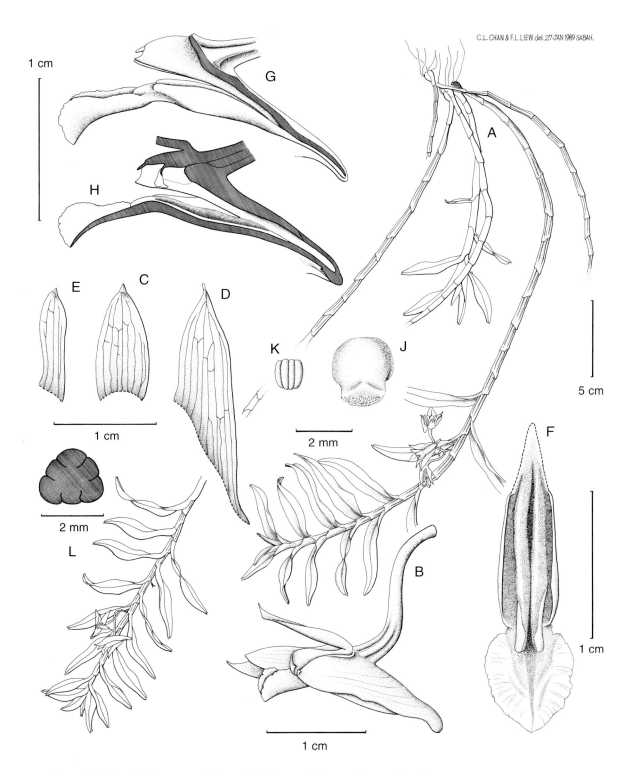

C.L.CHAN & F.L.LIEW del. 27 JAN 1989 SABAH.

Figure 39. Dendrobium metachilinum Rchb.f. - A: habit. - B: flower, side view. - C: dorsal sepal. - D: lateral sepal. - E: petal. - F: lip, flattened, front view. - G: pedicel with ovary, lip and column, side view. H: pedicel with ovary, lip and column, longitudinal section. - J: anther-cap, back view. - K: pollinia. – L: ovary, transverse section. All drawn from material cultivated at *Tenom Orchid Centre* by C.L. Chan and Liew Fui Ling.

sepal (0.8)1–1.4 × 0.4–0.51 cm, ovate, acutely apiculate. **Lateral sepals** 2.3 cm long, 5 mm wide distally, 1.4 cm wide at base, obliquely triangular-ovate, acutely apiculate. **Mentum** 0.9–1.4 cm long, narrowly conical, obtuse, held at an acute angle to 45° to pedicel and ovary. **Petals** 0.9 × 0.2–0.29 cm, narrowly oblong-ligulate, acutely apiculate. **Lip** 3-lobed, 1.7–2.2 cm long, 5 mm wide across side lobes, with a long narrow claw gradually widening to *c.* 4 mm; side lobes *c.* 9 mm long, *c.* 1.5 mm high, erect, narrowly oblong, free apical portion rounded, margin somewhat involute; mid-lobe 6–7 × 4–6 mm, ovate, acute, shallowly folded, fleshy, slightly crenulate, margin papillose; disc with 2 farinose fleshy keels terminating on base of mid-lobe. **Column** 4 mm long; foot 1–1.1 cm long; anther-cap *c.* 2.9 × *c.* 2.2 mm, oblong-ovate, minutely papillose. Plate 9C.

HABITAT AND ECOLOGY: Lowland forest beside streams. Alt. 200 m. Flowering observed in January and February.

DISTRIBUTION IN BORNEO: SABAH: Sandakan District, Telupid area.

GENERAL DISTRIBUTION: Peninsular Malaysia, Thailand, Sumatra, Borneo and Maluku (Ambon).

NOTES: Although placed in section *Distichophyllum*, *D. metachilinum* more closely resembles members of section *Conostalix*. The delimitation between these sections is not clear and future research may result in them being treated as one. Holttum (1964) comments that populations in Peninsular Malaysia are variable. Carr described a plant from Johor as var. *crenulatum* Carr, differing from the typical plant in the more slender, rather curved mentum and less fleshy, crenulate lip mid-lobe. The typical plant is known from only one location in Borneo and it is not known if similar variability exists in populations there.

DERIVATION OF NAME: The specific epithet is derived from the Greek *meta*, associated with, changed, substituted for, and the Latin *chilus*, lip. It is unclear as to what feature of the lip this refers.

40. DENDROBIUM PARTHENIUM Rchb.f.

Dendrobium parthenium *Rchb.f.* in Gard. Chron., ser. 2, 24: 489 (1885). Type: Borneo, locality unknown, *Bull* s.n. (holotype W).

Dendrobium sanderianum Rolfe in Kew Bull.: 155 (1894). Type: Borneo, locality unknown, imported by *Sander & Co.* (holotype K), **syn. nov.**

Epiphyte. **Roots** 2–2.5 mm in diameter, branching, smooth. **Stems** up to 20 or more per plant, 27–60 cm long, 4–5 mm in diameter, erect, orange-yellow, entirely clothed in leaf sheaths, internodes 1–3 cm long. **Leaf-blade** 1.5–6.3(–7) × 0.7–1.7 cm, ovate-oblong or oblong, minutely obtusely bidentate at apex, stiffly coriaceous; sheath (1–)2–2.5(–3) cm long, glabrous, yellowish green or brownish yellow. **Inflorescences** abbreviated, emerging through sheath base opposite

103

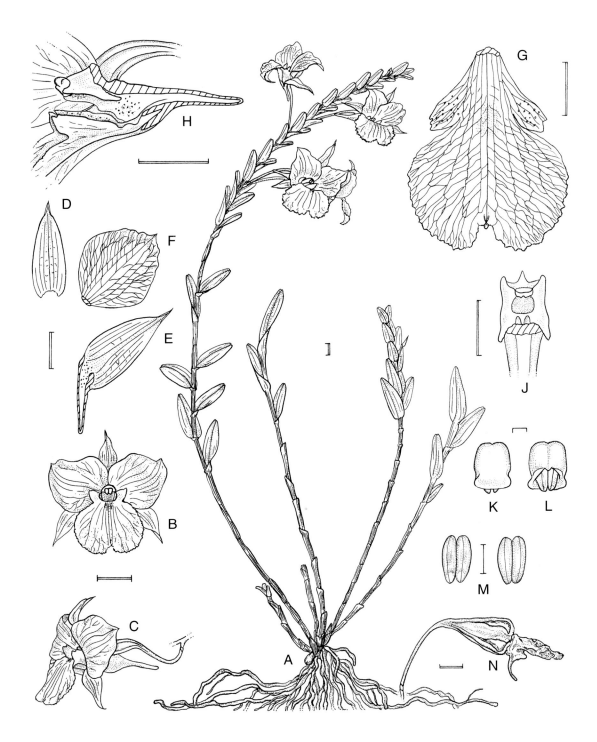

Figure 40. Dendrobium parthenium Rchb.f. - A: habit. - B: flower, front view. - C: flower, side view. - D: dorsal sepal. - E: lateral sepal. - F: petal. - G: lip, front view. - H: longitudinal section through mentum, with lower portion of lip and column. - J: column, front view. - K: anther-cap, back view. - L: anther-cap, interior view showing pollinia. - M: pollinia. - N: fruit. A drawn from miscellaneous material, B–J from *Lamb* K 33, K–M from *Lamb* K 33 *& Bailes & Cribb* 671 and N from *Lamb* SAN 91514, LKC 3164 by Linda Gurr. Scale: single bar = 1 mm; double bar = 1 cm.

blade, 2- to 3(–4)-flowered, usually 1 or 2 flowers open at a time; peduncle and rachis 5 mm long; floral bracts 2-3 mm long, ovate, obtuse to acute, glabrous. **Flowers** 2.5–3 cm across, pure white, ovary and apex of mentum pale green, lip with a yellowish green, pale orange-red or pale cerise-purple basal patch, ventral part of column below stigmatic cavity sometimes with purple lines, anther-cap sometimes with 3 purple basal blotches. **Pedicel** and **ovary** 2.5–3.5(–4.5) cm long, curved to somewhat sigmoid, ovary 3-keeled. **Sepals** spreading, to recurved distally. **Dorsal sepal** 2.1–2.3 × 0.7–0.9 cm, triangular-ovate, acuminate, dorsally somewhat carinate. **Lateral sepals**: upper margin 2.3–2.7 cm long, lower margin 4.1–4.2 cm long, 0.9–1.1 cm wide at base, obliquely triangular-ovate, acuminate, dorsally somewhat carinate. **Mentum** 1.6–1.9 cm long, 5–6 mm wide at base, narrowly conical, ovipositor-shaped, broadened and somewhat saccate at base, straight to hooked distally. **Petals** 2.5–2.6 × 1.9–2.1 cm, suborbicular-elliptic or oblong, obtuse and apiculate. **Lip** shallowly 3-lobed, (2.8–)3.2–3.5 cm long, 2.3–3 cm wide across mid-lobe, 1.7–2 cm wide across side lobes; side lobes 0.9–1.2 cm long, 3–4 mm high, oblong, obtuse; mid-lobe broadly obcordate, cuneate at base, crenulate, apiculate, disc smooth. **Column** 3–5 × 5 mm; foot 6–7 × 6–7 mm, concave, wings obtuse; anther-cap 3 × 2.8–3 mm, ovate to oblong, cucullate, smooth, minutely papillose at base. Plate 9D.

HABITAT AND ECOLOGY: Hill forest on ultramafic substrate, recorded as epiphytic on *Gymnostoma sumatranum*; lowland forest on extreme ultramafic substrate. Alt. sea level to 900 m. Flowering observed in January, February, May, June, August and September.

DISTRIBUTION IN BORNEO: SABAH: Mt. Kinabalu; Lahad Datu District, Baik Island.

GENERAL DISTRIBUTION: Endemic to Borneo.

NOTES: Reichenbach (1885) described *D. parthenium* thus: "take the stem of a very tall long-leaved *Dendrobium revolutum* Lindl., or of *Dendrobium dearei* Rchb.f., give it two flowered inflorescences of flowers like those of *Dendrobium radians* Rchb.f., and you have this lovely Bornese novelty, just introduced by Mr. W. Bull."

Rolfe (1894), when describing *D. sanderianum*, considered it to be "most like *D. dearei* Rchb.f." However, Rolfe later annotated the type sheet: "next to *D. parthenium* Rchb.f., but has a longer, narrower spur. (RAR.)" Examination of several collections preserved in alcohol shows the spur-like mentum to vary in length and width depending on the size and maturity of the flower. The two species appear identical in other respects and are treated here as conspecific.

DERIVATION OF NAME: Reichenbach states that "it is named *parthenium* in allusion to its white virginal flowers, which might, no doubt, have pleased Vesta herself, provided the Romans had known Dendrobia."

41. DENDROBIUM RADIANS Rchb.f.

Dendrobium radians *Rchb.f.* in Xenia Orch. 2: 130, t. 146, figs. I & II (1867). Type: Borneo, coll. and cult. *Low* s.n. (holotype W).

Epiphyte. **Stems** up to *c.* 60 cm long, erect, somewhat fractiflex distally, orange-yellow, entirely clothed in leaf sheaths, internodes 2–3.5 cm long. **Leaves** 2.8–5.5 × 1.2–2.5 cm, oblong-elliptic, apex unequally obtusely bilobed, nigrohirsute, particularly on reverse, sheaths persistently densely nigrohirsute. **Inflorescences** abbreviated, emerging through sheath base opposite blade, usually 4- to 5-flowered, often only 3 flowers open at a time; peduncle and rachis *c.* 1 cm long; floral bracts 1–2 cm long, ovate-elliptic, acuminate, nigrohirsute. **Flowers** up to 6 cm across, milky-white, claw of lip greenish, base of lip with an orange flush, sometimes with brownish cinnabar-red on disc and sometimes ochre spots at base of side lobes, column apex brownish, sometimes with a brownish cinnabar-red flush below stigmatic cavity, unscented. **Pedicel** and **ovary** 4.5–5 cm long, cylindrical, gently curved, glabrous. **Sepals** and **petals** spreading, 7-nerved. **Dorsal sepal** *c.* 2.7 × 1.3 cm, oblong-ovate, acute, mid-nerve strongly dorsally carinate, especially along distal half. **Lateral sepals** *c.* 5.7 cm long, free portion 3 cm long, 1.4 cm wide at base, obliquely triangular-ovate, acute, mid-nerve dorsally strongly carinate. **Mentum** 2.5–3 cm long, *c.* 7 mm wide at base, narrowly conical, infundibuliform, obtuse, straight or slightly curved. **Petals** 4 × 2.2–2.6 cm, broadly oblong-elliptic or ovate-elliptic, somewhat truncate, cuneate at base, margin rather uneven. **Lip** shallowly 3-lobed, main nerves 7, 3–3.1 cm long, 2.6–2.8 cm wide across mid-lobe, 1.5–1.7 cm wide across side lobes, claw *c.* 1.3 cm long, pandurate, with or without a short isthmus between side lobes and mid-lobe; side lobes obliquely rounded, margin entire, erect; mid-lobe transversely ovate to reniform, retuse to emarginate, becoming shallowly bilobulate, margin crenulate, slightly convex in centre, thicker-textured at base, otherwise thin-textured, surface, rough, minutely papillose. **Column** 6 × 4–5 mm; foot 5–6 mm long, 6 mm wide at entrance to mentum, wings rounded; anther-cap *c.* 3.5 × 3.1 mm, ovate, cucullate, sulcate on ventral surface, minutely papillose.

HABITAT AND ECOLOGY: Lower montane mossy forest. Alt. 1200 m. Flowering observed in September.

DISTRIBUTION IN BORNEO: SARAWAK: Mt. Dulit.

GENERAL DISTRIBUTION: Endemic to Borneo.

NOTES: Vegetatively *D. radians* (section Formosae) is virtually indistinguishable from the closely related *D. sculptum* Rchb.f. (see *Orchids of Borneo*, volume 3: fig. 30). However, *D. radians* appears to have a more distinctly three-lobed lip with a broader, shallowly bilobulate reniform mid-lobe. The mid-lobe also lacks the central dark orange blotch which seems to be characteristic of *D. sculptum*. Both species have only been collected on a couple of occasions since they were described in the 1860s and, consequently, the range of variability in the wild is unknown. I suspect that intermediates may occur suggesting, perhaps, that the two species should be united.

DERIVATION OF NAME: The Latin specific epithet *radians* means to radiate, spreading straight outwards from a common centre. This presumably refers to the wide-opening flower with spreading sepals and petals.

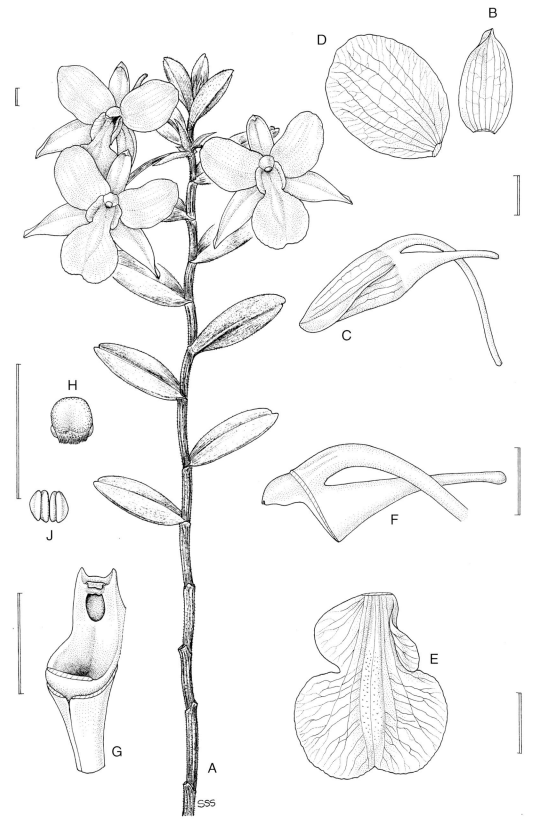

Figure 41. Dendrobium radians Rchb.f. - A: habit. - B: dorsal sepal. - C: pedicel with ovary, lateral sepal and mentum, side view. - D: petal. - E: lip, front view. - F: pedicel with ovary, mentum and column, side view. - G: column, oblique view. - H: anther-cap, back view. - J: pollinia. All drawn from *Richards* in *Synge* S. 421 by Susanna Stuart-Smith. Scale: single bar = 1 mm; double bar = 1 cm.

42. DENDROCHILUM ALPINUM Carr

Dendrochilum alpinum *Carr* in Gdns. Bull. Straits Settl. 8: 235 (1935). Type: Borneo, Sabah, Mt. Kinabalu, below Sayat Sayat, 3450 m, June 1933, *Carr* 3545, SFN 27624 (holotype SING, isotypes AMES, K).

Lithophyte or **epiphyte**. **Rhizome** to 5 cm long, branching. **Roots** 1–1.5 mm in diameter, minutely papillose. **Cataphylls** 3, 1.5–4 cm long, ovate-elliptic, acute to acuminate. **Pseudobulbs** crowded or borne up to 1 cm apart, 1.8–2.7 × 0.8–1 cm, ovoid, sometimes somewhat curved, golden yellow, orange or red. **Leaf-blade** (5–)7–14.5 × 0.7–1.5 cm, narrowly elliptic, narrowed above middle, apex conduplicate, acute, rigid-textured; petiole 0.8–3 cm long, sulcate. **Inflorescences** laxly 11- to 20-flowered, pendulous; flowers borne 5–6 mm apart; peduncle 3–10 cm long, slender, terete; non-floriferous bracts 1 or 2, 4–9 mm long, ovate, acute; rachis 7.5–12 cm long, quadrangular; floral bracts 5 × 4.5 mm, ovate, acute, margins erose towards apex. **Flowers** up to *c.* 1 cm across, sepals and petals either entirely salmon-pink or yellow, suffused salmon-pink, lip either brownish or salmon-pink with a brownish salmon-pink median line, keels and apex, column deep salmon-pink, stelidia brown, anther-cap cream suffused salmon-pink at base. **Pedicel** and **ovary** 2–5 mm long, narrowly clavate. **Dorsal sepal** 7.8 × 4.4 mm, oblong to oblong-elliptic, obtuse, curving forward. **Lateral sepals** 7.8 × 7.9 × 4 mm, ovate-oblong, obtuse to acute, apex slightly dorsally carinate on reverse, spreading. **Petals** 7.8 × 3.8 mm, oblong or obliquely-oblong, obtuse, twisted 90° from vertical. **Lip** entire, 6.4 × 5.8 mm, broadly ovate, shallowly retuse or subacute, rather incurved above middle, elevate along median nerve; disc with a thickened central area and provided with 2 tall curved, suberect or rather extrorse fleshy keels which form, with the central thickened area, 2 grooves with a low cushion at the base. **Column** 1.2–2 mm long; foot absent; apical hood dilated, obscurely 3-lobed to rounded; rostellum conspicuous, triangular, acute; stelidia 1.7–1.8 mm long, basal, oblong-oblanceolate, obtuse; anther-cap cucullate. Plate 9E & 10A.

HABITAT AND ECOLOGY: Upper montane forest; sides of granitic rocks sheltered by stunted trees. Alt. 2400 to 3700 m, abundant above 3200 m. Flowering observed in June, July, October and December.

DISTRIBUTION IN BORNEO: SABAH: Mt. Kinabalu.

GENERAL DISTRIBUTION: Endemic to Borneo.

NOTES: *Dendrochilum alpinum* has the largest pseudobulbs and flowers of any species in section *Eurybrachium* represented on Mt. Kinabalu. It is easily identified by the petals, which are twisted 90° from vertical, and the darkly pigmented lip contrasting with the pale sepals and petals.

DERIVATION OF NAME: The specific epithet refers to the virtually "alpine" habitat of this species.

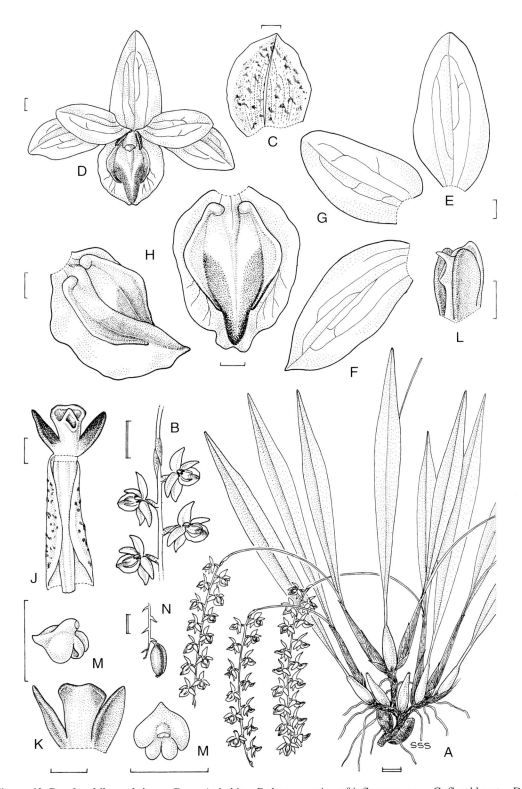

Figure 42. Dendrochilum alpinum Carr - A: habit. - B: lower portion of inflorescence . - C: floral bract. - D: flower, front view. E: dorsal sepal. - F: lateral sepal. - G: petal. - H: lip, front and oblique views. J: floral bract, pedicel with ovary and column, anther-cap removed, front view. - K: column, back view. - L: column, excluding stelidia, oblique view. - M: anther-cap, side and back views. - N: fruit. All drawn from *Carr* C. 3545, SFN 27624 (holotype) by Susanna Stuart-Smith. Scale: single bar = 1 mm; double bar = 1 cm.

43. DENDROCHILUM CORRUGATUM (Ridl.) J.J. Sm.

Dendrochilum corrugatum (*Ridl.*) *J.J. Sm.* in Recl. Trav. Bot. Néerl. 1: 65 (1904). Type: Borneo, Sabah, Mt. Kinabalu, Marai Parai Spur, 1700 m, *Haviland* 1814 (holotype SAR).

Platyclinis corrugata Ridl. in Stapf, Trans. Linn. Soc. Lond. 2, 5: 233 (1894).
Acoridium corrugatum (Ridl.) Rolfe in Orchid Rev. 12: 220 (1904).
Dendrochilum fimbriatum Ames, Orchidaceae 6: 51–52 (1920). Type: Borneo, Sabah, Mt.
 Kinabalu, Marai Parai Spur, 23 November 1915, *J. Clemens* 248 (holotype AMES).

Epiphyte, sometimes **lithophyte**. **Rhizome** abbreviated, or up to 6 cm long. **Roots** up to 25 (–45) cm long, 0.5–1 mm in diameter, smooth. **Cataphylls** 2–3, 0.5–4 cm long, acute to acuminate, pale brown, with darker brown speckles. **Pseudobulbs** crowded, often into large clumps, 0.5–1.6 × 0.5–0.7(–1.2) cm, ovoid-globose or broadly ellipsoid, rugulose, green suffused red, or entirely red. **Leaf-blade** 5.5–13.5 × 0.5–1 cm, linear to linear-lanceolate, acute, minutely apiculate, rigid; petiole 0.8–3 cm long, sulcate. **Inflorescences** subdense to densely many-flowered, curving; flowers borne 1.8–2 mm apart in 2 regular ranks; peduncle 5.5–10 cm long, terete, filiform; non-floriferous bracts (1–)3–4, 1–3.5 mm long, ovate-elliptic, acute; rachis 5–16 cm long, quadrangular, reddish; floral bracts (1.8–)2.5–3 × 1.5–2.5 mm, broadly ovate, subacute, rigid, reddish. **Flowers** unscented, sepals and petals creamy white, lip yellowish green, column deep pink or orange, anther-cap yellow. **Pedicel** and **ovary** 1.8–2 mm long, narrowly clavate, geniculate. **Dorsal sepal** 4–4.5 × 1.5 mm, lanceolate, acute, somewhat reflexed distally. **Lateral sepals** (3–)3.5 × 1.9–2 mm, ovate-lanceolate, falcate, apex acute or strongly reflexed. **Petals** 3.5 –4 × 1 mm, lanceolate, acute to acuminate, margin minutely erose-denticulate. **Lip** entire, or sometimes shallowly 3-lobed, 2 × 3–3.1 mm, broadly ovate-rotundate, somewhat concave at base, triangular-apiculate, margin shortly and irregularly fimbriate; disc bicallose, calli conspicuous, semi-orbicular, erect, fleshy, *c.* half as long a lip, one on each lateral nerve, confluent at base by an obscure transverse ridge. **Column** (0.5–)0.7–0.8 mm, fleshy; foot absent; apical hood rounded, entire; rostellum prominent, porrect, narrowly triangular, linguiform, acute; stelidia 0.3–0.4 mm long, basal, oblong, obtuse; anther-cap cordate, cucullate, glabrous.

HABITAT AND ECOLOGY: Mostly a twig epiphyte in lower montane forest and *Leptospermum* ridge scrub on ultramafic substrate; preferring open sites on extreme ultramafic substrate where it avoids the shade of a well-developed canopy. Alt. 1500 to 2100 m. Flowering observed from March to May, August, and from September to November.

DISTRIBUTION IN BORNEO: SABAH: Mt. Kinabalu.

GENERAL DISTRIBUTION: Endemic to Borneo.

NOTES: *Dendrochilum corrugatum* bears a close resemblance and is closely related to *D. alatum* Ames. It can be distinguished primarily by the broader, fimbriate lip, falcate lateral sepals, and petals which are not twisted and aligned 90° from vertical.

DERIVATION OF NAME: The specific epithet refers to the strongly wrinkled pseudobulbs which may, to some extent, be a misnomer since examination of dried material of related species shows that extensive shrinkage, distortion and wrinkling results in part from the process of pressing and

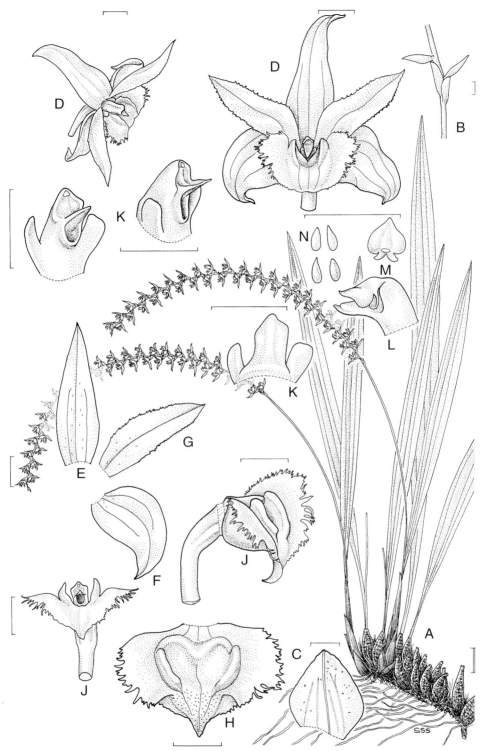

Figure 43. Dendrochilum corrugatum (Ridl.) J.J.Sm. - A: habit. - B: junction of rachis with peduncle showing two non-floriferous bracts and two floral bracts. - C: floral bract. - D: flower, front and top views. E: dorsal sepal. - F: lateral sepal. - G: petal. - H: lip, front view. - J: pedicel with ovary, lip and column, front and side views. - K: column with anther-cap removed, front, oblique and back views. L: column and anther-cap, side view. - M: anther-cap, back view. N: pollinia. A drawn from *Carr* C. 3128, SFN 24728, B from *J. & M.S. Clemens* 32244, and C–N from *Bailes & Cribb* 841 by Susanna Stuart-Smith. Scale: single bar = 1 mm; double bar = 1 cm.

drying. However, mature pseudobulbs of many orchids do become wrinkled to varying degrees and it is certainly not a character confined to this species.

44. DENDROCHILUM GRAVENHORSTII J.J. Sm.

Dendrochilum gravenhorstii *J.J. Sm.* in Bull. Jard. Bot. Buitenzorg, ser. 3, 2: 28–29 (1920). Type: Borneo, Kalimantan Barat, Upper Kapuas, Talaj River, 1916, *Gravenhorst* s.n., cult. hort. Bogor (holotype BO, isotype L).

Terrestrial or **epiphyte**. **Rhizome** up to 40 cm long, internodes 1–1.3 cm long, branching, clothed in numerous reddish-brown sheaths when young. **Roots** 0.2– 0.8 mm in diameter, branching distally, smooth. **Cataphylls** 4–6, 0.3–2.5 cm long, ovate-elliptic, acute to acuminate, glossy reddish brown, sometimes speckled. **Pseudobulbs** borne 1.3–4.5(–7) cm apart, often at an acute angle to rhizome, 1.3–2.5 × 0.55–0.75 cm, obliquely oblong-ovoid, becoming grooved, glossy, pale green to orange-ochre. **Leaf-blade** 4.5–11.5 × 0.5–0.8(–0.9) cm long, linear, ligulate, obtuse to acute, rigid, leathery; petiole 2–3 mm long, cuneate, canaliculate. **Inflorescences** subdensely many-flowered, gently decurved; flowers borne 2–3 mm apart; cataphylls 8–9, 0.3–3 cm long, imbricate, pale brown; peduncle 1–2 cm long, erect to porrect; rachis (4–)5–8 cm long, quadrangular, arcuate; floral bracts 0.8–1 × 1.75–1.8 mm, ovate-orbicular, obtuse to shortly acute. **Flowers** sweetly scented, very pale green to pale yellow, lip yellowish green to ochre with whitish margin, stelidia white. **Pedicel** and **ovary** 2–2.25 mm long, clavate. **Dorsal sepal** 4.4–4.5 × 1.3 mm, narrowly oblong, apiculate. **Lateral sepals** 4 × 1.6 mm, obliquely oblong-ovate, subfalcate, acute. **Petals** 3.7 × 0.8 mm, obliquely lanceolate to linear-lanceolate, acute. **Lip** 2.7 mm long, 1.25 mm wide at base, 0.8 mm wide at centre, subentire, shortly clawed, decurved, apex often recurved, lanceolate, dilated and auriculate below, auriculate basal portion with a few narrow basal teeth, erose above; disc with a raised, somewhat compressed, fleshy flange each side at the base, keels 3, arising on the dilated basal portion, thick, fleshy, outer keels terminating on distal portion of blade, median keel shorter and narrower, terminating on proximal portion of blade. **Column** 2 mm long; foot distinct, truncate; apical hood quadrangular, 2- to 4-toothed; rostellum prominent, acute; stelidia slightly longer than apical hood, arising opposite stigmatic cavity, obliquely linear-ligulate to slightly sigmoid, acute; anther-cap 0.5 mm wide, cucullate, triangular, acuminate or acute.

HABITAT AND ECOLOGY: Kerangas forest; primary mixed dipterocarp forest on steep slopes; podsolic dipterocarp/*Dacrydium* forest on very wet sandy soil. Alt. 100 to 1000 m. Flowering observed in June and July.

DISTRIBUTION IN BORNEO: KALIMANTAN BARAT: Upper Kapuas, Talaj River area. SABAH: Nabawan area; Sandakan District, Telupid area. SARAWAK: Marudi District, Bario area; Limbang District, Sipayang River.

GENERAL DISTRIBUTION: Endemic to Borneo.

Figure 44. Dendrochilum gravenhorstii J.J. Sm. - A: habit. - B: basal portion of leaf. - C: distal portion of inflorescence. - D: floral bract. - E: flower, side view. - F: dorsal sepal. - G: lateral sepal. H: petal. - J: lip, front view. - K: lip and column, side view. - L: pedicel with ovary, lip and column, oblique view. - M: column, back view. - N: anther-cap, side view. - O: pollinia. A & B drawn from *Awa & Lee* S. 40416 and C–O from *Gravenhorst* s.n. (holotype). Scale: single bar = 1 mm; double bar = 1 cm.

NOTES: *Dendrochilum gravenhorstii* belongs to subgenus *Dendrochilum* which includes *D. crassum* Ridl. (see *Orchids of Borneo*, volume 3: fig. 38) and the widespread *D. pallidiflavens* Blume. Lip ornamentation is generally simplest in subgenus *Dendrochilum* where the disc is provided with two free proximal longitudinal keels and sometimes a third shorter, median keel or swelling. *Dendrochilum gravenhorstii* is unusual in having a more complicated structure on the lip.

DERIVATION OF NAME: Named in honour of C.A. Gravenhorst (born 1884), the head of a Danish oil factory laboratory who travelled to the former Dutch East Indies to investigate the properties of oil-bearing seeds. He made several collections of living orchids in Kalimantan which were sent to J.J. Smith at Bogor (formerly Buitenzorg) Botanic Gardens in Java.

45. DENDROCHILUM JIEWHOEI J.J. Wood

Dendrochilum jiewhoei *J.J. Wood*, Dendrochilum of Borneo: 232, fig. 105 (2001). Type: Borneo, Sarawak, Kuching District, Mt. Penrissen, 1920's, *Mjöberg* s.n. (holotype AMES, isotype K).

Epiphyte. **Rhizome** up to *c.* 7 cm long, 2 mm in diameter. **Roots** elongate, sparsely to much-branched, smooth, 0.3–1 mm in diameter. **Cataphylls** 2–3, ovate-elliptic, obtuse to acute, finely nerved, pale brown, with obscure darker brown speckling, becoming fibrous. **Pseudobulbs** crowded on rhizome, ovoid to cylindrical, finely wrinkled in dried state, (0.6–)1–*c.* 1.8 × *c.* 0.4–0.8 cm, yellowish olive. **Leaf-blade** narrowly elliptic to oblong-elliptic, constricted *c.* 0.8–1.4 cm below apex, obtuse and mucronate, thin-textured, 6.5–11.5 × 0.9–1.8 cm, main nerves 5–7; petiole 1.4–2 cm long, slender, sulcate. **Inflorescences** gently curving, densely many-flowered, 1–1.4 cm in diameter, flowers borne 2–3 mm apart, lowermost opening last; peduncle filiform, (5–)7–8.5 cm long; non-floriferous bracts 1–4, 1.8–4 mm long, ovate, obtuse to acute; rachis quadrangular, (3.5–)13–20 cm long; floral bracts 2–3 × 2.1 mm long, ovate-oblong, obtuse and mucronate, pale brown. **Flowers** wide-opening, stellate, colour not recorded. **Pedicel** and **ovary** *c.* 2–2.4 mm long, slender. **Sepals** and **petals** 3-nerved. **Dorsal sepal** 4.5–4.6(–5.1) × 1.1–1.5 mm, narrowly elliptic, acute to acuminate. **Lateral sepals** 4.1–4.2(–5) × 1.3–1.5 mm, narrowly elliptic to somewhat ovate-elliptic, acute to acuminate. **Petals** 3.9–4.1 × 1.2 mm, narrowly elliptic, acute to acuminate, margins entire. **Lip** 3.2–4 mm long, *c.* 1.8–2.1 mm wide across side lobes, very shortly clawed, shortly 3-lobed, 3-nerved, very minutely papillose; side lobes triangular to narrowly triangular, minutely erose, free apical portion 0.4–0.8 mm long; mid-lobe *c.* 1.4–1.5(–2) × 1.2–1.3 mm, oblong-triangular to triangular-ovate, acute to acuminate, margin slightly irregular; disc with 2 keels joined at and forming a cavity at the base, extending *c.* 0.9–1.1 mm along blade, median nerve rather prominent for a short distance. **Column** 1.4–1.5 mm long; foot 0.1–0.2 mm long; apical hood rounded, slightly irregular to obscurely tridentate; stelidia 1–1.1 mm long, basal, falcate, acute, equalling or shorter than apical hood; rostellum small, ovate, obtuse; stigmatic cavity without a raised margin; anther-cap cucullate, smooth.

HABITAT AND ECOLOGY: Lower montane forest. Alt. 1320 m. Flowering time unknown.

DISTRIBUTION IN BORNEO: SARAWAK: Kuching District, Mt. Penrissen.

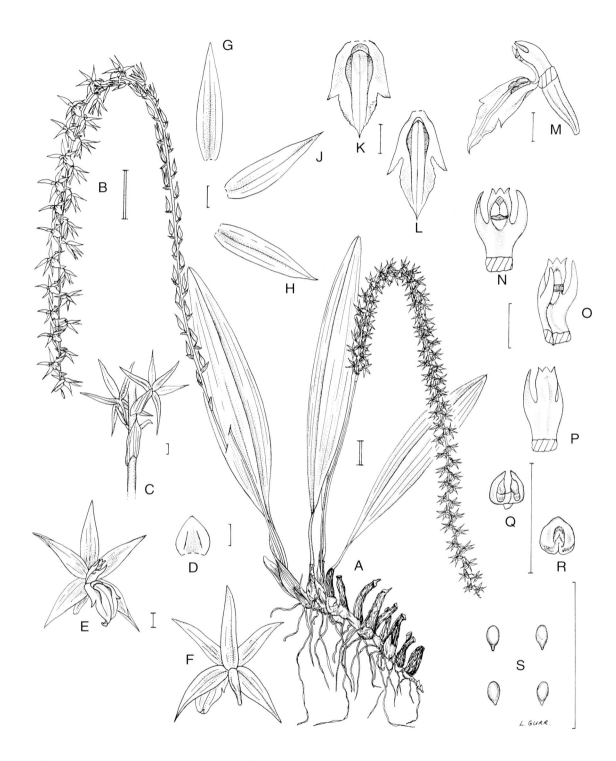

Figure 45. Dendrochilum jiewhoei J.J. Wood - A: habit. - B: inflorescence. - C: junction of rachis with peduncle showing one non-floriferous bract and two flowers. - D: floral bract. - E: flower, oblique view. - F: flower, back view. - G: dorsal sepal. - H: lateral sepal. - J: petal. - K & L: lips, front view. - M: pedicel with ovary, lip and column, side view. - N: column, front view. - O: column, oblique view. - P: column, back view. - Q: anther-cap, front view. R: anther-cap, back view. - S: pollinia. All drawn from *Mjöberg* s.n. (holotype) by Linda Gurr. Scale: single bar = 1 mm; double bar = 1 cm.

GENERAL DISTRIBUTION: Endemic to Borneo.

NOTES: Specimens of this recently described species were among a batch of previously unstudied material, collected in Sarawak in the 1920's by the Swedish zoologist Eric Mjöberg, on loan to Kew from the Orchid Herbarium of Oakes Ames, Harvard University in the United States (AMES) for determination. It is surprising that an undescribed species should have originated from Mt. Penrissen, which is relatively well collected compared to many areas of Sarawak. *D. jiewhoei* is known only from the type and it is curious that no subsequent collections are extant. Although possibly a rare endemic to Mt. Penrissen, its modest appearance suggests that it is more likely to have simply been overlooked elsewhere.

Dendrochilum jiewhoei is possibly most closely allied to the Kinabalu endemic *D. kamborangense* Ames. It is easily distinguished, however, by its shorter pseudobulbs and leaves, the lowermost buds on the inflorescence opening last, and the smaller flowers which have entire petals. The lip has triangular to narrowly triangular, minutely erose side lobes and a smaller, oblong-triangular to triangular-ovate mid-lobe. The column is shorter, with a less prominent foot, and basal stelidia which are equal to or shorter than the apical hood.

DERIVATION OF NAME: The specific epithet honours Mr Tan Jiew Hoe of Singapore, who has generously provided financial support to cover the printing costs of colour plates for several recent volumes on Bornean flora.

46. DENDROCHILUM JOCLEMENSII Ames

Dendrochilum joclemensii *Ames*, Orchidaceae 6: 55, plate 83, top (1920). Type: Borneo, Sabah, Mt. Kinabalu, Marai Parai Spur, 22 November 1915, *J. Clemens* 247 (holotype AMES).

Epiphyte, rarely **terrestrial**. **Rhizomes** up to 6 cm long, clump-forming. **Roots** 1 mm in diameter, minutely papillose. **Cataphylls** 3, 0.5–1.6 cm long, ovate-elliptic, acute to acuminate. **Pseudobulbs** 1–1.4 × 0.3–0.5 mm, crowded, subfusiform, smooth, becoming wrinkled. **Leaf-blade** 4.5–10 × 0.3–0.4 cm, linear, obtuse and apiculate to shortly acute, erect; petiole 3–9 mm long, sulcate. **Inflorescences** laxly few- to about 24-flowered, erect to gently curving; flowers borne 2–3 mm apart; peduncle 2.2–4 cm long, slender, greenish-orange; non-floriferous bracts absent; rachis 4–5.5 cm long, quadrangular, salmon-pink; floral bracts 2 × 0.5–0.6 mm, narrowly elliptic, acute, pale orange. **Flowers** opening from top of inflorescence (always?), wide opening, about 5.5 mm across, unscented, translucent salmon-pink, pale orange or yellow with pale salmon-pink at base of keels on lip, or very pale reddish ochre to brownish salmon-pink with very pale green petals, column cream, stelidia translucent white. **Pedicel** and **ovary** 2 mm long, narrowly clavate, curved. **Sepals** and **petals** spreading to reflexed. **Sepals** narrowly ovate-elliptic, acute to acuminate. **Dorsal sepal** 2.5–3 × 1 mm. **Lateral sepals** 2.5–3 × 1.1 mm, slightly carinate. **Petals** (2–)2.5–2.8 × (0.5–)0.9–1 mm, narrowly elliptic, acute. **Lip** 2–2.2 × 0.8–1(–1.3) mm, very obscurely auriculately lobed, lobes hidden by keels on disc, oblong, abruptly contracted near apex into an elongated acuminate tip, 3-nerved; disc with 2 suborbicular, laterally flattened, spreading keels confluent with outer nerves, each thickened and joined just above base of lip and terminating midway along, median nerve prominent and slightly raised. **Column** 1–1.5 mm long; apical hood

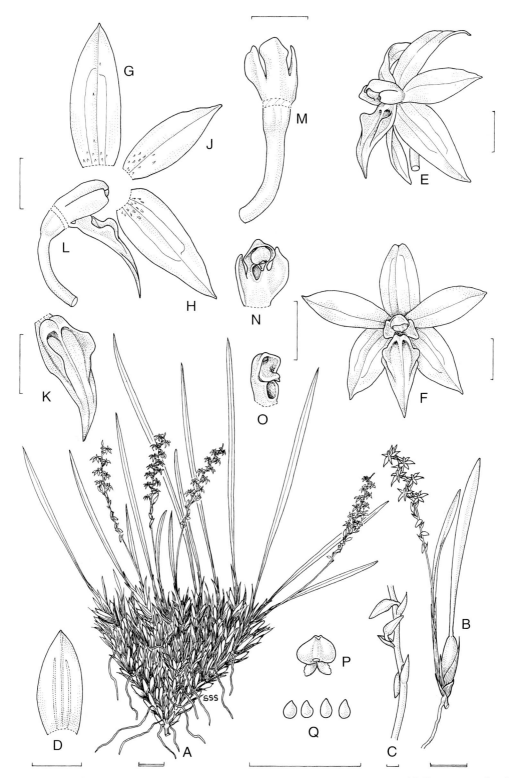

Figure 46. Dendrochilum joclemensii Ames - A & B: habits. - C: lower portion of inflorescence showing flower buds. - D: floral bract. - E: flower, oblique view. - F: flower, front view. - G: dorsal sepal. - H: lateral sepal. - J: petal. - K: lip, oblique view. - L: pedicel with ovary, lip and column, side view. - M: pedicel with ovary and column, back view. - N: column, oblique view. - O: column apex, anther-cap removed, oblique view. - P: anther-cap, back view. - Q: pollinia. A drawn from *Carr C.* 3710 and B–Q from *Gunsalam 10* by Susanna Stuart-Smith. Scale: single bar = 1 mm; double bar = 1 cm.

strongly concave, almost orbicular, entire; stelidia *c.* 0.8(–1) × 0.5 mm, basal, oblong, rounded to truncate, rather fleshy, erect; anther-cap minute, cucullate. Plate 10B & C.

HABITAT AND ECOLOGY: A twig epiphyte in lower montane ridge forest on ultramafic and sandstone substrates. Also recorded as a terrestrial on a mossy roadside bank by Lamb (pers. comm.). Alt. 1050 to 2000 m. Flowering observed in July and from October until December.

DISTRIBUTION IN BORNEO: SABAH: Crocker Range, Kimanis road; Mt. Kinabalu.

GENERAL DISTRIBUTION: Endemic to Borneo.

NOTE: *Dendrochilum joclemensii* belongs to section *Eurybrachium*.

DERIVATION OF NAME: Named to commemorate Joseph Clemens (1862–1936), a U.S. Army chaplain who, together with his wife Mary Strong Clemens and others, made extensive plant collections on Mt. Kinabalu.

47. DENDROCHILUM KELABITENSE J.J. Wood

Dendrochilum kelabitense *J.J. Wood* in Beaman *et al.,* Orchids of Sarawak : 276, fig. 57 (2001). Type: Borneo, Sarawak, Marudi District, trail from Bario to Pa Berang, 1000 m, March 1997, *Leiden* cult. (*Roelfsema, Schuiteman* & *Vogel*) 970198 (holotype L, isotype K, spirit material only).

Epiphyte. Rhizome not seen. **Roots** smooth. **Cataphylls** 4, 1–4.5 cm long. **Pseudobulbs** 2.2 × 1 cm, ellipsoid. **Leaf** (immature) 7.5 × 0.1 cm, linear, acicular, acute, somewhat fleshy, deeply canaliculate, rigid. **Inflorescences** subdensely many-flowered, decurved to pendulous; flowers borne 2.8–3 mm apart, lowermost up to 4 mm apart; peduncle 3–5 cm long, slender; non-floriferous bracts absent; rachis 9.5–13 cm long, quadrangular; floral bracts 2–3 mm long, narrowly triangular, acuminate, 3-nerved. **Flowers** opening from bottom of inflorescence, colour not recorded, but probably white or greenish. **Pedicel** and **ovary** 1.8–1.9 mm long, narrowly clavate. **Sepals** and **petals** 3-nerved. **Dorsal sepal** 4.4–4.5 × 1 mm, narrowly elliptic, acuminate. **Lateral sepals** 4–4.2 × 1.1 mm, narrowly elliptic, acuminate. **Petals** 4 × 0.8–0.9 mm, linear-lanceolate, acuminate. **Lip** very shortly stipitate to column-foot, 3-lobed, 3 mm long, 1 mm wide across mid-lobe, decurved, 3-nerved; side lobes *c.* 6 × 0.1 mm, narrowly triangular, acute, entire, outer margin slightly incurved at base, tooth-like, erect; mid-lobe *c.* 2.1–2.2 mm long, narrowly elliptic from a 0.5–0.6 mm wide base, acute, entire; disc with 3 low keels along nerves and joined at base, extending from base and terminating above base of mid-lobe, median less pronounced. **Column** *c.* 2 mm long, curving forward; foot *c.* 0.2 mm long; apical hood shallowly irregularly tridentate, median tooth sometimes obscurely bidentate; rostellum prominent, ovate, acute; stelidia borne just below stigmatic cavity, equalling apical hood, *c.* 1.2 mm long, sigmoid, falcate, acute; stigmatic cavity narrow; anther-cap *c.* 0.6–0.7 × 0.6–0.7 mm, cucullate, smooth.

HABITAT AND ECOLOGY: Peaty undisturbed kerangas forest. Alt. 1000 m. Flowering observed in March.

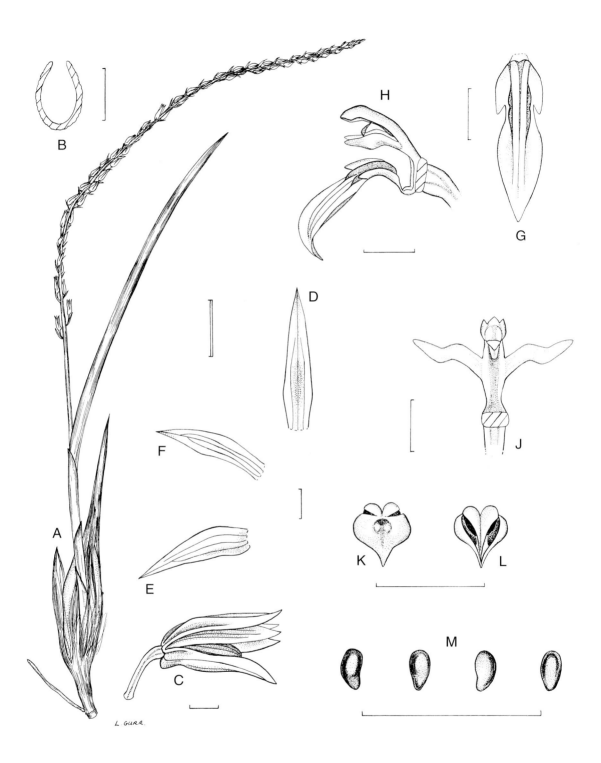

Figure 47. Dendrochilum kelabitense J.J. Wood - A: habit. - B: leaf, transverse section. - C: flower, side view. - D: dorsal sepal. E: lateral sepal. - F: petal. - G: lip, front view. - H: lip and column, side view. - J: column, with stelidia opened out, front view. - K: anther-cap, back view. - L: anther-cap, interior view. - M: pollinia. All drawn from *Roelfsema, Schuiteman & Vogel* LC 970198 (holotype) by Linda Gurr. Scale: single bar = 1 mm; double bar = 1 cm.

DISTRIBUTION IN BORNEO: SARAWAK: Marudi District, Kelabit Highlands.

GENERAL DISTRIBUTION: Endemic to Borneo.

NOTES: Wood (2001) enumerated 81 species of *Dendrochilum* from Borneo and predicted that additional undescribed taxa almost certainly occur, particularly in little investigated montane areas of the island. This rather undistinguished species from the Kelabit Highlands of Sarawak is one of these. Its floral morphology tentatively places it closest to *D. angustipetalum* Ames, a species so far known only from between 1200 and 2000 m on Mt. Kinabalu and Mt. Tembuyuken in Sabah. *Dendrochilum kelabitense* can be distinguished by its linear, acicular, somewhat fleshy, deeply channelled leaves, lack of non-floriferous bracts, broader petals, and lip with entire side lobes and a less prominent, not elevate median keel on the disc. The type collection does not include rhizomes or mature leaves. Living material is held at the Hortus Botanicus of Leiden University in The Netherlands.

DERIVATION OF NAME: Named after the Kelabit Highland region of Sarawak from where the type was collected.

48. DENDROCHILUM LONGIPES J.J. Sm.

Dendrochilum longipes *J.J. Sm.* in Bull. Jard. Bot. Buitenzorg, ser. 2, 3: 55 (1912). Type: Borneo, Sarawak, Limbang District, Mt. Batu Lawi, Ulu Limbang, May 1911, *Moulton* 15 (holotype BO, possibly destroyed, isotypes L, SAR).

Dendrochilum mantis J.J. Sm. in Bull. Jard Bot. Buitenzorg, ser. 3, 11: 108 (1931). Type: Borneo, Kalimantan Timur, West Kutai (Koetai), summit of Mt. Kemul (Kemoel), 1800 m, 13 October 1925, *Endert* 3991 (holotype L).

Epiphyte, terrestrial, sometimes a **lithophyte**. **Rhizome** elongate, internodes 0.9–3 cm long, branching. **Roots** 0.5–2 mm in diameter, smooth. **Cataphylls** on rhizome 1–2 cm long, tubular, ovate, obtuse; at base of pseudobulbs 3–4, 0.5–9 cm long, ovate-elliptic, obtuse or subacute, accrescent, speckled. **Pseudobulbs** 1.5–6.5 cm apart, often borne at an acute angle to rhizome, 5–13 × 0.3–0.4 cm, narrowly cylindrical, cauliform. **Leaf-blade** 5.5–18 × 1–3.5 cm, narrowly elliptic, acute or shortly obtuse and somewhat conduplicate, often constricted below apex, stiff-textured, leathery, main nerves 5–9, with numerous small transverse nerves, pale brownish green or pale green; petiole 0.4–2 cm long, sulcate. **Inflorescences** erect, rigid, densely many-flowered, flowers borne 2–3.2 mm apart; peduncle 1–4 cm long, greenish brown or pinkish; non-floriferous bract solitary, or rarely 2, 2–3.5 mm long, ovate, obtuse, adpressed; rachis 17–28 cm long, quadrangular; floral bracts 2–3.7 × 2.7–2.8 mm, broadly ovate to semi-orbicular, obtuse, erose, spreading. **Flowers** fragrant, variously described as yellowish green, tinged brownish, lip light salmon-pink, or sepals and petals translucent greenish yellow, light brown, glistening flesh-pink or very pale green, flushed pink, lip pink, dull deep red, peach, pale orange or with red mid-lobe, yellow side lobes, calli yellow or creamy white, column yellow or pale flesh-pink. **Pedicel** and **ovary** 3 mm long, clavate. **Sepals** 3-nerved. **Dorsal sepal** 3–3.7 × 1.2–1.5 mm, ovate-oblong to oblong-elliptic, shortly acute to acuminate, concave. **Lateral sepals** 3–3.6 × 1.4–1.7 mm,

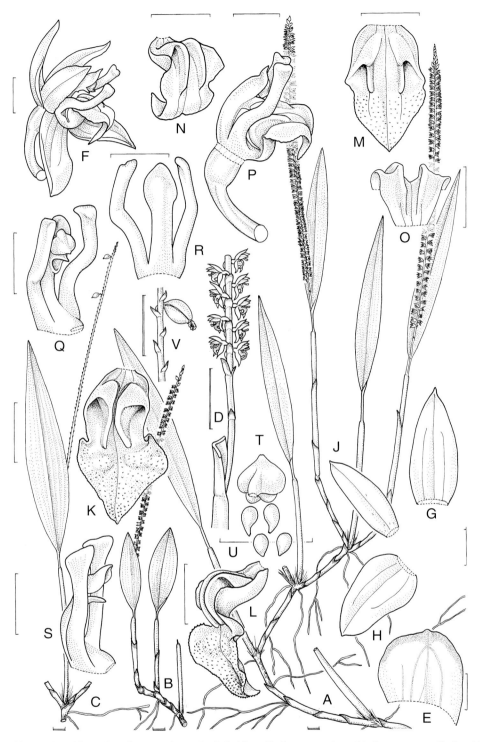

Figure 48. Dendrochilum longipes J.J. Sm. - A–C: habits. - D: lower portion of inflorescence. - E: floral bract. - F: flower, side view. - G: dorsal sepal. - H: lateral sepal. - J: petal. - K: lip, front view. - L: lip, side view. - M: lip, front view. - N: lip, oblique view. O: basal portion of lip showing keels, front view. - P: pedicel with ovary, lip and column, side view. - Q: column, oblique view. - R: column, back view. - S: column, side view. - T: anther-cap, back view. - U: pollinia. - V: fruit. A, K, L & S drawn from *Endert* 3991 (holotype of *D. mantis*), B from Yii S. 44432, C from *Kato et al.* B.11071, D–J, M–R and T–V from *Lewis* 345 by Susanna Stuart-Smith. Scale: single bar = 1 mm; double bar = 1 cm.

obliquely oblong-ovate, shortly acute to acuminate, concave, slightly keeled. **Petals** 2.8–3.5 × 0.8–1 mm, obliquely narrowly elliptic, sometimes somewhat falcate, acute, concave, 1- to obscurely 3-nerved. **Lip** subentire, strongly decurved at middle, 3-nerved, 2–2.6 mm long, 1.4 mm wide across side lobes when flattened; side lobes 1 mm long, obscure, erect, quadrangular or semi-orbicular and rounded, slightly undulate; mid-lobe 1.5–1.6 × 1.5–1.8 mm, broadly ovate-triangular, rounded at base, acute, undulate, papillose; disc with 2 prominent, tall, semi-orbicular keels adnate above base to margin of side lobes and terminating at or just beyond junction of side lobes and mid-lobe, appearing as a flange each side running to margin of side lobes and forming a concave area. **Column** 2 mm long, curved; foot distinct; apical hood obtuse and rounded, or obscurely 3-lobed; rostellum shortly ovate-triangular, acute, convex; stelidia 2.2–3.1 mm long, basal, linear-ligulate, dilated and obliquely obtuse, spathulate-truncate and convex at apex, often subfalcate; anther-cap 0.5–0.6 mm wide, cucullate, broadly ovate-triangular, apex narrowly truncate.

HABITAT AND ECOLOGY: Lower and upper montane ridge forest; mossy forest; exposed places in cloud forest associated with *Nepenthes* spp., *Rhododendron* spp. and *Vaccinium* spp.; montane ericaceous scrub; on boulders. Alt. 1400 to 2400 m. Flowering observed from April to June, August to October, and December.

DISTRIBUTION IN BORNEO: KALIMANTAN BARAT: Serawai, Raya Hill. KALIMANTAN TENGAH: Raya Hill, Katingan and Samba River areas. KALIMANTAN TIMUR: Mt. Kemul; Mt. Buduk Rakik, Apo Kayan area. SARAWAK: Mt. Batu Lawi; Batu Tiban Hill; Mt. Mulu; Mt. Murud.

GENERAL DISTRIBUTION: Endemic to Borneo.

NOTES: J.J. Smith described *D. mantis* from Mt. Kemul as late as 1931 stating that it "must be a near ally of *D. subintegrum* Ames." From examination of the type it is clear that *D. mantis* is conspecific with *D. longipes* which Smith had described much earlier in 1912. He apparently either overlooked or had forgotten *D. longipes* when studying the material from Mt. Kemul.

DERIVATION OF NAME: The specific epithet is derived from the Latin *longus*, long, elongate and refers to the elongated rhizomes and pseudobulbs.

49. DENDROCHILUM LONGIRACHIS Ames

Dendrochilum longirachis *Ames*, Orchidaceae 6: 60–62 (1920). Type: Borneo, Sabah, Mt. Kinabalu, Kiau, 29 Nov. 1915, *J. Clemens* 332 (holotype AMES, isotypes BO, K).

Epiphyte. **Rhizome** to 3 cm long, branching. **Roots** 0.2–1.5 mm in diameter, branching, smooth, forming a dense mass. **Cataphylls** 3–4, 0.5–5 cm long, ovate-elliptic, obtuse to acute, speckled. **Pseudobulbs** (1.5–)5–9.5 × 0.5–0.7 cm, crowded, cylindrical to narrowly fusiform, broadest proximally, shiny green. **Leaf-blade** (4–)8–15(–19) × (1.5–)3–4.3 cm, oblong-elliptic to elliptic, rounded to acute, thin-textured, main nerves (7–)9, with numerous tiny transverse nerves; petiole (2–)6–11 mm long, sulcate. **Inflorescences** pendulous, laxly up to 130 or more-flowered,

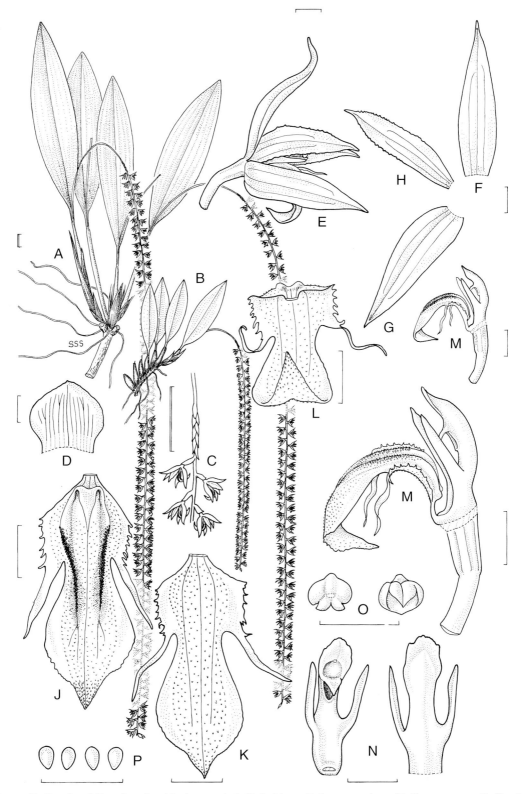

Figure 49. Dendrochilum longirachis Ames - A & B: habits. - C: lower portion of inflorescence. - D: floral bract. - E: flower, side view. - F: dorsal sepal. - G: lateral sepal. - H: petal. - J: lip, flattened, front view. - K: lip, flattened, back view. - L: lip, natural position, back view. - M: pedicel with ovary, lip and column, side view. - N: column, anther-cap removed, front and back views. - O: anther-cap, back and interior views. - P: pollinia. A drawn from *Nooteboom & Chai* 01890, B from *Richards S.* 483, and C–P from *Wood* 694 by Susanna Stuart-Smith. Scale: single bar = 1 mm; double bar = 1 cm.

flowers borne (2.5–)4 mm apart; peduncle 3–8.5 cm long, filiform, slightly dilated distally, brownish pink; non-floriferous bracts (3–)5–9, 1.5–2 mm long, ovate, obtuse, erose, pinkish brown, rachis 20–45 cm long, quadrangular, brownish pink; floral bracts 2–2.5 mm long, broadly ovate, rounded, somewhat erose, salmon-pink. **Flowers** slightly sweet-scented, sepals and petals greenish yellow, palest green, translucent yellow, dull greenish cream or white, lip yellow or pale green, side lobes often white, keels brown, column pale yellow. **Pedicel** and **ovary** 2–2.2 mm long, narrowly clavate. **Dorsal sepal** 4.5–5.6 × 1.6–1.7 mm, oblong-elliptic, acute. **Lateral sepals** 4.5–5.6 × 1.6–1.7 mm, slightly obliquely oblong-elliptic, acute. **Petals** 4–5.5 × 1 mm, linear-elliptic, acute, entire or minutely erose. **Lip** stipitate to column-foot by a narrowly oblong claw 0.1–0.2 mm long, 3-lobed, curved, 3-nerved, (3.5–)4.5–4.6 mm long, 1.8–1.9 mm wide across side lobes, minutely papillose; side lobes narrowly triangular with a narrow setaceous free apical portion 1.5–1.6 mm long, reaching to middle of mid-lobe, outer margin irregularly serrate; mid-lobe 2.5–2.9 × 1.5–2 mm, cuneate-obovate, apiculate, margin entire or minutely erose; disc with 2 erect-spreading, fleshy, partially papillose keels joined at base with thickened or elevate median nerve, extending from apex of claw and terminating on proximal portion of mid-lobe, elevate portion of median nerve terminating about halfway along disc. **Column** 2.5–2.6 mm long; foot short; apical hood oblong, rounded to truncate, entire or obscurely erose; rostellum ovate-triangular, rostrate; stelidia 1.4–1.5 mm long, arising just below level of stigmatic cavity, linear-subulate, acute; anther-cap cucullate, minutely papillose.

HABITAT AND ECOLOGY: Riverine lower montane forest with *Rhododendron* understorey; dense primary forest to 30 metres high dominated by *Agathis* spp. and *Lithocarpus* spp. on poor sandy soil over sandstone. Alt. 1100 to 1500(–2000) m. Flowering observed in January, April, and from September until December.

DISTRIBUTION IN BORNEO: KALIMANTAN TIMUR: Apo Kayan. SABAH: Crocker Range, Sinsuron road; Mt. Kinabalu; Sipitang District. SARAWAK: Mt. Dulit; Mt. Murud.

GENERAL DISTRIBUTION: Endemic to Borneo.

NOTE: *Dendrochilum longirachis* belongs to section *Platyclinis*.

DERIVATION OF NAME: The specific epithet is derived from the Latin *longus*, long, and *rachis*, the rachis, ie. the axis of the inflorescence above the peduncle, which is very elongate in this species.

50. DENDROCHILUM SIMPLEX J.J. Sm.

Dendrochilum simplex *J.J. Sm.* in Bull. Dép. Agric. Indes Néerl. 22: 13–14 (1909). Type: Borneo, Kalimantan Tengah, Liangangang (Liang Gagang), *Hallier* 2646 (holotype BO, isotypes K, L).

Dendrochilum remotum J.J. Sm. in Bull. Jard. Bot. Buitenzorg, ser. 2, 3: 57–58 (1912). Type: Borneo, Sarawak, Limbang District, Mt. Batu Lawi, Ulu Limbang, *Moulton* 10 (holotype SAR).

Figure 50. Dendrochilum simplex J.J. Sm. - A: habit. - B: floral bract. - C: flower, side view. - D: lip, front view. - E: lip and column, side view. - F: column, front view. - G: pollinia. - H: floral bract. - J: flower, side view. - K: dorsal sepal. - L: lateral sepal. - M: petal. - N: lip, front view. - O: pedicel with ovary, lip and column with part of lateral sepal and petal, side view. - P: column, oblique view. - Q: column, side view. - R: column, foot and anther-cap removed, oblique view. - S: column, back view. - T: anther-cap, front and back views. - U: pollinia. A–G drawn from *Hallier* 2646 (holotype), H–U from *Vermeulen & Lamb* 324 by Susanna Stuart-Smith. Scale: single bar = 1 mm; double bar = 1 cm.

Terrestrial, epiphytic or **lithophytic**. **Rhizome** up to *c.* 40 cm long, branching, covered when young with persistent reddish brown sheaths. **Roots** elongate, 0.1–0.2 mm in diameter, primary roots usually minutely papillose, producing a tangle of branching smooth secondary roots distally. **Cataphylls** 5–6, 0.3–3.6 cm long, ovate-elliptic, acute to acuminate, persistent, reddish brown. **Pseudobulbs** 1–5.5 cm apart, 0.8–2.75 × 0.3–0.5 cm, narrowly oblong to narrowly fusiform, smooth, spreading. **Leaf-blade** 2.5–16 × 0.5–1.8 cm, ovate-elliptic to narrowly elliptic, often constricted below apex, acute to shortly acuminate, sometimes narrowly obtuse, thin-textured, main nerves 5–7; petiole 0.15–1.5 cm long, sulcate to canaliculate. **Inflorescences** gently curving, laxly to subdensely many-flowered, flowers borne 1.5–2.2(–2.5) mm apart; peduncle 2–7.5 cm long, filiform; non-floriferous bracts 1–6, 2–3 mm long, ovate, apiculate to acuminate; rachis 4–14 cm long, quadrangular, slightly compressed; floral bracts 2–2.9 mm long, ovate-orbicular or broadly ovate, apiculate, acute, sometimes obtuse, slightly erose, concave. **Flowers** fragrant, creamy white to white, pale green, yellowish green, creamy gold or yellowish cream, lip often greenish at base. **Pedicel** and **ovary** 0.75–2 mm long, slender, geniculate or straight. **Dorsal sepal** 3.7–4 × 0.7–1.1 mm, narrowly oblong to linear-lanceolate, apex acute and often recurved. **Lateral sepals** 3–3.9 × 0.7–1 mm, obliquely narrowly lanceolate, falcate, acute. **Petals** 2.9–3.8 × 0.6–1 mm, narrowly lanceolate, falcate, acute, minutely erose-crenulate. **Lip** entire or shallowly 3-lobed, shortly clawed, decurved, 3-nerved, 1.1–1.6 × 0.6–1 mm, oblong, shortly acute; side lobes (when present) erect, small, shallowly rounded, somewhat crenulate; mid-lobe (when present) *c.* 0.4 mm long, semi-orbicular to ovate; disc with 2 low keels united below to form a U-shape. **Column** 1.2–1.75 mm long; foot 0.2–0.4 mm long; apical hood oblong-quadrangular, obtuse to acute, or variously toothed; rostellum recurved, shortly triangular, acute, convex; stelidia *c.* 0.3–1 mm long, arising either side of stigmatic cavity, lanceolate, subulate, subfalcate, acute to acuminate; anther-cap minute, cucullate.

HABITAT AND ECOLOGY: Lower montane forest, on sandstone and ultramafic substrates; podsol forest with *Dacrydium* spp. and dipterocarps on very wet sandy soil, often associated with *Bromheadia finlaysoniana*, *Bulbophyllum nabawanense*, *Nepenthes ampullaria* and aroids; rocky places; riparian forest. Alt. 400 to 2000 m. Flowering observed in March, and from May until December.

DISTRIBUTION IN BORNEO: KALIMANTAN TENGAH: Liangangang. SABAH: Crocker Range, Mt. Alab, Sinsuron road; Mt. Kinabalu; Mt. Monkobo; Nabawan area; Penampang District, Tunggul Togudon; Pig Hill. SARAWAK: Mt. Batu Lawi; Mt. Murud; Mt. Penrissen; Hose Mountains; Mengiong/Balleh Rivers, Upper Entulu.

GENERAL DISTRIBUTION: Endemic to Borneo.

NOTES: *Dendrochilum simplex*, belonging to section *Platyclinis*, varies in habit according to habitat. Robust terrestrial populations growing in shady podsol forest near Nabawan in Sabah have leaves up to 16 cm long. Others from Sarawak, referred to *D. remotum* by J.J. Smith, have smaller leaves and flowers. Shallow lobing of the lip, reported in *D. remotum*, also occurs in larger flowered populations.

DERIVATION OF NAME: The Latin specific epithet refers to the simple, undivided lip found in many populations.

51. DENDROCHILUM SUBULIBRACHIUM J.J. Sm.

Dendrochilum subulibrachium *J.J. Sm.* in Bull. Jard. Bot. Buitenzorg, ser. 3, 11: 109–110 (1931). Type: Borneo, Kalimantan Timur, Long Petak, 800 m, 12 September 1925, *Endert* 3221 (holotype L, not located).

Epiphyte. **Rhizome** to *c.* 3 cm long. **Roots** elongate, with very few branches, minutely papillose, 0.2–0.5 mm in diameter. **Cataphylls** 3, 0.4–2 cm long, ovate-elliptic, obtuse to acute, speckled. **Pseudobulbs** 0.5–1.1 × 0.4–0.5 cm, grouped close together, oblong-ovoid. **Leaf-blade** 2.5–6.5 × 0.1–0.4 cm, linear, obtuse, rigid, fleshy, main nerves 3; petiole 0.3–1 cm long, slender, sulcate. **Inflorescences** gently curving, produced with the leaf nearly fully expanded, laxly 9- to *c.* 28-flowered, flowers borne 2.5–3.5 mm apart; peduncle 5–9 cm long, filiform; non-floriferous bracts absent or rarely solitary, 3 mm long; rachis 3–9 cm long, quadrangular, often flexuose; floral bracts 2(–3.5) × 1 mm, obtuse, erose distally. **Flowers** pale greenish to yellow, with 2 pale brown longitudinal central streaks on lip. **Pedicel** and **ovary** 1.5–2 mm long, narrowly clavate. **Sepals** and **petals** incurved. **Dorsal sepal** 3.7–4.1 × 1.4–1.5 mm, oblong-ovate, obtuse to acute. **Lateral sepals** 3.7–4 × 1.5 mm, obliquely oblong-elliptic or subovate-oblong, acute. **Petals** 3.5–3.9 × 1 mm, narrowly ovate-elliptic, acute to acuminate, sometimes slightly erose **Lip** stipitate to column-foot by a short claw, 3-lobed, 3-nerved, gently undulate, 3.5–4 mm long; side lobes *c.* 1.4 × 2–2.3 mm, spreading, transversely quadrangular, truncate at base, apex acuminate, outer margin irregularly lacinulate; mid-lobe 2–2.5 × 2.7–3 mm, obovate-truncate from a narrow cuneate base, minutely erose distally; disc with 2 prominent, smooth keels united at base, terminating above base of mid-lobe, median nerve slightly elevate. **Column** 2–2.5 mm long, incurved, fleshy at base, slender above; foot prominent; apical hood entire, obtuse, recurved; rostellum ovate-triangular, obtuse, convex, obscurely carinate; stelidia 1 mm long, borne above the base, linear-subulate, acute, reaching to base of stigmatic cavity; anther-cap cucullate, with an apical wart.

HABITAT AND ECOLOGY: Lower montane ridge forest; hill forest. Alt. 800 to 1730 m. Flowering observed from August until October.

DISTRIBUTION IN BORNEO: KALIMANTAN TIMUR: Long Petak. SARAWAK: Limbang District, Bario, Pa Mario River; Mt. Mulu.

GENERAL DISTRIBUTION: Endemic to Borneo.

NOTES: This graceful species belongs to section *Platyclinis*. A closely related species, *D. magaense* J.J. Wood, was figured in *Orchids of Borneo*, volume 3: fig. 57.

DERIVATION OF NAME: The specific epithet is derived from the Latin *subula*, a fine sharp point, and *brachium*, an arm, in reference to the subulate stelidia on the column.

Figure 51. Dendrochilum subulibrachium J.J. Sm. - A & B: habits. - C: floral bract. - D: flower, side view. - E: dorsal sepal. - F: lateral sepal. - G: petal. - H: lip, front view. - J: pedicel with ovary and column, anther-cap removed, side view. - K: column, oblique view. - L: column, back view. - M: anther-cap, back view. - N: pollinia. A & C–N drawn from *Lewis* 349 and B from *Awa & Lee* S. 50756 by Susanna Stuart-Smith. Scale: single bar = 1 mm: double bar = 1 cm.

52. DENDROCHILUM TRANSVERSUM Carr

Dendrochilum transversum *Carr* in Gard. Bull. Straits Settl. 8: 233–234 (1935). Type: Borneo, Sabah, Mt. Kinabalu, Marai Parai Spur, May 1933, *Carr* 3477, SFN 27431 (holotype SING, isotypes AMES, K, LAE).

Epiphyte. **Rhizome** up to 6 cm long, branching. **Roots** 1 mm in diameter, branching, smooth. **Cataphylls** 3–4, 0.5–5.5 cm long, ovate-elliptic, acute, pale brown, speckled darker brown. **Pseudobulbs** 2.5–4.5 × 0.5–0.7 cm, crowded, narrowly cylindric, green, suffused red, or entirely red. **Leaf-blade** 9–23 × 0.6–1.2 cm, linear-lanceolate or linear-oblanceolate, abruptly narrowed below an acute, shortly cuspidate apex, thin-textured, main nerves 5, deep olive-green; petiole 1.5–4 cm long, sulcate. **Inflorescences** erect to curving, laxly many-flowered, flowers borne 2–3 mm apart; peduncle (9–)13–20 cm long, very slender; non-floriferous bracts 1–2, 3.5–5 mm long, ovate-elliptic, acute; rachis 7–9 cm long, quadrangular; floral bracts 2.8–3.8 mm long, ovate, apiculate. **Flowers** 7–8 mm across, sepals and petals rose-pink or salmon-pink, lip deep ochre with darker apex and margins, column pinkish brown, stelidia ochre tipped dark ochre-brown, occasionally entire flower may be orange-yellow. **Pedicel** and **ovary** 2.8–4.5 mm long, narrowly clavate, slightly curved. **Dorsal sepal** 4.7–5 × 2.1 mm, ovate-elliptic, acute. **Lateral sepals** 4.5–4.7 × 2–2.5 mm, obliquely ovate-elliptic, acute. **Petals** 4.5–5 × 1.6–1.7 mm, oblong-elliptic, acute, sometimes twisted and aligned up to 90° from vertical. **Lip** entire, 2.6–3 × 4–4.4 mm, transversely oblong, subapiculate, recurved distally, margins recurved in upper half, minutely erose; disc provided with a tall, broadly horseshoe-shaped keel enclosing a concave area, extending 3/4 the length of the lip; surface minutely papillose, especially distally. **Column** 1(–1.7) mm long; foot absent; apical hood slightly dilated and shallowly 3-lobed or entire and transversely elliptic; stelidia 1.1–1.2 mm long, basal, oblong, obtuse, minutely papillose distally; margins of stigmatic cavity prominent, elevated; anther-cap not seen.

HABITAT AND ECOLOGY: Twig epiphyte in upper montane mossy forest on ultramafic and non-ultramafic substrates. Alt. 2400 to 2800 m. Flowering observed in February and May.

DISTRIBUTION IN BORNEO: SABAH: Mt. Kinabalu.

GENERAL DISTRIBUTION: Endemic to Borneo.

NOTES: *Dendrochilum transversum* is one of nine species of section *Eurybrachium* that are endemic to the higher slopes of Mt. Kinabalu. Many of these are restricted to ultramafic localities which are isolated and, rather like islands, act as barriers to dispersal. In addition, there are no neighbouring mountains of comparable elevation upon which new populations could establish via wind-blown seed.

DERIVATION OF NAME: The specific epithet refers to the broad transversely oblong lip.

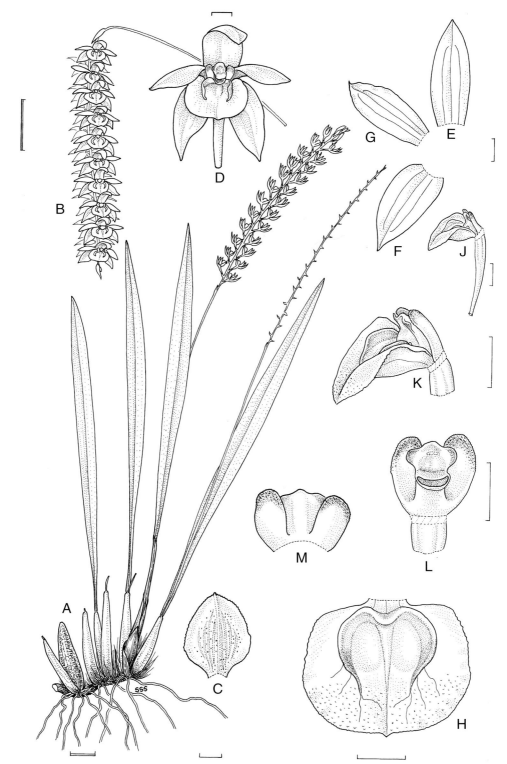

Figure 52. Dendrochilum transversum Carr. - A: habit. - B: inflorescence. - C: floral bract. - D: flower, front view. - E: dorsal sepal. - F: lateral sepal. - G: petal. - H: lip, front view. - J: pedicel with ovary, lip and column, side view. - K: upper portion of ovary, lip and column, side view. - L: upper portion of ovary and column, anther-cap lost, front view. - M: column, back view. A, C & E–M drawn from *Carr* C. 3477, SFN 27431(holotype), and B & D from *Barkman et al.* 95 by Susanna Stuart-Smith. Scale: single bar = 1 mm; double bar = 1 cm.

53. DILOCHIA PARVIFLORA J.J. Sm.

Dilochia parviflora *J.J. Sm.* in Bull. Jard. Bot. Buitenzorg, ser. 3, 11: 111 (1931). Type: Borneo, Kalimantan Timur, Kutai (West Koetai), Mt. Kemal (Kemoel, Kemul), 1800 m, Oct. 1925, *Endert 4262* (holotype L, isotype BO).

Arundina parviflora (J.J. Sm.) Masam., Enum. Phan. Born.: 120 (1942).

Terrestrial or **epiphytic**. **Stems** several, leafy, up to 1 m long, internodes 0.8–2 cm long, uppermost shorter. **Leaves** bifarius; blade (3–)4.5–8 × (1–)1.5–2.6 cm, oblong-ovate, long-acuminate or acute, shortly contracted at base, coriaceous, glossy above, dull beneath, main nerves *c.* 11; sheath 0.8–2 cm long, tubular, apex truncate. **Inflorescence** terminal, paniculate, 8–12 cm long; peduncle 1.5–3.5 cm long; branches 5–7, 4–5 cm long, spreading, slightly curved, with up to 10 flowers, rachis angular, internodes 3–7 mm long; bracts subtending branches 1–2 cm long, clasping branch, cymbiform, concave, acute to shortly acuminate, persistent, brown; floral bracts 5–8 × 9 mm, cymbiform, concave, semi-ovate, shortly acuminate or conical-apiculate, densely punctate, many-nerved, cream coloured to brown. **Flowers** *c.* 1.3 cm across, sepals and petals cream to greenish, finely spotted magenta-purple on exterior, interior cream to pale pinkish, lip bright magenta-purple with cream side lobes and keels, column cream, purplish at base, anther-cap purple. **Pedicel** and **ovary** 5–7 mm long, pedicel very slender, ovary obconical. **Dorsal sepal** 9.5 × 4 mm, oblong-elliptic, obtuse, apiculate, concave, incurved, dorsally slightly carinate, 5-nerved. **Lateral sepals** 1 × 0.4 cm, obliquely subovate-oblong, apex narrowed, acute, concave, suberect, conspicuously carinate, especially at apex. **Petals** 0.9 × 0.27 cm, narrowly obliquely ovate-oblong, obtuse, erect to spreading, recurved above, abruptly shortly clawed at base, claw 1 mm long, 3-nerved. **Lip** 3-lobed, aligned parallel with column, *c.* 9 mm long, *c.* 5 mm wide across side lobes, concave, especially at base, recurved at apex, longitudinally sulcate below; side lobes *c.* 1.7 × 1.7 mm, erect, obliquely triangular-ovate, obtuse, margin irregular; mid-lobe *c.* 4 × 4.4 mm, suborbicular, broadly rounded to truncate, shortly obtusely trilobulate, margin irregular, convex; disc with 3 longitudinal parallel keels, rarely verruculose below, usually verruculose on mid-lobe, lateral ones much thickened in the lower part. **Column** *c.* 7.6 mm long, slender, margins alate, especially in the middle, foot 2 mm long; anther-cap transversely somewhat ovate, truncate; pollinia 8.

HABITAT AND ECOLOGY: Hill and lower montane forest on limestone; padang vegetation in shallow peat overlying fine sand; kerangas. Alt. 600 to 1800 m. Flowering observed in June, August and October.

DISTRIBUTION IN BORNEO: KALIMANTAN TIMUR: Mt. Kemal. SABAH: Mt. Kinabalu. SARAWAK: Gunung Mulu National Park; Belaga District, Batu Laga; Marudi District, Pa Main/Marauro River.

GENERAL DISTRIBUTION: Endemic to Borneo.

NOTES: *Dilochia* (tribe *Arethuseae*) is a small genus of about six or seven species distributed from Myanmar (Burma), Thailand and Peninsular Malaysia though Indonesia, eastwards to the

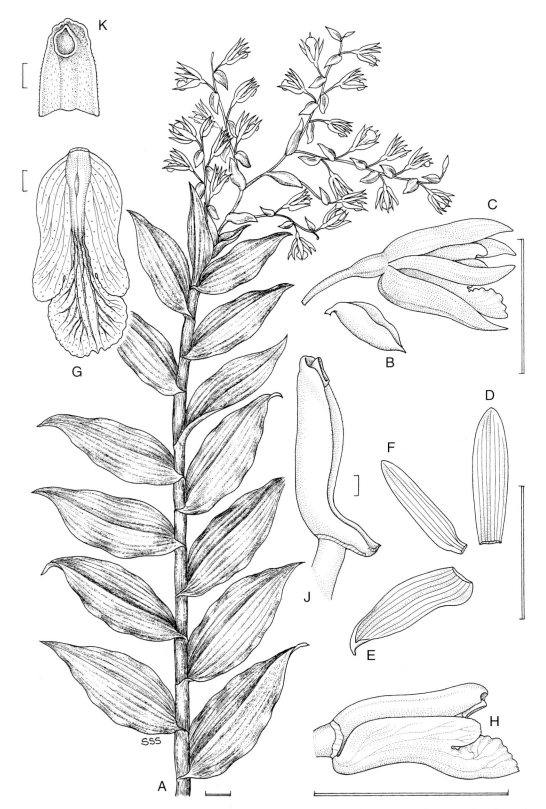

Figure 53. Dilochia parviflora J.J. Sm. - A: habit. - B: floral bract. - C: flower, side view. - D: dorsal sepal. - E: lateral sepal. - F: petal. - G: lip, flattened, front view. - H: column and lip, side view. - J: column, side view. - K: column apex, anther-cap lost, front view. A–K drawn from *Mohtar* S. 48082 by Susanna Stuart-Smith. Scale: single bar = 1 mm; double bar = 1 cm.

Philippines and New Guinea. Five species are represented in Borneo. Although closely related to *Arundina*, the inflorescences are usually branched, the flowers are always much smaller, are subtended by large, concave, deciduous bracts and have narrower petals.

DERIVATION OF NAME: The generic name is derived from the Greek *di-*, double, and *lochos*, row, rank, in reference to the distichous disposition of the leaves and, in some species, bracts. The specific epithet is derived from the Latin *parvus*, small, and *floralis*, floral, relating to the flower, referring to the rather small flowers of this species.

54. DILOCHIA RIGIDA (Ridl.) J.J. Wood

Dilochia rigida (*Ridl.*) *J.J. Wood* in Wood *et al.*, Plants of Mt. Kinabalu 2, Orchids: 206 (1993). Type: Borneo, Sabah, Mt. Kinabalu, *Haviland* 1251 (holotype K).

Bromheadia rigida Ridl. in Stapf in Trans. Linn. Soc. London, Bot. 4: 239 (1894).
Arundina gracilis Ames & C. Schweinf., Orchidaceae 6: 96 (1920). Type: Borneo, Sabah, Mt. Kinabalu, Marai Parai Spur, *J. Clemens* 370 (holotype AMES, isotype BM).
Dilochia gracilis (Ames & C. Schweinf.) Carr in Gard. Bull. Straits Settlem. 8: 91 (1935).

Epiphyte or **terrestrial. Rhizome** elongate, branching, creeping, rooting profusely at nodes. **Stem** 25–60 cm or more long, slender, erect, rigid, rooting at lower nodes, sometimes with a few short branches, enclosed by persistent leaf sheaths, internodes (1–)1.4–3(–3.5) cm long. **Leaf-blade** (3–)10–13.5 × (0.4–)0.9–1.1(–1.3) cm, distichous, spreading, linear, narrowed at base, apex obtuse or minutely unequally bilobed, minutely mucronate and minutely erose, rigid, thick, coriaceous, margins somewhat revolute, mid-nerve sulcate above, carinate beneath, uppermost leaves usually abruptly much shorter; sheaths (1–)1.4–3(–3.5) cm long, coriaceous, striate, turning brownish green or sometimes crimson-brown. **Inflorescence** terminal, unbranched, pendent or reflexed, 3- to 6(-7)-flowered; peduncle abbreviated; rachis 0.5–1 cm long, fractiflex; floral bracts 1.4–1.5 cm long, broadly ovate, cymbiform, conduplicate, dorsally carinate near apex, subacute to acute, lowermost often with long foliaceous tips, enveloping pedicel and ovary. **Flowers** often slightly sweetly scented, sepals cream or white, sometimes flushed green inside, exterior usually stained crimson or purplish down the centre, sometimes stained greenish purple on reverse, petals cream or white, flushed purple at base on reverse, lip white with a yellow patch on base of mid-lobe, sometimes pale lemon-yellow distally, sometimes with some purple staining between keels, column cream or white flushed red or purplish above and at base of foot. **Pedicel** and **ovary** 1–1.5 cm long, cylindrical to narrowly clavate. **Dorsal sepal** 1.8–2.25 × 0.65–0.75 cm, oblong-elliptic to narrowly elliptic, apex subacute to acute, slightly cucullate, main nerves 7(-9). **Lateral sepals** 1.6–2 × 0.51–0.65 cm, oblong-elliptic, subfalcate, acute, dorsally carinate, especially distally, main nerves 7. **Petals** 1.7–2.1 × 0.33–0.45 cm, linear-ligulate, subfalcate, obtuse to acute, main nerves 5. **Lip** 3-lobed, 1.6–1.7 cm long, 0.95–1 cm wide across side lobes, ovate-oblong in outline, thickened at base; side lobes erect, free part *c.* 2.7 mm long, 2.2–2.5 mm high, triangular-oblong, apex broadly rounded, lower margin sometimes irregular; mid-lobe 7–8 × 5.5–7 mm, oblong-subquadrate, broadly truncate, apiculate, convex at centre below apex; disc with 3 fleshy, irregular, undulate keels extending from base of lip to just below apex of mid-lobe, with an additional shorter lateral keel each side extending from middle of side lobes and terminating at varying points from

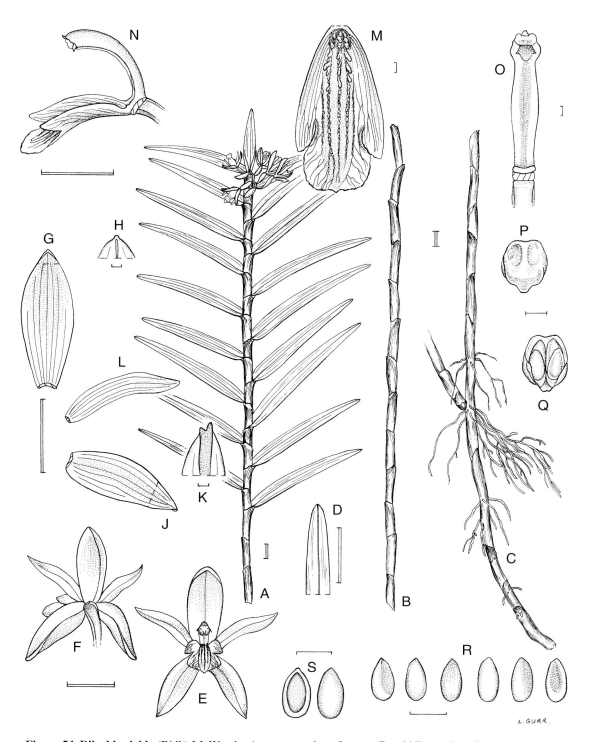

Figure 54. Dilochia rigida (Ridl.) J.J. Wood. - A: upper portion of stem. - B: middle portion of stem. - C: lower rooting portion of stem. - D: leaf apex, adaxial surface. - E: flower, front view. - F: flower, back view. - G: dorsal sepal. - H: close-up of apex of dorsal sepal, abaxial surface. - J: lateral sepal. - K: close-up of apex of lateral sepal, abaxial surface. - L: petal. - M: lip, spread out, front view. - N: lip and column, side view. - O: column, front view. - P: anther-cap, back view. - Q: anther-cap, interior view. - R: pollinia. - S: close-up of two pollinia. A–C drawn from *J. & M.S.Clemens* 50421 & *Comber* 147, D from *Comber* 147, and E–S from *Wood* 769 by Linda Gurr. Scale: single bar = 1 mm; double bar = 1 cm.

halfway or just beyond on mid-lobe. **Column** 1.4 × *c*. 0.29–0.3 cm, curved, flattened on lower surface, winged above, apical wing shortly toothed, foot obscure; rostellum ovate, obtuse; anther-cap *c*. 2.4 × 2.4 mm, cucullate; pollinia 8. Plate 10D.

HABITAT AND ECOLOGY: Lower and upper montane mossy forest, often on ridges, growing in mossy, often exposed sites, often associated with *Chelonistele sulphurea* var. *sulphurea* and *Epigeneium tricallosum*; montane heath forest of *Dacrydium, Ericaceae* and *Schima*, etc. with field layer of *Dipteris conjugata, Epigeneium tricallosum* and *Gleichenia* spp.; oak-laurel forest; often on ultramafic substrates; recorded as a common terrestrial in lower montane ridge forest composed of *Dacrydium, Leptospermum, Phyllocladus, Rhododendron* and *Tristania*, etc., with an understorey of *Gahnia*, bamboo and rattans. Alt. 1300 to 2400 m. Flowering observed throughout the year.

DISTRIBUTION IN BORNEO: SABAH: Crocker Range, eg. Mt. Alab, Mt. Lumaku; Mt. Kinabalu; Bukit Monkobo; Mt. Trus Madi; Sipitang District, Maga River area. SARAWAK: Mt. Dulit; Gunung Mulu National Park; Mt. Murud; Tama Abu Range.

GENERAL DISTRIBUTION: Endemic to Borneo.

NOTES: *Dilochia rigida* was originally placed in *Bromheadia* by Ridley. *Bromheadia*, however, apart from belonging to a different tribe (*Cymbidieae*), has regularly alternate, distichous bracts, flowers borne one or a few at a time in succession, and only two pollinia.

This is a characteristic plant of mossy montane forest where it is most frequently found growing as a terrestrial among mosses and other vegetation along exposed ridges. It spreads rapidly vegetatively, often forming large colonies, but seems not to be very free flowering.

DERIVATION OF NAME: The specific epithet is derived from the Latin *rigidus*, rigid, which describes the habit of the plant.

55. DIPLOCAULOBIUM BREVICOLLE (J.J. Sm.) Kraenzl.

Diplocaulobium brevicolle (*J.J. Sm.*) *Kraenzl.* in Engler, Pflanzenr. Orch. Monandr. IV. 50 II. B. 21: 335 (1910). Type: Origin unknown, cult. Bogor (holotype BO).

Dendrobium brevicolle J.J. Sm. in Ic. Bogor. 2: 82, t. 115B (1903).
Diplocaulobium vanleeuwenii sensu Wood & Cribb, Checklist Orch. Borneo: 273, plate 9F, (1994), *non* (J.J. Sm.) P.F. Hunt & Summerh. (1961).

Epiphyte. **Rhizome** up to 6 cm or more long, covered with brown sheaths, rooting profusely. **Pseudobulbs** up to 20 or more, congested, clump-forming, of one internode only, 2–5.5 × 0.3–0.6 cm, narrowly ovate, narrowed distally, laterally flattened, longitudinally sulcate, dark green or straw-coloured, partially in a papery sheath, 1-leaved. **Leaves** 6–9 × 0.3–0.5 cm, linear, apex equally acutely bidentate, attenuate at base, sessile, erect, 1-nerved. **Inflorescences** terminal, 1-flowered, emerging from a 0.9–1.5 cm long, straw-coloured, acute sheath, flowers borne in succession; peduncle (0.6–)1–1.5 cm long; floral bract *c*. 9 mm long, papery, with a few black hairs. **Flowers** *c*. 3.3 cm across, ephemeral, pedicel cream to yellowish white, ovary yellowish green, sepals and petals translucent pinkish white grading to purple-maroon distally, with darker

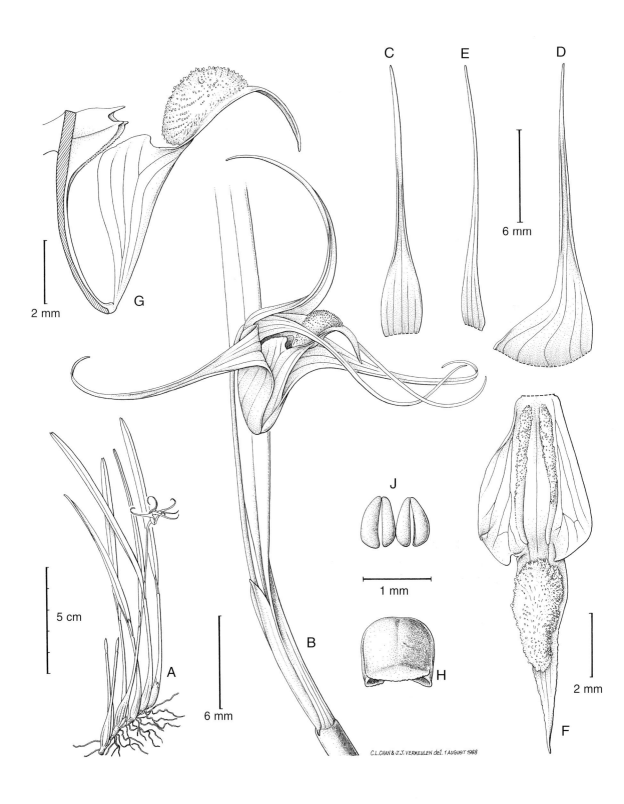

Figure 55. Diplocaulobium brevicolle (J.J. Sm.) Kraenzl. - A: habit. - B: flowering shoot. - C: dorsal sepal. - D: lateral sepal. - E: petal - F: lip, front view. - G: lip and column, side view. - H: anther-cap, back view. - J: pollinia. All drawn from material cultivated at *Tenom Orchid Centre* by C.L. Chan and Jaap Vermeulen.

purple edges, lip side lobes purple-maroon, mid-lobe translucent cream, dark purple at base, central cushion white, column white, foot stained purple-maroon, anther-cap cream. **Pedicel** and **ovary** (2–)2.7–2.8 cm long, slender. **Sepals** spreading. **Dorsal sepal** 1.75–1.8 × 0.25–0.35 cm, narrowly ovate, caudate, acute, 5-nerved. **Lateral sepals** 1.8 × 0.3 cm, 0.7 cm wide at base, triangular, caudate, acute, 5-nerved. **Mentum** 5.5–7 mm long, obtuse. **Petals** 1.1–1.7 × 0.15–0.2 mm, linear, acute, 3-nerved. **Lip** 1–1.2 cm long, 0.4–0.45 cm wide across side lobes; side lobes 5 mm long, *c.* 1–2 mm high, semi-obovate, rounded, erect; mid-lobe 5.5–6 × 1.3–2 mm, narrowly triangular, acuminate, deflexed, bearing a prominent, oblong, farinose-pilose cushion; disc between side lobes bearing 2 almost parallel glandular, papillose-hairy keels terminating at junction of side lobes and mid-lobe. **Column** 2 mm long; foot 5.5–7 mm long; anther-cap 1.8 × 1–1.1 mm, quadrate, cucullate; pollinia 4. Plate 10E.

HABITAT AND ECOLOGY: Podsol forest on sandstone; recorded as epiphytic on trunks of *Dacrydium pectinatum*; mixed dipterocarp forest. Alt. 50 to 700 m. Flowering observed in April, June and August.

DISTRIBUTION IN BORNEO: BRUNEI: Locality unknown. SABAH: Nabawan area; Telupid area. SARAWAK: Kuching District, near Kuching; Lubok Antu District, Lanjak Entimau Protected Forest; Lundu District, Upper Sematan River; Marudi District, Melinau River; Serian District, Majang Hill.

GENERAL DISTRIBUTION: Endemic to Borneo.

NOTES: *Diplocaulobium* was formerly treated as a section of *Dendrobium*. It is distinguished by having one-noded, 1-leaved pseudobulbs which taper from a fleshy, flask-shaped base into a slender neck from which one- or two-flowered, successive, terminal inflorescences are produced. The ephemeral flowers are borne on elongated pedicels and most commonly have narrowly caudate sepals and petals producing a rather spidery appearance.

About 100 species are currently recognised, the majority of which are endemic to New Guinea, with three only recorded from Borneo. One species extends west to Peninsular Malaysia while others are found in Australia and the Pacific islands. In New Guinea the numerous species are found in habitats as diverse as coastal mangrove and upper montane forest.

The plant depicted by Wood & Cribb (1994, plate 9F) and referred to *D. vanleeuwenii* (J.J. Sm.) P.F. Hunt & Summerh., is this species. This name has been in use in Sabah for the last ten years.

DERIVATION OF NAME: The generic name is derived from the Greek *diplous*, double, and *kaulos*, stem, and refers to the dimorphic appearance of the pseudobulbs. The specific epithet is derived from the Latin *brevis*, short, and *collum*, neck, neck-like prolongation, collar, in reference to the shape of the pseudobulbs.

56. DIPLOCAULOBIUM LONGICOLLE (Lindl.) Kraenzl.

Diplocaulobium longicolle (*Lindl.*) *Kraenzl.* in Engler, Pflanzenr. Orch. Monandr. IV. 50 II. B. 21: 340 (1910). Type: Singapore, *Cuming* s.n., cult. Loddiges (holotype K-LINDL).

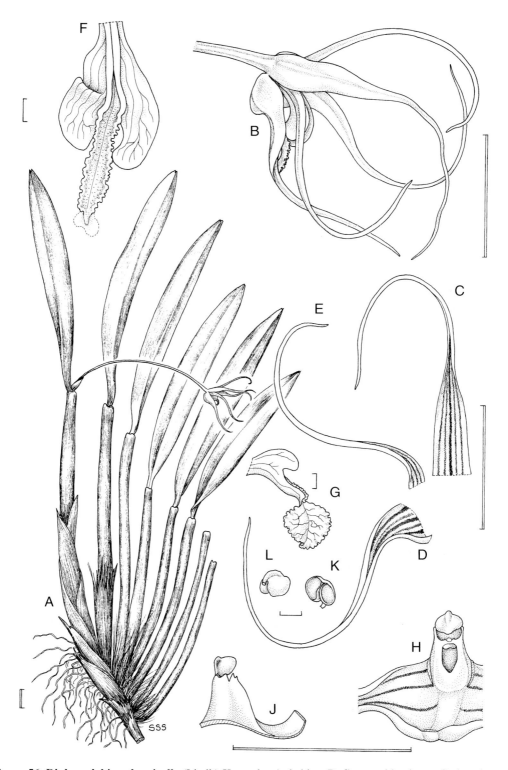

Figure 56. Diplocaulobium longicolle (Lindl.) Kraenzl. - A: habit. - B: flower, side view. - C: dorsal sepal. - D: lateral sepal. - E: petal. - F: lip, front view, epichile lost. - G: lip, epichile spread out, side view. - H: column and lower portion of lateral sepals, front view. - J: column with anther-cap, side view. - K: anther-cap, interior view. - L: anther-cap, side view. A–L drawn from *Richards* 2334 by Susanna Stuart-Smith. Scale: single bar = 1 mm; double bar = 1 cm.

Dendrobium longicolle Lindl. in Bot. Reg. 26: 74, misc. 172 (1840).

Diplocaulobium malayanum Carr in Gard. Bull. Straits Settlem. 7: 6 (1932). Type: Peninsular Malaysia, Johor, *Laycock* 90 (holotype SING, isotype K).

Epiphyte. **Rhizome** *c*. 7 cm long, covered in papery sheaths, rooting profusely. **Pseudobulbs** 6–25 cm long, up to *c*. 8.5 mm wide just above base, 4–5 mm wide at apex, clump-forming, cylindrical above a fusiform base, laterally compressed, tapering towards a dilated apex, becoming longitudinally wrinkled, erect, yellowish green, one-leaved. **Leaves** 6–16 × 0.9–2.1(–2.4) cm, narrowly elliptic, narrowed, cuneate, slightly twisted and conduplicate at base, apex equally or slightly unequally acutely bidentate, main nerves 4–5. **Inflorescences** terminal, 1- to 2-flowered, emerging from a 2–3 cm long, acuminate sheath, flowers borne in succession at intervals of some weeks; peduncle up to *c*. 2 cm long; floral bract 5–9 mm long, papery. **Flowers** open for *c*. 2 days, sepals and petals cream at base grading to yellow, distal portion dull red to deep purple, lip pale yellow, mid-lobe brighter yellow, column pale yellow. **Pedicel** and **ovary** up to 8 cm long, slender. **Sepals** spreading. **Dorsal sepal** 2.7–3.5 × 0.3 cm, narrowly linear-triangular, long subulate-acuminate, concave, margins incurved, furfuraceous-punctate, 5-nerved. **Lateral sepals** 2.7–3.5 × 0.4 cm, narrowly linear-triangular, long subulate-acuminate, falcate, concave, margins incurved, anterior margin dilate above base into an oblong, slightly rounded obtuse lobe, furfuraceous-punctate, 5-nerved. **Mentum** 3-4 mm long, incurved, oblong, obtuse. **Petals** 2.55–2.8 × 0.15 cm, linear, long subulate-acuminate from below middle, concave, margins incurved, very sparsely furfuraceous-punctate, 3-nerved. **Lip** *c*. 1.2 cm long, *c*. 5 mm wide across side lobes, decurved above; side lobes 4 mm long, triangular, obtuse, curved; mid-lobe spathulate-flabellate, claw *c*. 3 mm long, oblong, blade *c*. 5 × 5 mm, abruptly dilate, triangular to ovate, obtuse, margin uneven, undulate; disc with 2 elevate, elongate keels extending to apex of claw of mid-lobe, and sometimes a lower, straight median one. **Column** *c*. 2.8 mm long; foot 3–4 mm long, incurved; clinandrium excavate. Plate 10F.

HABITAT AND ECOLOGY: Lowland and hill forest. Alt. below 300 m. Flowering observed in October.

DISTRIBUTION IN BORNEO: BRUNEI: Locality unknown (A. Lamb, pers. comm.). SARAWAK: Kuching District, Matang Road; Marudi District, Mt. Dulit.

GENERAL DISTRIBUTION: Peninsular Malaysia, Singapore and Borneo.

NOTE: This species occurs as an epiphyte in mature mangrove in the southern half of Peninsular Malaysia and Singapore.

DERIVATION OF NAME: The specific epithet is derived from the Latin *longus*, long, and *collum*, neck, in reference to the shape of the pseudobulbs.

57. DOSSINIA MARMORATA E. Morren

Dossinia marmorata *E. Morren* in Ann. Soc. Roy. Agric. Gand 4: 171, t. 193 (1848). Type: 'Java', cult. *Verschaffelt* (holotype BR).

Cheirostylis marmorata (E. Morren) Lindl. ex Lem. in Fl. des Serres, ser. 1, 4: t. 370 (1848).
Anoectochilus lowii E.J. Lowe, Beautiful Leaved Plants: 81–82, fig. 40 (1868).

Terrestrial. Rhizome fleshy, producing woolly roots at the nodes. **Stem** 1–4 cm long, fleshy, producing woolly roots from lowermost nodes, leafy. **Leaves** 3–5(–7), grouped at base; blade 4–11 × 3.5–7 cm, ovate or broadly elliptic, rather fleshy, main nerves 5–7, dark, almost blackish green with iridescent pink, greenish yellow or golden nerves forming a reticulate pattern, reverse pink to purple; petiole 1.5–4.5 cm long, sheathing, whitish or pinkish green. **Inflorescence** erect, laxly to densely many-flowered (often between 24 and 35); peduncle 10–28 cm long, densely shortly pubescent, salmon-pink; non-floriferous bracts 2–4, 1–3 cm long, acute to acuminate, lowermost sometimes leafy, densely shortly pubescent, remote, pale pink; rachis *c.* 13–31 cm long, densely shortly pubescent, pink to purple; floral bracts 4–8(–10) mm long, narrowly elliptic, acute to acuminate, densely shortly pubescent, membranous, pink. **Flowers** resupinate, sepals and petals translucent white, stained brownish-pink, lip white, column white, red above, flushed pink below, pollinia yellow. **Pedicel** and **ovary** 1 cm long, densely shortly pubescent, green. **Sepals** densely shortly pubescent on reverse. **Dorsal sepal** 5–7 × 3 mm, oblong to ovate, obtuse, 1-nerved. **Lateral sepals** 6.2–8 × 3–4 mm, obliquely ovate, obtuse, 1-nerved. **Petals** 5.9–7 × 1.8–1.9 mm, connivent with dorsal sepal, obliquely subfalcate, subacute, glabrous, 2-nerved. **Lip** 7–9 mm long, porrect to horizontal; hypochile 2–2.1 mm wide, saccate, with 2 large, sessile basal glands and a fleshy, sulcate median callus, margin of sac extended each side into a narrow, toothed flange, surface minutely papillose, especially below junction with epichile; epichile *c.* 5.5 mm wide, abruptly expanded into a transverse, obtusely bilobed, minutely papillose blade. **Column** 2 mm long, extended below into a minutely papillose, concave, sulcate, shortly bidentate, lamellate projection *c.* 2.5 mm long, which has 2 short, triangular, acute basal teeth and a lateral flange on either side; rostellum *c.* 1.1 mm long, horizontal, fleshy, clavate, sulcate and shallowly v-shaped in cross-section distally, bearing 2 tiny, acute, tooth-like projections; anther-cap *c.* 2.1–2.2 × *c.* 1.9 mm, cordate, acuminate; pollinia 2. Plate 11A–B.

HABITAT AND ECOLOGY: Lowland and hill forest on limestone, growing in leaf litter or moss on or between rocks and on ledges. Alt. sea level to 400 m. Flowering observed in March, April, June, August, October and December.

DISTRIBUTION IN BORNEO: KALIMANTAN: Locality unknown. SABAH: Pensiangan District, Batu Ponggol; Kunak District, Batu Baturong west of Kunak. SARAWAK: Bau District, Bau area, Bidi; Kuching District, Braang, Mt. Serapi, Padawan/Manok Hill and Mentawa Hill; Lundu District, Mt. Gading; Marudi District, Gunung Mulu National Park; Miri District, Baram, Gunung Subis.

GENERAL DISTRIBUTION: Endemic to Borneo.

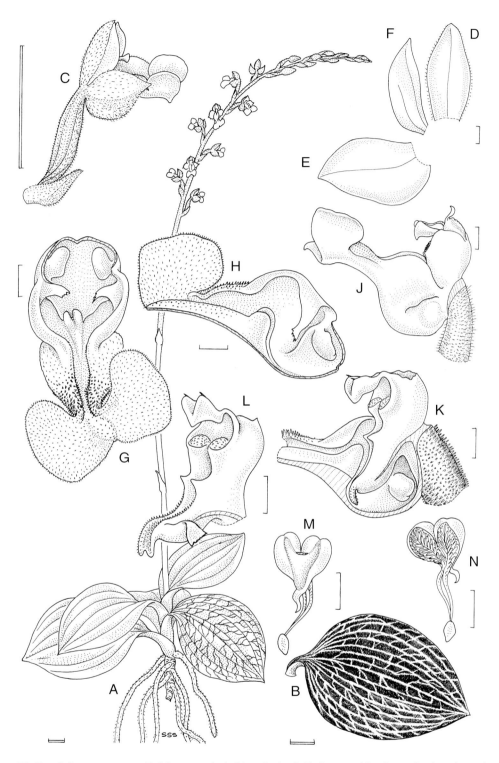

Figure 57. Dossinia marmorata E. Morren. - A: habit. - B: leaf. C: flower, side view. - D: dorsal sepal. - E: lateral sepal. - F: petal. - G: lip, front view. - H: lip, longitudinal section. - J: lip, column and portion of ovary, side view. - K: basal portion of lip, longitudinal section, column and portion of ovary, side view. - L: column, oblique view. - M: pollinarium, front view. - N: pollinarium, thecae open to show sectile pollinia, front view. A drawn from *Haviland* 838, and B–N from material cultivated at *Royal Botanic Gardens, Kew*, by Susanna Stuart-Smith. Scale: single bar = 1 mm; double bar = 1 cm.

NOTES: This attractive jewel orchid was introduced into cultivation in Belgium in 1847 by Ambroise Verschaffelt from material said to have been discovered in Java by M. Low. A beautiful illustration was published, under the name *Cheirostylis marmorata*, in the Belgian periodical *Flore des Serres et des Jardins de l'Europe* the following year. Similar in content to Curtis's Botanical Magazine and Botanical Register, this high quality 19th century journal described and figured new introductions of rare and horticulturally desirable plants to Europe.

Dossinia would appear to be monospecific and endemic to Borneo, the original provenance of Java having proven to be false. When not in flower, *D. marmorata* could easily be mistaken for a species of *Macodes*. In *Macodes*, however, the flowers are non-resupinate and have a lip epichile that is twisted to one side and a twisted column with a less prominent, cleft rostellum.

DERIVATION OF NAME: The generic name honours the Belgian botanist M.P.E. Dossin (1777–1852). The specific epithet is derived from the Latin *marmoratus*, marbled, irregularly striped or nerved, and refers to the colourful leaves.

58. ENTOMOPHOBIA KINABALUENSIS (Ames) de Vogel

Entomophobia kinabaluensis (*Ames*) *de Vogel* in Blumea 30: 199 (1984). Type: Borneo, Sabah, Mt. Kinabalu, Marai Parai Spur, *J. Clemens* 279 (holotype AMES).

Pholidota kinabaluensis Ames, Orchidaceae 6: 68 (1920).

Epiphyte or **lithophyte**. **Rhizome** short, creeping. **Roots** 1.5–3mm in diameter, branching. **Rhizome** short, creeping. **Cataphylls** 7–8, 1–14 cm long, ovate-elliptic, acute, imbricate. **Pseudobulbs** borne close together, 2–5 × 1–1.5 cm, oblong or ovate oblong, all turned to one side of rhizome, bifoliate, purple. **Leaf-blade** 16.5–65 × 1–1.9 cm, linear, acute, stiffly and thinly coriaceous, main nerves 3–7; petiole (3.5–)6–13(–17) cm long, sulcate. **Inflorescences** proteranthous to synanthous with the newly emerging young leaves, 15- to 42-flowered, flowers borne 2.5–5 mm apart; peduncle 4.5–11 cm long, erect, of 1 internode, rather flattened, elongating up to 30 cm long after anthesis; rachis 6–14 cm long, straight to slightly zigzag, angular; floral bracts15–25 × 8–16 mm, ovate, acuminate to truncate, folded along midrib, stiffly herbaceous, many-nerved, glabrous, persistent at anthesis, or at least remaining clasped around flower, shed during fruit setting, pale green. **Flowers** white or cream, lip greenish, sometimes with a pink spot, distichous but all turned to one side (secund), almost entirely closed, glabrous. **Pedicel** and **ovary** 8.5–13.5 mm long, angular. **Dorsal sepal** 7–10 × 2.8–4.5 mm, ovate-oblong, truncate to acuminate, deeply concave, 5-nerved. **Lateral sepals** 7–10 × 3.5–4.5, ovate-oblong, slightly asymmetric, acute, shallowly concave, 5-nerved, mid-nerve a ± distinct, rounded keel. **Petals** 7–9.7 × 4–6 mm, broadly spathulate, obtuse to ± apiculate, nerves 3–5. **Lip** 5–8 mm long, 1.7–3.5 mm wide, 2–3.7 mm high, not clearly divided into a hypochile and epichile, not exserted from perianth, laterally ± flattened; base deeply saccate, separated from front part by a transverse, high, slightly bent, fleshy callus which fits into stigmatic cavity, thus blocking entrance to backside of lip and stigma, lateral margins at base adnate to column for *c.* 2 mm; front part recurved, acute to

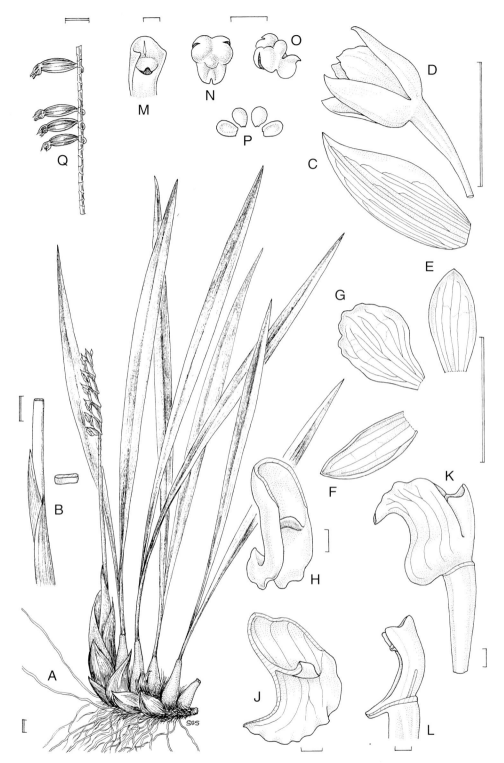

Figure 58. Entomophobia kinabaluensis (Ames) deVogel - A: habit. - B: cataphylls and base of peduncle with close-up of transverse section through flattened peduncle. - C: floral bract, abaxial surface. - D: flower, side view. - E: dorsal sepal. - F: lateral sepal. - G: petal. - H: lip, oblique view. - J: lip, longitudinal section. - K: pedicel with ovary, lip and column, anther-cap removed, side view. - L: column, anther-cap removed, side view. - M: column apex, anther-cap removed, oblique view. - N: anther-cap, back view. - O: anther-cap, side view. - P: pollinia. - Q: portion of infructescence. A & B drawn from *Lamb* AL 859/87, C–P from *Argent & Jermy* 1007 and Q from *J. & M.S.Clemens* 30607 by Susanna Stuart-Smith. Scale: single bar = 1 mm; double bar = 1 cm.

rounded, margins slightly undulate. **Column** 4–5 mm long, laterally flattened; foot absent; apical hood 3-lobed; rostellum broadly triangular, notched; anther-cap 1.2–1.5 × 0.7–1.3 mm, cordate, acute, top distinctly recurved; pollinia 4. Plate 11C & D.

HABITAT AND ECOLOGY: Very low and open podsol forest; wet kerangas over sandstone; low scrub on limestone; lower montane ridge forest on ultramafic substrate; limestone rocks. Alt. 900 to 2300 m. Flowering observed from January until July, September, October and December, but probably throughout the year.

DISTRIBUTION IN BORNEO: BRUNEI: Bukit Retak. KALIMANTAN TIMUR: fide de Vogel (1986). SABAH: Mt. Kinabalu; Crocker Range, Kimanis road; Sipitang District, Ulu Padas. SARAWAK: Gunung Mulu National Park; Mt. Penrissen; Kelabit Highlands.

GENERAL DISTRIBUTION: Endemic to Borneo.

NOTES: De Vogel (1986) proposed the monospecific genus *Entomophobia* to accommodate *Pholidota kinabaluensis*, on account of its unique morphological characters which all aim at obstructing entrance to the flower by pollinators. The flower remains almost completely closed during anthesis, enclosed by the floral bract which by then has already been detached from the rachis.

DERIVATION OF NAME: The generic name is derived from the Greek *entomos*, insect, and *phobia*, a fear, in reference to the structure of the flowers which deters potential pollinating insects. The specific epithet refers to Mt. Kinabalu, the type locality.

59. EPIGENEIUM ZEBRINUM (J.J. Sm.) Summerh.

Epigeneium zebrinum (*J.J. Sm.*) *Summerh.* in Kew Bull. 12: 266 (1957). Type: Borneo, Kalimantan Barat, Kelam, *Hallier* s.n. (holotype BO).

Dendrobium zebrinum J.J. Sm. in Icones Bogor. 2: 72, t. 113c (1903).
D. sulphuratum Ridl. in J. Straits Branch Roy. Asiat. Soc. 49: 28 (1908). Type: Borneo, Sarawak, Kuching District, Sijingkat (Sajingkat), *Hewitt* s.n. (holotype SING).
Sarcopodium sulphuratum (Ridl.) Rolfe in Orchid Rev. 18: 239 (1910).
S. zebrinum (J.J. Sm.) Kraenzl. in Engl., Pflanzenr. Orch. Monandr. IV. 50. II. B. 21: 324 (1910).
Dendrobium citrinocastaneum Burkill in Gard. Bull. Straits Settlem. 3: 12 (1923). Type: Peninsular Malaysia, Johor, near Johor Baharu, *Feddersen* s.n. (holotype SING).
Sarcopodium citrinocastaneum (Burkill) Ridl., Fl. Mal. Pen. 4: 29 (1924).
Katherinea citrinocastanea (Burkill) A.D. Hawkes in Lloydia 19: 95 (1956).
K. zebrina (J.J. Sm.) A.D. Hawkes in Lloydia 19: 98 (1956).

Epiphyte. **Rhizome** 20 or more cm long, *c.* 0.4–0.5 cm in diameter, creeping, vigorous, branching, covered in imbricate sheaths 0.8–1.2 cm long, internodes 2.5–8 cm long, rooting

144

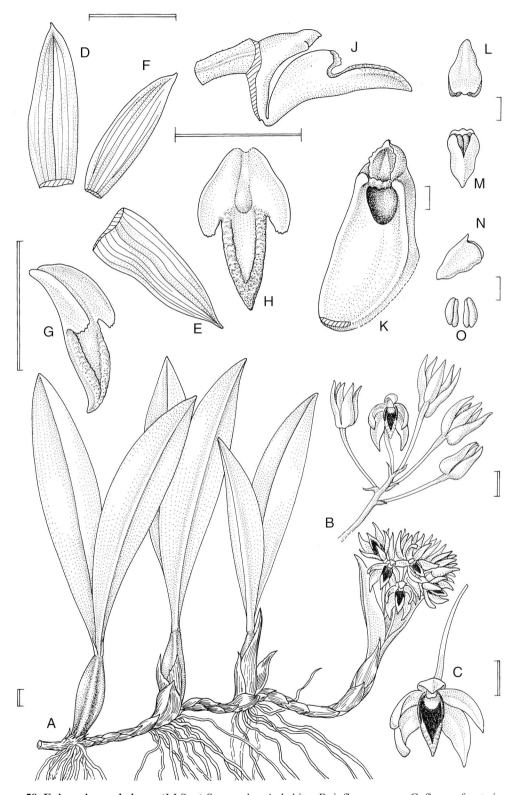

Figure 59. Epigeneium zebrinum (J.J.Sm.) Summerh. - A: habit. - B: inflorescence. - C: flower, front view. - D: dorsal sepal. E: lateral sepal. - F: petal. - G: lip, oblique view. - H: lip, front view. - J: pedicel with ovary, lip and column, side view. - K: column, anther-cap removed, oblique view. - L: anther-cap, back view. - M: anther-cap, interior view showing pollinia. - N: anther-cap, side view. - O: pollinia. A drawn from *Carr* 256 and B–O from *Lewis* 102 by Susanna Stuart-Smith. Scale: single bar = 1 mm; double bar = 1 cm.

profusely at nodes. **Cataphylls** subtending pseudobulbs and leaves 4–5(–9), 1.2–5 cm long, ovate-elliptic, acute, dark brown. **Pseudobulbs** 3–c.5.7 × 1.5–2(–3) cm, bifoliate, ovate, quadrangular, shiny yellowish green. **Leaf-blade** 10.5–18 × 2.5–4.6 cm, narrowly elliptic, obtuse to subacute, margin somewhat revolute, coriaceous, bright green, flushed dark brownish when young, main nerves 9–12; petiole 0.5–1 cm long, conduplicate. **Inflorescences** synanthous, two emerging with developing leaves, each (3)4- to 7-flowered; peduncle 3.5–6 cm long, naked, pale green; rachis 1.3–1.8 cm long; floral bracts 4–7 mm long, triangular-ovate, acuminate, pale green, speckled brown. **Flowers** *c.* 3 cm across, unscented, sepals and petals sulphur- or lemon-yellow, lip lemon-yellow below, blood-red above, side lobes dark chestnut-brown, apex of mid-lobe lemon-yellow, or dark chestnut-brown above, mid-lobe with a yellow margin, column yellow, foot pale yellow, dark yellow to brown at apex, anther-cap yellowish white. **Pedicel** and **ovary** 2.8–3.5 cm long, cylindrical, very minutely ramentaceous. **Sepals** very sparsely ramentaceous on reverse. **Dorsal sepal** 1.8–1.85 × 0.6–0.65 cm, oblong-elliptic, acute, nerves *c.* 9. **Lateral sepals** 1.8–2 × 0.65–0.7(–2) cm, slightly obliquely ovate-elliptic, acute, nerves 9–10. **Mentum** (4.5–)5 × 3 mm, obtuse, retuse. **Petals** 1.6–1.7 × 0.49(–0.5) cm, narrowly elliptic, acute, glabrous, main nerves 5. **Lip** 3-lobed, (1.25–)1.3–1.4 cm long, *c.* 8 mm wide when flattened, gently decurved, glabrous; side lobes 7 mm long, *c.* 2 mm high, erect, oblong, rounded, slightly rugulose on inner surface; mid-lobe 6.5–7 × 3.1–3.5(–3.7) mm, narrowly ovate-elliptic, acute or subacute, somewhat carinate at apex on reverse, shallowly concave, fleshy, margins thickened, slightly rugulose; disc with 2 obscure, low, rounded basal ridges and a central thickened area between side lobes. **Column** 6–6.5 × 2.5–4 mm, apex minutely erose; foot 4–5 × 2.5–3.5 mm, shallowly concave; stigmatic cavity oblong to rounded; anther-cap *c.* 2.5–2.9 × 1.3–1.5 mm, triangular-ovate, acute, base truncate, glabrous; pollinia 4. Plate 12A.

HABITAT AND ECOLOGY: Not recorded for Borneo; recorded from lowland dipterocarp forest in Peninsular Malaysia. Alt. lowlands. Flowering observed in August, September and November in Peninsular Malaysia.

DISTRIBUTION IN BORNEO: KALIMANTAN BARAT: Kelam Hill. SARAWAK: Kuching District.

GENERAL DISTRIBUTION: Peninsular Malaysia, Sumatra and Borneo.

NOTES: Ten species of *Epigeneium* are recorded from Borneo, five of which have already appeared in earlier volumes in this series. See volume one (*E. kinabaluense,* fig. 36; *E. longerepens*, fig. 37; *E. treacherianum*, fig. 38) and volume three (*E. speculum,* fig. 72; *E. tricallosum*, fig. 73). *Epigeneium zebrinum* is a lowland species subsequently described on two occasions by Ridley (1908) and Burkill (1923). Burkill's epithet *citrinocastaneum* aptly describes the attractive lemon-yellow and chestnut-brown coloration on the lip.

DERIVATION OF NAME: The generic name is derived from the Greek *epi*, upon, and *geneion*, a chin, alluding to the position of the lateral sepals and petals on the column-foot. The specific epithet is derived from the Latin *zebrinus*, striped fairly regularly with white or yellow which, in my view, is less descriptive of the flowers than either Burkill's or Ridley's epithet.

60. ERIA AURANTIA J.J. Sm.

Eria aurantia *J.J. Sm.* in Bull. Jard. Bot. Buitenzorg, ser. 2, 3: 10 (1912). Type: Borneo, Kalimantan, *A.W. Nieuwenhuis* s.n., cult. Bogor, n. 2148 (holotype BO).

E. aurantiaca J.J. Sm. in Bull. Dép. Agric. Indes Néerl., 45: 19 (1911), *non* Ridl. (1910).

Robust clump-forming **terrestrial** or **epiphyte**. **Stems** to over 3 m high, 1.5–2 cm in diameter, internodes 2.5–4.5 cm long, terete below, cylindrical above, leafy throughout, entirely enclosed by sheathing leaf bases, erect to spreading. **Leaf-blade** (14–)18–28.5 × (2.8–)4–5.5 cm, oblong-elliptic to elliptic, apex asymmetrical, acute, coriaceous, shiny; sheath 2.5–4.5 cm long, tubular. **Inflorescences** borne in a recess above a node, sessile, 3- to 4-flowered; rachis 1.8 cm long, golden yellow ochre to orange; floral bracts 5–6, 1.9–2.5 × 1–1.8 cm, alternate, ovate-elliptic, apex acute to acuminate and slightly upcurved, convex, fleshy, orange ochre. **Flowers** not opening widely, fleshy, sepals yellow ochre to orange, cream distally, petals yellowish, cream distally, apical margins dark brownish ochre, lip cream, dark brownish ochre distally, disc spotted brownish ochre with ochre keels, column yellowish cream. **Ovary** sessile, 2.3–2.5 cm long, yellow ochre. **Dorsal sepal** 1.3–1.6 × 0.6–0.7 cm, oblong, obtuse, concave. **Lateral sepals** 1.3–1.5 × 0.9 cm, obliquely oblong-triangular, obtuse, somewhat falcate, concave, adnate to column-foot to form an obtuse mentum. **Petals** 1.1–1.4 × 0.6 cm, oblong, obtuse, slightly falcate and concave, margins undulate distally. **Lip** 3-lobed, 1.1–1.2 cm long, 0.7 cm wide (measured across middle), gently curved; free portion of side lobes 2–2.2 mm long, ovate-triangular, acute to acuminate, erect, velutinous; mid-lobe 2–3 × 4 mm, oblong, apex truncate and shallowly bilobed, lobules curved, obtuse and separated by a 1–1.5 mm wide sinus, velutinous; disc with 3 villous hairy keels, median keel broader and extending to centre of mid-lobe, outer keels narrower, higher and glabrous distally and extending to base of mid-lobe. **Column** 7–8 mm long; foot 5–6 mm long; stelidia short, obtuse; anther-cap cucullate, obtuse.

HABITAT AND ECOLOGY: Lower montane podsol forest composed of *Eugenia, Leptospermum, Podocarpus,* etc., with an understorey of *Pandanus* spp., *Rhododendron durionifolia* var. *lumakuensis,* and a field layer including *Nepenthes reinwardtiana, Appendicula* spp., *Dilochia* spp., *Araceae* and ferns; often epiphytic in forked boles of *Eugenia.* Alt. 1200 to 1300 m. Flowering observed in October.

DISTRIBUTION IN BORNEO: KALIMANTAN: province unspecified. SABAH: Sipitang District, Long Pa Sia area.

GENERAL DISTRIBUTION: Endemic to Borneo.

NOTES: *Eria aurantia* is one of the most robust members of section *Cylindrolobus* in Borneo. *Cylindrolobus* superficially resembles certain species of *Dendrobium* section *Grastidium* in habit and is distinctive in having long, often cane-like leafy stems bearing short lateral inflorescences from or opposite the leaf axils. The flowers are subtended by conspicuous, often brightly coloured persistent bracts. Unlike the majority of *Eria,* all parts of the plant are usually glabrous and future research may necessitate the removal of *Cylindrolobus* from *Eria.*

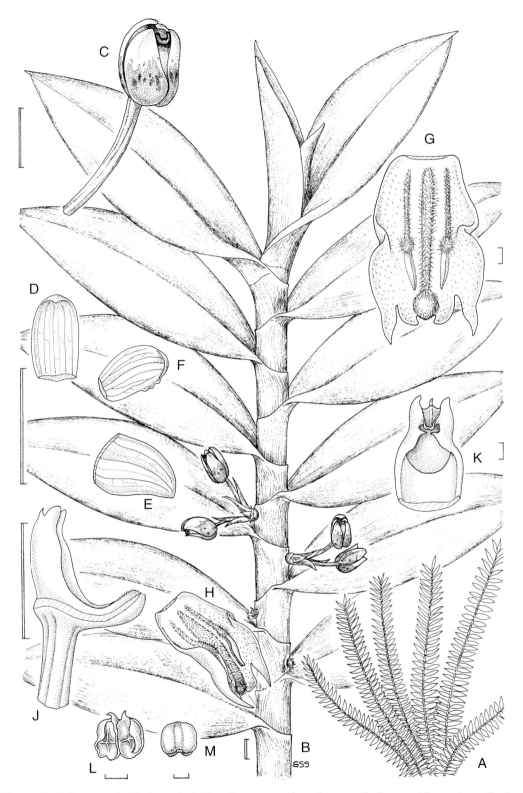

Figure 60. Eria aurantia J.J. Sm. - A: habit. - B: upper portion of stem. - C: flower, oblique view. - D: dorsal sepal. - E: lateral sepal. - F: petal. - G: lip, flattened, front view. - H: lip, oblique view. - J: pedicel with ovary and column, anther-cap detached, side view. - K: column, anther-cap detached, front view. - L: anther-cap, interior view. - M: anther-cap, back view. All drawn from *Wood* 701 by Susanna Stuart-Smith. Scale: single bar = 1 cm; double bar = 1 mm.

I have provisionally refered the plant figured here to J.J. Smith's *Eria aurantia*. The flowers of my collection, however, differ in a few details from the figure provided by Smith (1934, t. 56 I). In particular, the mid-lobe of the lip is shallowly bilobed, with a more apically swollen median keel, and the dorsal sepal is rather broader. Until further collections are made, it will remain unclear whether these differences fall within the natural variation of the species.

DERIVATION OF NAME: The specific epithet is derived from the Latin *aurantiacus*, orange, and refers to the flower colour.

61. ERIA CYMBIDIFOLIA Ridl. var. CYMBIDIFOLIA

Eria cymbidifolia *Ridl.* in Journ. Bot. 36: 212 (1898). Types: Borneo, Kalimantan Barat, Pontianak, *Ridley* s.n., cult. Singapore Botanic Garden 1893 (lectotype SING); Sumatra, Deli Baros, cult. 1910 (syntype SING).

Eria cymbiformis J.J. Sm. in Rec. Trav. Bot. Néerl. 1: 152 (1904). Type: Sumatra, Singalang, 700 m, *Korthals* (holotype BO).

var. **cymbidifolia**

Clump-forming **epiphyte** or **lithophyte**. **Pseudobulbs** absent. **Cataphylls** 3–4, 1.5–11 cm long, spotted and stained dull purple, ageing brown, margins papery. **Stem** 3–5(–8) cm long, entirely enclosed by leaf sheaths. **Leaves** 5–7(–14), distichous, *Cymbidium*-like; blade (12–)15–32(–45) × 0.8–1.2(–2) cm, ligulate, apex unequally acutely bilobed, articulated to a sheathing, imbricate base, curved; sheath 5–9 cm long, margins pale brown and papery, similar to those of the genus *Agrostophyllum*. **Inflorescences** erect, somewhat curved at apex, subdensely many-flowered; peduncle and rachis pale green, sparsely puberulus; peduncle 8–10 cm long, hidden by leaf sheath below; non-floriferous bracts 5–6, 4–6 mm long, oblong-ovate, obtuse; rachis 7–10 cm long; floral bracts 3–4.5 mm long, triangular-ovate, acute, reflexed, minutely puberulus or ramentaceous, particularly along margins. **Flowers** non-resupinate, resembling those of the genus *Polystachya*, sepals whitish to palest straw, petals with 2 purple longitudinal stripes, lip palest lemon-yellow, column purplish, slightly scented. **Pedicel** and **ovary** 6–7(–9) mm long, narrowly clavate, puberulus. **Dorsal sepal** 5–5.5 × 3–3.2 mm, oblong, obtuse, glabrous. **Lateral sepals** 5 × 5.5 mm, broadly ovate, oblique, obtuse, glabrous. **Mentum** 4 mm long, obtuse. **Petals** 4–4.5 × 2.1–2.2 mm, oblong-elliptic, obtuse to subacute. **Lip** 4.5 mm long, 4.5–5 mm wide when flattened, 3-lobed, concave, cymbiform, saccate at base, minutely papillose, nectariferous; side lobes 0.9–1.8 mm long, rounded or triangular, acute, margins often recurved, each with a papillose-hairy spathulate flange-like basal callus extending on to mid-lobe; mid-lobe 0.5–0.9 mm long, triangular, subacute; disc with a prominent glabrous, thickened median ridge-like nectary attenuated distally and terminating just below apex of mid-lobe, sometimes thickened at apex. **Column** 1.8 mm long, glabrous; foot 4 mm long, central portion hirsute; anther-cap 0.8–0.9 × 1 mm, ovate, cucullate. Plate 12B.

HABITAT AND ECOLOGY: Lower montane oak-chestnut forest; lower montane oak-laurel forest; hill forest; rocky slopes; sometimes on ultramafic substrate. Alt. 900 to 2400 m. Flowering observed throughout the year.

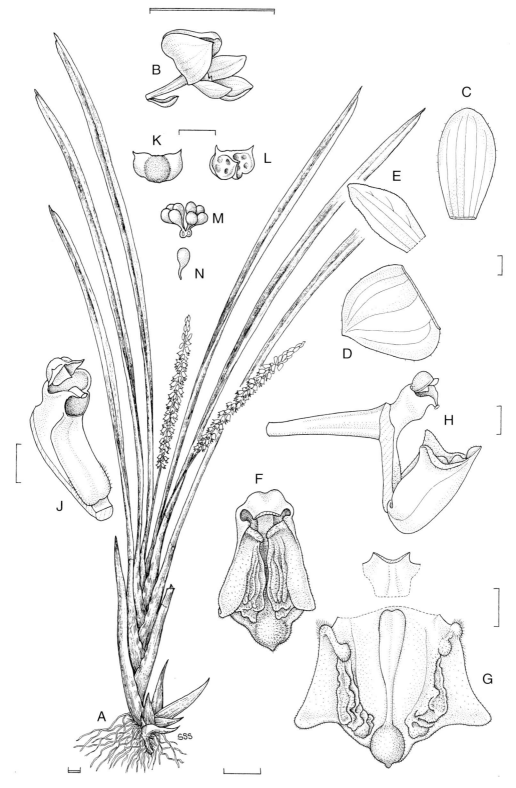

Figure 61. Eria cymbidifolia Ridl. var. **cymbidifolia** - A: habit. - B: flower, side view. - C: dorsal sepal. - D: lateral sepal. - E: petal. - F: lip, natural position, front view. - G: lip, flattened, base detached, front view. - H : pedicel with ovary, lip and column, side view. - J: column, oblique view. - K: anther-cap, back view. - L: anther-cap, interior view. - M: pollinarium. - N: pollinium. A drawn from *Beaman* 9571 and B–N from *Bailes & Cribb* 506 by Susanna Stuart-Smith. Scale: single bar = 1 mm; double bar = 1 cm.

DISTRIBUTION IN BORNEO: KALIMANTAN BARAT: Pontianak area. SABAH: Mt. Kinabalu; Mt. Trus Madi. SARAWAK: Mt. Dulit.

GENERAL DISTRIBUTION: Sumatra, Borneo, Sulawesi and the Philippines.

NOTES: *Eria cymbidifolia* belongs to section *Cymboglossum* J.J.Sm which frequently have a habit recalling *Cymbidium* and usually non-resupinate flowers. Non-flowering specimens of *E. cymbidifolia* closely resemble *Cymbidium bicolor* Lindl. subsp. *pubescens* (Lindl.) Du Puy and P.J. Cribb. In section *Cymboglossum* the lip is distinctly concave, boat-shaped, variously three-lobed and provided with a thickened median ridge-like nectary and two lateral, papillose, flange-like calli.

Although stature and leaf size are variable in *E. cymbidifolia,* most Bornean material examined is of modest size with leaves measuring around 15–30 × 0.8–1.2 cm. Rather larger plants with leaves up to 1.6 cm wide and a hairy inflorescence have been collected on Mt. Kinabalu. Material of *E. cymbidifolia* from Sumatra and the Philippines in the Kew herbarium have a densely hairy inflorescence and sepals reminiscent of *E. pseudocymbiformis* var. *hirsuta* (see fig. 62a). Most collections from Borneo examined have only a sparsely puberulus inflorescence and glabrous sepals.

DERIVATION OF NAME: The specific epithet is Latin and means boat-shaped, in reference to the lip.

61a. ERIA CYMBIDIFOLIA Ridl. var. PANDANIFOLIA J.J. Wood

Eria cymbidifolia *Ridl.* var. **pandanifolia** *J.J. Wood* in Lindleyana 5, 2: 95 (1990). Type: Borneo, Sarawak, Mt. Dulit, Ulu Tinjar, near Long Kapa, 9 August 1932, *Synge* S.154 (holotype K).

Robust **epiphyte. Leaves** 3–6; blade 45–55 × (2–)3–3.5 cm, ensiform, obliquely subacutely bilobed, or obliquely acute; sheath 10–19 cm long. **Inflorescence** peduncle and rachis densely tomentose-hirsute. **Pedicel** and **ovary** densely tomentose-hirsute. **Sepals** glabrous or puberulus.

HABITAT AND ECOLOGY: Hill forest; lower montane forest; upper montane forest; mostly on ultramafic substrate. Alt. 300 to 2700 m. Flowering observed in February, April, November and December.

DISTRIBUTION IN BORNEO: SABAH: Mt. Kinabalu. SARAWAK: Mt. Dulit; Kapit District.

GENERAL DISTRIBUTION: Endemic to Borneo.

NOTES: This variant is distinguished by its robust habit and broad leaves reminiscent of the genus *Pandanus* (Pandanaceae). The flowers of the type collection are described as having a slightly sweet scent.

DERIVATION OF NAME: The specific epithet means *Pandanus* (screw pine) - leaved.

62. ERIA PSEUDOCYMBIFORMIS J.J. Wood

Eria pseudocymbiformis *J.J. Wood* in Kew Bull. 39: 84, fig. 9 (1984). Type: Borneo, Sarawak, Gunung Mulu National Park, Mt. Api, below Pinnacle Camp, 24 February 1978, *I. Nielsen* 486 (AAU, holotype, isotype K).

Eria cymbidifolia Ridl. var. *longipes* Carr in Gard. Bull. Straits Settlements 8, 2: 99 (1935), nom. nud.

var. **pseudocymbiformis**

Epiphyte. **Stems** elongate, (8–)14–30 × 1–2 cm, covered with numerous imbricate, acute, scarious-edged green sheaths 3–18 cm long, lowermost densely mottled and spotted purple, uppermost less so. **Leaves** 3–5, apical; blade (15–) 23–30(–43) × (0.8–)1–4 cm, linear-ligulate to ligulate-ensiform, obliquely acute, conduplicate at base, coriaceous, spreading or arcuate; sheath (10–)15–20 cm long, unspotted, margin pale brown, scarious. **Inflorescences** erect to porrect, subdense to dense, many-flowered; peduncle and rachis glabrous or puberulus; peduncle 5–10 cm long; non-floriferous bracts 0.3–10 cm long, ovate, acute; rachis (5–)7–16 cm long; floral bracts 0.5–1 mm long, triangular-ovate, acute, glabrous, reflexed. **Flowers** non-resupinate, cream, white or pale translucent lemon, with crimson or purple markings on petals, lip side lobes and column, reported as having a slight bitter scent, or unscented. **Pedicel** and **ovary** 5–7 mm long, glabrous. **Sepals** and **petals** glabrous. **Dorsal sepal** 5–7 × 2.5–3.5 mm, oblong, obtuse to subacute, concave. **Lateral sepals** 4.5–6 × 5–8 mm, obliquely ovate, obtuse. **Mentum** 3–6 mm long, obtuse. **Petals** 4.5–6 × 1.5–3 mm, obliquely oblong or oblong-elliptic, obtuse. **Lip** 5–6 mm long, 4.5–6 mm wide when flattened, erect, shallowly 3-lobed distally, concave-cymbiform, saccate at base, minutely papillose, with a prominent, glabrous raised median ridge-like nectary which is attenuated distally and terminates in a fleshy, oval callus or rugose area at base of mid-lobe; side lobes 1 mm long, triangular, obtuse, margin often reflexed, each with a papillose hairy, fleshy basal keel, margin of keels connivent; mid-lobe 1–2 mm long, triangular-ovate, obtuse to subacute. **Column** 1.5–2 mm long, with short curved stelidia; foot 3–5 mm long, puberulus; anther-cap 1 × 1 mm, ovate.

HABITAT AND ECOLOGY: Lower montane forest on limestone, oak-laurel forest; mossy forest; sometimes on ultramafic substrate. Alt. 800 to 2100 m. Flowering observed in February, April, May, and from July until September.

DISTRIBUTION IN BORNEO: SABAH: Mt. Kinabalu; Crocker Range, Kimanis road. SARAWAK: Gunung Mulu National Park; Mt. Dulit.

GENERAL DISTRIBUTION: Endemic to Borneo.

NOTE: Although Carr (1935) considered this plant to be a variety of *E. cymbidifolia* Ridl., the elongated stems, often up to 30 cm long, immediately distinguish it from the imbricate, fan-shaped cymbidium-like growth-form of the former.

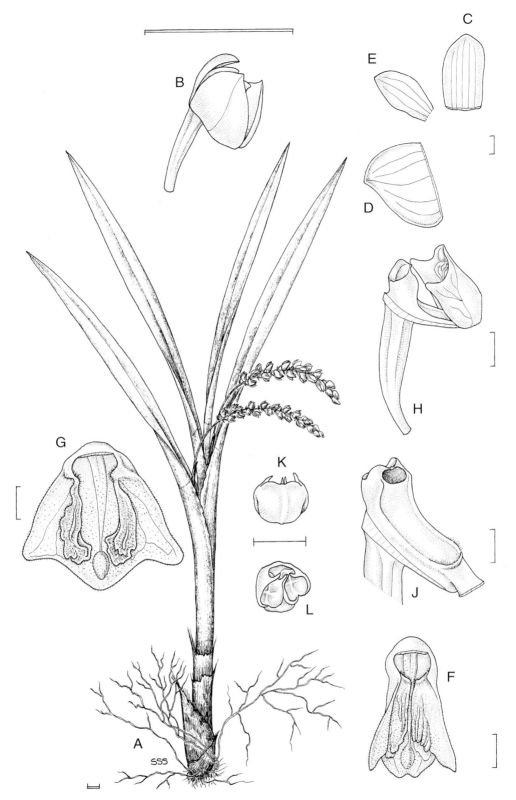

Figure 62. Eria pseudocymbiformis J.J. Wood var. **pseudocymbiformis** - A: habit. - B: flower, side view. - C: dorsal sepal. - D: lateral sepal. - E: petal. - F: lip, natural position. - G: lip, flattened. - H: pedicel with ovary, lip and column, side view. - J: column, anther-cap removed, oblique view. - K: anther-cap. back view. - L: anther-cap, interior view. A drawn from *Synge* S. 368 & B–L from *Nielsen* 486 by Susanna Stuart-Smith. Scale : single bar = 1 mm; double bar = 1 cm.

62a. ERIA PSEUDOCYMBIFORMIS J.J. Wood
var. HIRSUTA J.J. Wood

Eria pseudocymbiformis *J.J. Wood* var. **hirsuta** *J.J. Wood* in Lindleyana 5 (2): 97 (1990). Type: Borneo, Sabah, Mt. Kinabalu, head of Columbon River, 11 July 1933, *J. & M.S. Clemens* 33938 (holotype BM, isotype K).

Epiphyte or **lithophyte**. **Inflorescence** peduncle and rachis densely greyish-tomentose hirsute. **Sepals** densely hirsute on reverse. **Flowers** resupinate, variously described as pink to white, cream and purple, pinkish white with purple stripes on petals and lip.

HABITAT AND ECOLOGY: Lower montane forest; upper montane mossy forest and scrub forest, frequently on ultramafic substrate. Alt. 900 to 2900 m. Flowering observed from April to July, November and December.

DISTRIBUTION IN BORNEO: SABAH: Mt. Kinabalu; Mt. Trus Madi; Penampang District.

GENERAL DISTRIBUTION: Endemic to Borneo.

NOTES: This variant differs from var. *pseudocymbiformis* by the dense greyish-hairy tomentose covering on the inflorescence, and the resupinate flowers. The floral bracts are only tomentose along the margins. *Beaman* 10484, collected along the Kota Kinabalu to Tambunan road, has rather short, narrowly elliptic leaves, but in all other respects is identical with material from Mt. Kinabalu.

DERIVATION OF NAME: The Latin varietal epithet means hairy and refers to the greyish-hairy inflorescence.

63. ERIA SACCIFERA Hook.f.

Eria saccifera *Hook.f.*, Fl. Brit. Ind. 5: 797 (1890). Type: Peninsular Malaysia, Mt. Batu Pateh, *Wray* 1215 (holotype K).

Eria brookesii Ridl. in J. Straits Branch Roy. Asiat. Soc. 50: 136 (1908). Type: Borneo, Sarawak, Bidi, *Brookes* s.n. (holotype SING, isotype K).
Eria saccata Ridl. in J. Straits Branch Roy. Asiat. Soc. 61: 39 (1912). Type: Peninsular Malaysia, Perak, Mt. Korbu (Kerbau), *Haniff* s.n., cult. Singapore (holotype SING, isotype K).

Epiphyte. **Pseudobulbs** 4–21 × 2–4.5 cm, cylindrical, broadly cylindrical or ellipsoid, clump-forming, erect, when young with conspicuous pale green vertical nerves, bearing several 5.5–12 cm long, dark reddish brown, often brown-speckled, parchment-textured, triangular-ovate, acute sheaths. **Leaves** 2–6, distal; blade (5–)12–39 × (2.5–)3–8 cm, oblong-elliptic, acute to acuminate, cuneate below, coriaceous, main nerves 6–8, mid-green above, dull grey-green below; petiole 3–8 cm long, sulcate-conduplicate. **Inflorescences** borne from upper portion of pseudobulbs, laxly to

Figure 63. Eria saccifera Hook. f. - A: habit. - B: portion of leaf, adaxial surface. - C: inflorescence. - D: flower, front view. - E: floral bract and flower, side view. - F: dorsal sepal. - G: lateral sepal. - H: petal. - J: lip, flattened, front view. - K: pedicel with ovary, lip and column, side view. - L: column, side view. - M: column, oblique view. - N: apex of column, back view. - O: anther-cap, back view. - P: anther-cap, interior view. - Q: pollinia. - R: stellate hair from ovary. A drawn from *Wood* 813 and B–R from *Bailes & Cribb 592* by Judi Stone. Scale: single bar = 1 mm; double bar = 1 cm

densely many-flowered, gently decurved; peduncle 2–4 cm long, densely shortly pubescent, hairs dull cerise-purple; non-floriferous bracts 2–3, 0.5–1 cm long, ovate, obtuse; rachis 7–12 cm long, densely shortly pubescent, hairs dull cerise-purple; floral bracts 0.7–1.2 × 0.4–0.7 cm, oblong or ovate-oblong, obtuse to acute, strongly concave, lowermost less so, very sparsely hairy, especially at base, reflexed, pale green or lemon. **Flowers** unscented, sepals pale creamy lemon, greenish, straw-coloured or dull lilac, petals similar, often stained pale lilac-mauve or flesh-coloured, sometimes tipped lemon-yellow, mentum dull cerise-purple, lip pale lemon-yellow with pale lilac-mauve markings or cerise-purple to reddish brown, edged lemon-yellow at apex, creamy at centre, column and anther-cap flesh-coloured. **Pedicel** and **ovary** 0.7–1.1 cm long, slender, densely shortly pubescent, hairs dull cerise-purple. **Dorsal sepal** 0.7 × 0.4 cm, oblong-elliptic or ovate-elliptic, obtuse, concave, sparsely hirsute on reverse, main nerves 3, curving forward. **Lateral sepals** 8–1.1 × 0.7–0.71 cm, obliquely triangular-ovate, anterior portion rounded, concave, obtuse to subacute, cucullate, sparsely hirsute on reverse, main nerves 3. **Mentum** 0.5–0.7 cm long; saccate, rounded. **Petals** 0.5–0.7 × 0.3–0.4 cm, ovate-elliptic, sometimes subfalcate, obtuse to subacute, main nerves 3. **Lip** 0.6–0.7 × 0.8–0.9 cm when flattened, with a long claw, expanding into a flabelliform or subreniform, entire or obscurely 3-lobed blade, its lower sides erect, its margins unevenly undulate, obtuse, surface glabrous, rugose to rugulose; disc with 2 fleshy elevate flanges which coalesce at centre of blade as a fleshy, rugose or rugulose mass. **Column** 0.1–0.38 cm long; foot 0.6–0.7 cm long, apex curving upward, hamate, with a thickened transverse ridge or pronounced ovate, acute, concave flange at junction with lip; anther-cap 1–1.2 × 1.5–2 mm, broadly triangular-ovate, variously 3-toothed. Plate 12C.

HABITAT AND ECOLOGY: Kerangas forest; riverine forest; lower montane forest; ridge forest with epiphytic *Rhododendron*; recorded from sandstone and ultramafic substrates. Alt. *c.* 200 to 1900 m. Flowering observed from March until June.

DISTRIBUTION IN BORNEO: SABAH: Crocker Range, Kimanis road; Mt. Kinabalu; Sipitang District, Mt. Lumaku; Tenom District, above Kallang Waterfall. SARAWAK: Bau District, Bidi; Kapit District, Hose Mountains, Ulu Temiai, Mujong; Marudi District, Gunung Mulu National Park, Mt. Murud.

GENERAL DISTRIBUTION: Peninsular Malaysia and Borneo.

NOTES: The identity of this species, which is probably widespread in Borneo, has proven problematic. Examination of extant material suggests, however, that it falls within the circumscription of *E. saccifera*, described in 1890 by Joseph Hooker from material collected in Peninsular Malaysia by Leonard Wray. Although the type material is badly preserved, the general morphology suggests that it is best placed in section *Pinalia* alongside species such as *E. apertiflora* Summerh. and *E. bipunctata* Lindl. *Eria brookesii*, described from Sarawak by Ridley, would seem to be conspecific. Ridley, incorrectly compared this with *E. densa* Ridl., a species belonging to section *Urostachya* which it superficially resembles. Four years later Ridley described yet another species as *E. saccata* Ridl., which also appears to be conspecific. The Bornean populations of *E. saccifera* vary somewhat in flower colour and in some individuals the thickened transverse ridge at the apex of the column-foot is developed into a distinct flange. However, the flowers examined seem to match well with dissection drawings of *E. saccifera* provided by J.J. Smith (1930, t. 48 IV). The rather similar *E. suaveolens* Ridl., also described from

Peninsular Malaysia, was placed by Seidenfaden & Wood (1992) in section *Hymeneria*. This also seems to be closely related to *E. saccifera*, but is described as having a lip with small side lobes each provided with a short keel.

DERIVATION OF NAME: The specific epithet is derived from the Latin *sacciformis*, bag-shaped, sac-shaped, and refers to the shape of the mentum.

64. GEESINKORCHIS ALATICALLOSA de Vogel

Geesinkorchis alaticallosa *de Vogel* in Blumea 30: 201, pl.1d–i (1984). Type: Borneo, Kalimantan Timur, Apo Kayan, *Geesink* 8965 (holotype L, isotype BO).

Terrestrial. **Rhizome** short, creeping. **Roots** elongate, often over 25 cm long, 1–3.5 mm in diameter. **Cataphylls** of young shoot 4–6, 2.5–8 cm long. **Pseudobulbs** all turned to one side on rhizome, (2.5–)7.5–9.5(–12) × 2–3 cm, cylindrical, rather flattened, bifoliate. **Leaf-blade** 6.5–30 × (1.3–)2–3.3(–4.2) cm, linear-lanceolate, acute, rather stiffly coriaceous, main nerves 3–7; petiole 3–9 cm long, deeply canaliculate. **Inflorescences** starting proteranthous, continuing to produce up to 75 or more flowers successively, one or two at a time, long after pseudobulb and leaves are fully developed, erect, rigid, flowers borne 3–6 mm apart; peduncle 27–43 cm long, somewhat flattened; rachis 17–27 cm long, straight to slightly zigzag, much elongating during anthesis; non-floriferous bracts at base of rachis 7–8, 1–2 cm long, ovate-elliptic, concave, acute; floral bracts 13.5–15.5 × 9.9–11.5 mm, ovate, acute, closely imbricate, distichous, folded along midrib, caducous at anthesis, parchment-textured, with groups of stellately arranged, uniseriate hairs inside. **Flowers** resupinate, sepals translucent gold, yellowish brown or apricot, petals golden brownish orange to apricot-buff, hypochile cinnamon-brown, edged translucent white, epichile cinnamon-orange to ochre, edged translucent white, callus whitish or pale brownish, column dark chocolate-brown to blackish purple. **Pedicel** and **ovary** 4–5 mm long. **Dorsal sepal** 7–8 × 3.5–4 mm, ovate, acute to acuminate, deeply concave, with a few stellately arranged hairs inside. **Lateral sepals** 7–8.2 × 2.5–3.5 mm, ovate-oblong, acute to acuminate, similarly hairy. **Petals** *c.* 7.7 × 1 mm, linear, truncate, curved or rolled backwards, glabrous. **Lip** 7–8 mm long, pandurate; hypochile 3.2–4 mm long, saccate part *c.* 0.9 mm long, side lobes 1.5–2 × 1.5–2 mm, width across spread out lobes 5.5–6.5 mm, ± triangular, rounded, decurved, disc with a central, erect, sturdy, laterally compressed callus 1–1.5 mm long, 0.3 mm wide, 1–1.5 mm high, which blocks opening to stigma, on either side of central callus a low, thin, laterally slanting, 0.5–1 mm high, short keel; epichile 3.5–4 × 5.5 mm, basal portion 2–3.4 mm wide, its margin densely covered with thin, minute lacinulae, side lobes 2.5–4 × 1.5–2.3 mm, broadly rounded, margins smooth and upturned. **Column** *c.* 4 mm long, deeply concave; foot absent; apex truncate, rather irregular; margins smooth, curved to the front; rostellum triangular, 2-lobed; anther-cap 2–2.2 × *c.* 1.5 mm, long-cordate, acuminate; pollinia 4, on a slender stipes. Plate 13A & B.

HABITAT AND ECOLOGY: Kerangas forest on podsolic soil, composed of *Eugenia* spp., *Dacrydium* spp., etc., with *Rhododendron malayanum, Nepenthes* spp., rattans, etc.; often growing in large clumps among the roots of shrubs and trees; ridge scrub. Alt. 400 to 1000 m. Flowering observed in March, May, July and September, but probably throughout the year.

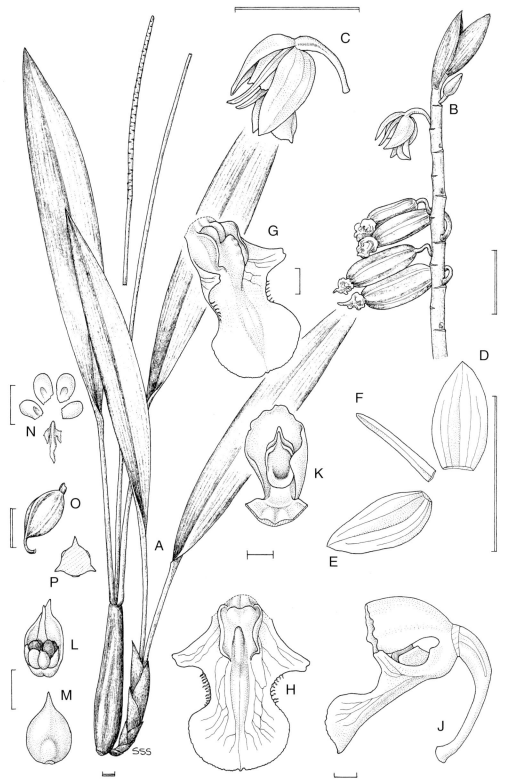

Figure 64. Geesinkorchis alaticallosa deVogel - A. habit. - B: upper part of inflorescence. - C: flower, side view. - D: dorsal sepal. - E: lateral sepal. - F: petal. - G: lip, natural position, oblique view. - H: lip, flattened, front view. - J: pedicel with ovary, lip and column, side view. - K: column, front view. - L: anther-cap, interior view. - M: anther-cap, back view. - N: stipes and pollinia. - O: fruit capsule. - P: fruit capsule, transverse section. All drawn from *Wood* 751 by Susanna Stuart-Smith. Scale: single bar = 1 mm; double bar = 1 cm.

DISTRIBUTION IN BORNEO: KALIMANTAN TIMUR: Apo Kayan. SABAH: Nabawan area; Sipitang District, Long Pa Sia. SARAWAK: Kelabit Highlands, Bario area.

GENERAL DISTRIBUTION: Endemic to Borneo.

NOTES: *Geesinkorchis* was established by de Vogel in 1984. It is unique among genera in subtribe *Coelogyninae* in having pollinia borne on a minute, thin stipes, rather than on separate caudicles. In other small flowered *Coelogyninae* the flowers are never produced in succession on an elongating rachis. Larger flowered species with a similar way of flowering to *Geesinkorchis* have a different floral morphology and bract arrangement.

Two species are currently recognised, although de Vogel (pers. comm.) has indicated that a further two, one from Brunei, the other from the Hose Mountains in Sarawak, remain undescribed.

DERIVATION OF NAME: The generic name honours the late Rob Geesink of the Rijksherbarium, Leiden in The Netherlands who collected the type material. The specific epithet is derived from the Latin *alatus*, winged, and *callosus*, callose, bearing a callus or hardened thickening, in reference to the wing-like keels on the lip.

65. GEESINKORCHIS PHAIOSTELE (Ridl.) de Vogel

Geesinkorchis phaiostele (*Ridl.*) *de Vogel* in Blumea 30: 201, pl. 1c (1984). Type: Borneo, Sarawak, Mt. Pueh, *Lewis* s.n. (lectotype K).

Coelogyne phaiostele Ridl. in J. Straits Branch Roy. Asiat. Soc. 54: 51 (1910).
Coelogyne ridleyana Schltr. in Feddes Repert. 8: 561 (1910). Type: Borneo, Sarawak, Mt. Santubong, *Brooks* s.n. (lectotype BM, isolectotype K).
Pholidota triloba J.J. Sm. in Brittonia 1: 105 (1931). Type: Borneo, Sarawak, Mt. Pueh, *J. & M.S. Clemens* 20398 (holotype L, isotype NY).

Epiphyte. **Rhizome** short, creeping. **Roots** elongate, often over 25 cm long, 1–3 mm in diameter. **Cataphylls** of young shoot 5–11, 0.5–12.5 cm long. **Pseudobulbs** all turned to one side on rhizome, (2.5–)3–8(–24) × (0.8–)1–2 cm, cylindrical, rather flattened, bifoliate. **Leaf-blade** 6.5–30 × (1.3–)2–3.3(–4.2) cm, linear-lanceolate, acute, rather stiffly coriaceous, main nerves 3–5; petiole 3–13(–17.5) cm long, deeply canaliculate. **Inflorescences** starting proteranthous, continuing to produce up to between 50 and 100 flowers successively, one or two at a time, erect, rigid, flowers borne 3–5 mm apart; peduncle (8–)25–33 cm long, somewhat flattened; rachis 20–35 cm long, straight to slightly zigzag; non-floriferous bracts at base of rachis (0–)2–6 (scars only seen); floral bracts 12–17 × 8–13 mm, ovate, acute, closely imbricate, distichous, folded along midrib, caducous at anthesis, hairy inside, papery-textured. **Flowers** resupinate, sepals pale greenish white to dull ochre or pale orange-yellow, petals cream, lip white with a brown blotch in front of keels, and a brown line on front margins of side lobes of hypochile and margin of narrow basal part of epichile, column brown, anther-cap ochre. **Pedicel** and **ovary** 5.5–8.5 mm long. **Dorsal sepal** 5.5–7.8 × 3–4.2 mm, ovate, acute to acuminate, deeply concave, sparsely hairy inside. **Lateral sepals** 5.2–8 × 2.3–4 mm, ovate to ovate-oblong, acute to acuminate, similarly

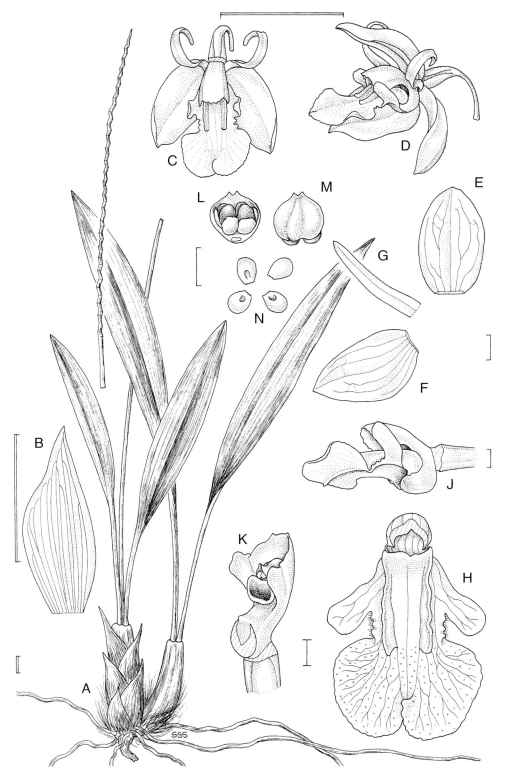

Figure 65. Geesinkorchis phaiostele (Ridl.) deVogel - A: habit. - B: folded floral bract. - C: flower, front view. - D: flower, oblique view. - E: dorsal sepal. - F: lateral sepal. - G: petal. - H: lip, flattened, front view. - J: upper portion of pedicel with ovary, lip and column, side view. - K: upper portion of pedicel with ovary and column, oblique view. - L: anther-cap, interior view. - M: anther-cap, back view. - N: pollinia. A drawn from *Brooks* 14 (isolectotype of *Coelogyne ridleyana*) and B–N from *Burtt & Woods* 2667 by Susanna Stuart-Smith. Scale: single bar = 1 mm; double bar = 1 cm.

hairy. **Petals** 5–8 × 0.5–1.5 mm, linear, obtuse, curved or rolled backwards, glabrous. **Lip** (4.5–)7–8 mm long, pandurate; hypochile 3–4 mm long, saccate part 1–2 mm long, 1–7 mm wide, 1–5 mm high, side lobes (0.5–)1–1.8 × (0.5–)0.8–1.2 mm, width across flattened lobes (2–)4.4–6 mm, erect, triangular to ligulate or obliquely spathulate, disc with 2 parallel, rather long, ± fleshy low keels, back of hypochile sac either or not with 3 short keels; epichile (2.7–)3.5–4.9 × 4–5 mm, sometimes retuse, basal part 1.8–3 mm wide, margins usually with a row of densely placed, minute, uniseriate hairs, side lobes 1.2–2 × 2–3 mm, margin irregular. **Column** 3–4.2 mm long, deeply concave; foot absent; apex broadly rounded with irregularly serrate to entire top margin, side lobes ± triangular, rounded at apex, turned to the front; rostellum broadly triangular or broadly rounded, incised at top; anther-cap 1–1.2 × 1–1.2 mm, cordate, acute.

HABITAT AND ECOLOGY: Cloud forest on sandstone substrate; epiphytic on tree trunks in vegetation on the summits of hills and mountains. Alt. 700 to 2000 m. Flowering observed in January, June, September to November, but probably throughout the year

DISTRIBUTION IN BORNEO: KALIMANTAN TIMUR: Locality unknown. SABAH: Sipitang District, Meligan Range. SARAWAK: Mt. Penrissen, Mt. Pueh, Mt. Santubong.

GENERAL DISTRIBUTION: Endemic to Borneo.

NOTES: De Vogel (1986) comments that a collection by *Boden-Kloss* (SFN 12290), labelled "Sumatra, Siberut", is of dubious provenance. In his opinion it is very unlikely that *G. phaiostele* occurs on this low elevation island, the correct provenance probably being Borneo, from where Boden-Kloss is also known to have made collections.

DERIVATION OF NAME: The specific epithet is derived from the Greek *phaeo*, dark, and *stele*, column, referring to the dark brown-coloured column.

66. GOODYERA CONDENSATA Ormerod & J.J. Wood

Goodyera condensata *Ormerod & J.J. Wood* in Orchid Rev. 109 (1242): 370, fig. 315 (2001). Type: Borneo, Sabah, Mt. Kinabalu, Mesilau River, 1500m, 21 January 1964, *Chew & Corner* RSNB 4052 (K).

Terrestrial. Rhizome creeping, terete, of 2 or more internodes each 1.5–4 × 0.4 cm. **Stem** 20–40 cm long, 6–8 mm in diameter, succulent. **Leaves** 3–6, blade 5–10(–12) × 3–4.7 cm, broadly ovate-elliptic, acuminate, thin-textured, green with fine white reticulatenerves on upper surface; petiole 0.5–2.5 cm long, with a 1.5–2.5 cm long sheathing portion; uppermost leaves grading into 4.5–6 × 1.5–2.5 cm foliaceous bracts. **Inflorescence** terminal, conical to oblong, 4.5–8.5 cm long, *c.* 3–3.3 cm wide; peduncle 1.5–2 mm long; rachis 2.5–5.5 cm long, densely many-flowered; floral bracts 1.5 cm long, *c.* 0.3–0.35 cm wide at base, uppremost smaller, slightly surpassing flowers, narrowly lanceolate, acuminate, margin minutely erose, particularly towards base, 3-nerved. **Flowers** resupinate, not opening widely, having a musky scent, white, lip yellow-orange with a white tip, column pale yellow, anther-cap brown. **Pedicel** and **ovary** 0.8–1 × 0.25–0.3 cm, glabrous. **Sepals** deeply concave, connate for 1mm at base, broadly ovate, subacute to obtuse,

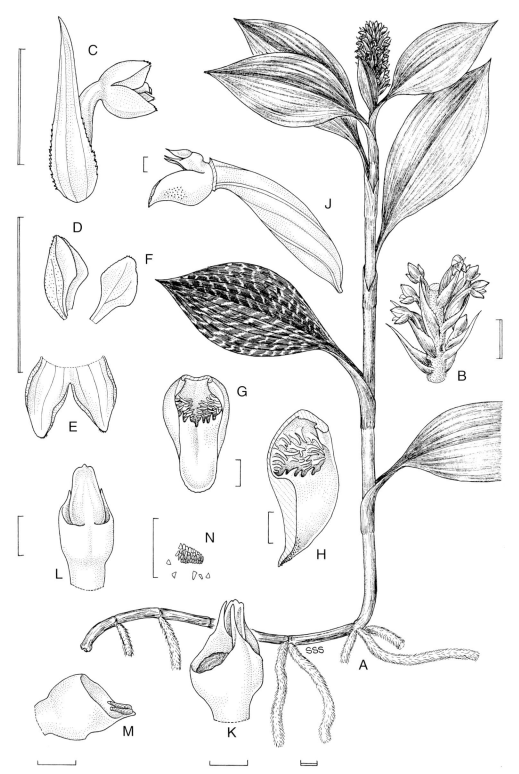

Figure 66. Goodyera condensata Ormerod & J.J. Wood. - A: habit. - B: portion of inflorescence. - C: floral bract and flower, side view. - D: dorsal sepal and connivent petals, side view. - E: synsepalum, adaxial surface. - F: petal. - G: lip, front view. - H: lip, longitudinal section. - J: pedicel with ovary, lip and column, side view. - K: column, oblique view. L: column, back view. - M: column, side view. - N: portion of sectile pollinium. A drawn from *Carr* 3437 and B–N from *Lamb* AL 1350/91 by Susanna Stuart-Smith. Scale: single bar = 1 cm; double bar = 1 mm.

exterior minutely papillose at apex. **Dorsal sepal** 5 × 2.1–2.2 mm. **Lateral sepals** 5 × 2.9–3 mm. **Petals** 4.5–4.6 × 2.5–2.6 mm, rhombic, obtuse, broadly clawed. **Lip** 5 mm long, 2–2.1 mm wide at base, *c.* 1.5 mm wide at apex, broadly ovate, obtuse, fleshy; hypochile semiglobose, containing many fleshy papillae; epichile ovate-lingulate, 1 mm thick; interior of lip fleshiest on epichile. **Column** 2–2.5 × 1.5 mm. Plate 14A.

HABITAT AND ECOLOGY: Rocky areas with alluvial soil near waterfalls in mixed lower montane *Castanopsis*/*Lithocarpus* forest on sandstone, in deep shade; on the boles of mossy tree trunks. Alt. 1400–2700 m. Flowering observed in January, May and June.

DISTRIBUTION IN BORNEO: SABAH: Mt. Kinabalu. Crocker Range, Mt. Alab.

GENERAL DISTRIBUTION: Endemic to Borneo.

NOTES: In 1935 Cedric Carr described *G. ustulata* from Mt. Kinabalu. It is clear, after study of material of *G. ustulata* and comparison with the type of *G. rostellata* Ames & C. Schweinf., that the two are conspecific. Prior to this, in 1933, Carr had misapplied the name *G. rostellata* when determining two of his Kinabalu collections (Carr 3437 & SFN 27592). This and several other collections from the mountain and elsewhere, clearly represented a hitherto undescribed species related to *G. rostellata*. This was described in 2001 by Wood and Ormerod as *G. condensata*. The two collections originally assigned by Carr to *G. rostellata* clearly belong to *G. condensata* and were never cited in his 1935 paper. *Goodyera condensata* is distinguished from all other Bornean *Goodyera* by the character combination of much broader, ovate-elliptic leaves which have only fine, white reticulate veins on the upper surface, a shorter, condensed conical to oblong inflorescence borne on a rather short peduncle, and flowers with a shorter, stouter column.

    Goodyera rostellata is a much less robust species with distinctly narrower leaves than *G. condensata*. These bear a bolder and broader pattern of white, pink or gold reticulate nerves on their upper surface. The inflorescence of *G. rostellata* is longer, narrower and much less dense and is borne on a longer peduncle. The flowers are hirsute and have a more slender column.

DERIVATION OF NAME: The generic name commemorates the English botanist John Goodyer (1592–1664). The specific epithet refers to the condensed, many-flowered inflorescences.

67. HABENARIA LOBBII Rchb.f.

Habenaria lobbii *Rchb.f.* in Linnaea 41: 50 (1877). Type: Borneo, Sarawak, *Lobb* s.n. (holotype K-LINDL, isotype W).

Habenaria marmorophylla Ridl. in J. Linn. Soc., Bot. 31: 304 (1895). Type: Borneo, Sarawak, Bau District, Buso, *Haviland* s.n. (holotype SING).
Habenaria havilandii Kraenzl., Orch. Gen. Sp. Pl. 1: 427 (1898). Type: Borneo, Sarawak, Bau District, Buso, *Haviland* s.n. (holotype B, destroyed, isotypes K, SAR).

Terrestrial. **Roots** up to 15 × *c.* 0.5 cm, thick, fleshy, woolly-hairy. **Leaves** 4–6, grouped towards base, sometimes one on lower portion of peduncle, *c.* 12–38 × 2.5–5.8 cm, *c.* 1.5 cm wide near base, lanceolate, acuminate, attenuate to a sheathing base, somewhat fleshy, margin undulate,

Figure 67. Habenaria lobbii Rchb.f. - A: habit. - B: floral bract. C: flower bud. - D: flower, oblique view. - E: dorsal sepal. - F: lateral sepal. - G: petal. - H: lip, front view. - J: lip and column, back view. - K: column, front view. - L: column, front view, showing protruding caudicles and viscidia. - M: pollinarium. - N: sectile structure of a portion of a pollinium. A drawn from *Lobb* s.n. (holotype) and B–N from *Rickards 90* by Susanna Stuart-Smith. Scale: single bar = 1 mm; double bar = 1 cm.

midrib dorsally sharply carinate on reverse, sheathing base shallowly v-shaped in cross-section, main nerves *c.* 7, with numerous small, transverse nerves. **Inflorescence** erect, laxly *c.* 12- to 22-flowered, flowers borne 1–3 cm apart; peduncle 25–55 cm long, *c.* 0.8 cm in diameter, obscurely ribbed; non-floriferous bracts 5–6, 4–7 × 1–2 cm, narrowly elliptic, acuminate, leafy; rachis 12–30 cm long, ribbed and very narrowly winged from base of floral bracts; floral bracts 1.5–2.5(–3) × *c.* 1 cm, ovate-elliptic, long-acuminate. **Flowers** *c.* 3 cm across, fleshy, green to greenish yellow. **Pedicel** and **ovary** 2.5–3 cm long, the pedicel 4–5 mm long, gently sigmoid, distinctly ribbed. **Dorsal sepal** and **petals** connivent to form a porrect, cucullate hood. **Dorsal sepal** 1–1.1 × 0.4–0.5 cm (unflattened), ovate, obtuse, strongly concave-cymbiform, cucullate, main nerves 3, each raised and strongly carinate on reverse. **Lateral sepals** 1.2 × 0.8–0.9 cm, obliquely ovate-elliptic, obtuse, strongly deflexed and parallel with each other, lower margins slightly revolute, main nerves 3, prominent and raised on reverse. **Petals** 1–1.1 cm long, 2.5 mm wide at base, 1.5 mm wide above, entire, linear, subfalcate, subacute, margins somewhat revolute, especially the outer, main nerve 1. **Lip** 2 cm long, cruciform; side lobes 12 × 3.1–3.2 mm, narrowly oblong, subacute, thinly fleshy, shallowly angled up from horizontal; mid-lobe 15 × 3–5 mm, narrowly oblong to linear, obtuse to subacute, thickly fleshy, vertical, pendent; spur 2.3–2.5 cm long, clavate, narrowly conical at base, dilated distally to *c.* 4 mm wide, laterally flattened, gently upcurved. **Column** 8 mm long; rostellar arms and stigmatic canals 3–4 mm long. Plate 13C & D.

HABITAT AND ECOLOGY: Lowland limestone hills; hill mixed dipterocarp forest; lower montane mossy forest. Alt. lowlands to 1100 m. Flowering observed in January, May, June, September and October, but probably throughout the year.

DISTRIBUTION IN BORNEO: KALIMANTAN: Locality unknown. SARAWAK: Bau District, Bau, Mt. Setiak, Taiton Hill, Bungo Range, Jambusan Hill, Krian Hill, Tiang River; Belaga District, Pantu Hill, Tasu Hill; Kuching District, Mt. Penrissen, Mt. Siburan, Padawan, Tiang Bekap; Gunung Mulu National Park, path to Deer Cave.

GENERAL DISTRIBUTION: Endemic to Borneo.

NOTES: The genus *Habenaria* contains between 600 and 800 species having a cosmopolitan distribution, but being particularly well represented in tropical Africa. Areas of great diversity are also found in mainland and South East Asia, as well as Central and South America. Ten species have been identified from Borneo, of which *H. setifolia* Carr was depicted in *Orchids of Borneo*, volume 1: fig. 45.

Seidenfaden (pers. comm.) studied Reichenbach's drawing of *H. lobbii* deposited in Vienna herbarium (W) while noting that the type material was probably at Kew. Wood & Ormerod (1998) relate that a sheet was found in the Lindley Herbarium at Kew bearing the annotation "Sarawak, on limestone rocks" which Reichenbach had also written on his drawing in Vienna. This has been designated as the holotype of *H. lobbii*, the study of which shows it to be conspecific with *H. marmorophylla*. This later and more widely used epithet becomes a synonym.

DERIVATION OF NAME: The generic name is derived from the Latin *habena*, reins, from the long, strap-like divisions of the petals (in many species) and the lip. The specific epithet honours Thomas Lobb (1820–1894), a professional orchid collector who, in 1857, was probably the second earliest collector of Sarawak orchids.

68. HETAERIA ANOMALA (Lindl.) Rchb.f.

Hetaeria anomala (*Lindl.*) *Rchb.f.* in Trans. Linn. Soc. London, Bot. 30 : 142 (1874).Type: India, Assam, Tingree, *Griffith* s.n. (holotype K–LINDL).

Aetheria anomala Lindl. in J. Linn. Soc., Bot., 1: 185 (1857).
Hetaeria biloba (Ridl.) Seidenf. & J.J. Wood in Orch. Pen. Mal. & Sing.: 95 (1992). Type: Peninsular Malaysia, Telom, on ridge above Batang Padang Valley, *Ridley* s.n. (holotype SING).
Zeuxine biloba Ridl. in J. Fed. Malay States Mus. 4: 73 (1909).
Hetaeria grandiflora Ridl. in J. Straits Branch Roy. Asiat. Soc. 87: 98 (1923). Type: Sumatra, Berastagi hill woods on the way to Sibayak, *Ridley* s.n. (holotype SING, isotype K).
H. rotundiloba J.J. Sm. in Svensk Bot. Tidskr. 20: 470 (1926). Type: Sulawesi, Balaang Mongondou, *Kaudern* 129 (holotype L).

Terrestrial. **Rhizome** fleshy, producing woolly roots at the nodes. **Stem** 30–50 cm long, decumbent and rooting at the nodes below, fleshy, glabrous, often producing a new growth 7–10 cm from the base, purple green. **Leaves** 3–5, uppermost bract-like; blade (1.5–)3–7 × (1–)2–3 cm, elliptic, sometimes asymmetrical, acute, abruptly narrowed to a sheathing petiole, glabrous, green, with 3 main nerves and faintly tessellated reticulate nerves; petiole (1.5–)2–3 cm long, sheath amplexicaul, dull red or pale pink. **Inflorescence** 8–12-flowered, lax; peduncle 18–24 cm long, reddish pink, white-pubescent; non-floriferous bracts 2–3, 1–2 cm long, adpressed, acute, glabrous, reddish pink; rachis 2.5–12 cm long, greenish pink, white-pubescent; floral bracts 6–10 × 2–3 mm, narrowly elliptic, acute to acuminate, salmon-pink, white-pubescent below and along margin. **Flowers** non-resupinate, sepals reddish pink, petals pink, lip and column white. **Pedicel** and **ovary** 9–11 mm long, salmon-pink, densely white-pubescent. **Sepals** pubescent. **Dorsal sepal** 5 × 2.5 mm, oblong to oblong-ovate, obtuse. **Lateral sepals** 5–6 × 3–4 mm, oblong to oblong-ovate, obtuse. **Petals** 4–5 × 1–1.1 mm, linear-ligulate, obtuse, glabrous. **Lip** glabrous, basal portion saccate, 2–2.5 × 2 mm, with 5 prominent nerves, each of the outer 4 nerves with a tall erose keel; claw 1 × 1 mm; terminal blade consisting of 2 strongly recurved obliquely oblong, obtuse lobes each 3.5 × 3 mm. **Column** 2 mm long, with short, acute lateral projections; anther-cap 3 mm long, ovate-acuminate; pollinia 2. Plate 14B.

HABITAT AND ECOLOGY: Lower and upper montane forest; oak-laurel forest; in humus among rocks; beside rivers; old landslide areas on ultramafic substrate. Alt. 1600 to 2100 m. Flowering observed from January until March, and December.

DISTRIBUTION IN BORNEO: SABAH: Mt. Kinabalu.

GENERAL DISTRIBUTION: Peninsular Malaysia, Thailand, Sumatra and Sulawesi.

NOTES: *Hetaeria* contains around 20 species distributed in the Old World tropics, extending from India to Fiji. The majority of species are found in Malaysia, with two, including *H. anomala*, occuring in Borneo. The second species, *H. hylophiloides* (Carr) Ormerod & J.J. Wood, was transferred from *Goodyera* in 2001. A third, *H. angustifolia* Carr, has since been transferred to *Rhomboda* by Ormerod (1995).

DERIVATION OF NAME: The generic name is derived from the Greek *hetaireia*, companionship, and refers to the intimate association of species of this genus with those of *Goodyera* and related genera of subtribe Goodyerinae.

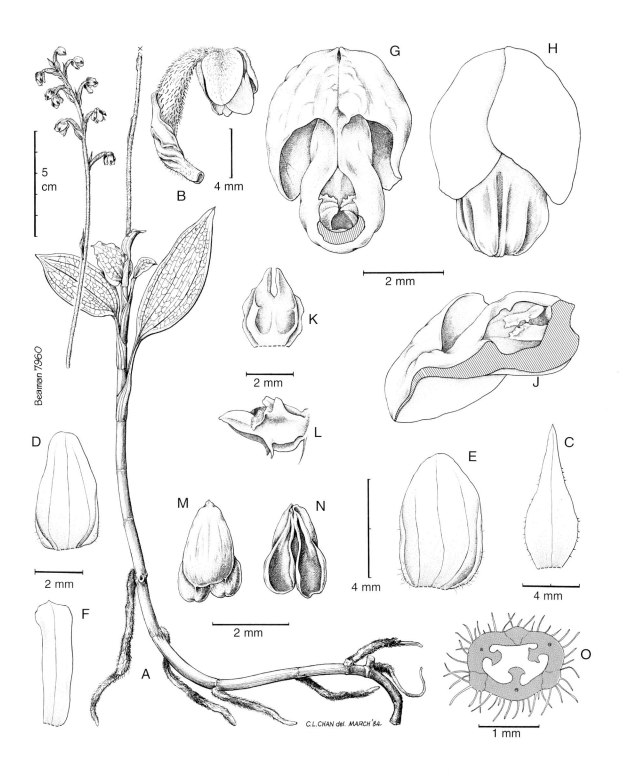

Figure 68. Hetaeria anomala (Lindl.) Rchb. f. - A: habit. - B: flower, side view. - C: floral bract. - D: dorsal sepal. - E: lateral sepal. - F: petal. - G: lip, front view. - H: lip, back view. - J: lip, longitudinal section. - K: column, front view. - L: column, side view. - M: anther-cap, back view. - N: anther-cap, interior view. - O: ovary, transverse section. All drawn from *Beaman 7960* by C.L. Chan.

69. LIPARIS ANOPHELES J.J. Wood

Liparis anopheles *J.J. Wood* in Nord. J. Bot. 11 (1): 85, fig. 1 (1991). Type: Borneo, Sabah, Mt. Trus Madi, above Kidukarok, *c.* 1560 m, 15 June 1988, *Surat* in *Wood* 781 (holotype K).

Epiphyte. **Rhizome** elongate, 2 mm in diameter, creeping, tough, rooting profusely at nodes, clothed with short, imbricate, scarious sheaths. **Cataphylls** 2–3, 3–6.5 cm long, ovate-elliptic, acuminate. **Pseudobulbs** 3.5–5(–6) cm long, 0.8 cm wide at base, 0.4–0.5 cm wide above, borne 2–3 cm apart, unifoliate, basal 1–1.5 cm narrowly pyriform, remainder narrowly attenuate and 3 winged. **Leaves** 27–36 × 1.5–2 cm, ligulate, ensiform, acute, gradually attenuated into a conduplicate petiole sheathing base of peduncle, dark olive-green above, paler below. **Inflorescences** erect; peduncle and rachis always shorter than leaf; peduncle 10–17.5 cm long, 1–1.5 mm wide at middle, 2–2.1 mm wide just below rachis, wiry, 2 winged, naked; rachis 1–4 cm long, curving, slightly zigzag; floral bracts 5–6 mm long, alternately distichous, conduplicate, acute, ascending, somewhat imbricate. **Flowers** successive, non-resupinate, pale tan, column straw-coloured, scented of cucumber. **Pedicel** and **ovary** 1–1.2 cm long, slender. **Sepals** slightly carinate, margins revolute. **Dorsal sepal** 10 × 2.2 mm, narrowly ovate-elliptic, acute, spreading to reflexed. **Lateral sepals** 10 × 2.2 mm, as dorsal, but slightly oblique. **Petals** 10 mm long, *c.* 0.6 mm wide at base, *c.* 0.2 mm wide above, linear, acute. **Lip** 12 mm long, 2.2 mm wide above base and distally, 2 mm wide at middle, ligulate, concave and fleshy at base, apex shortly 3-toothed, disc with a glabrous central raised fleshy keel arising *c.* 4.5–5 mm from base and merging into blade at apex, margin minutely papillose distally, basal callus glabrous, swollen, broad and shallowly crescent-shaped. **Column** 5 mm long, *c.* 1.5 mm wide at base, narrower above, curving; anther-cap 0.9 mm long, cucullate; pollinia 4. Plate 14C.

HABITAT AND ECOLOGY: Lower to upper montane forest, sometimes with *Agathis*. Alt. 1560 to 2400 m. Flowering observed in June and September.

DISTRIBUTION IN BORNEO: SABAH: Mt. Trus Madi. SARAWAK: Marudi District, Tama Abu Range; Mt. Murud.

GENERAL DISTRIBUTION: Endemic to Borneo.

NOTES: *Liparis anopheles* belongs to section *Distichae* which contains between 25 and 30 species centred on Indonesia and New Guinea. About 10 species, including *L. anopheles*, have been recorded from Borneo. Other members of this section illustrated in this series are *L. lobongensis* Ames (*Orchids of Borneo*, volume 3: fig. 80) and *L. pandurata* Ames (volume 3: fig. 81).

DERIVATION OF NAME: The generic name is derived from the Greek *liparos*, fat, greasy or shining, referring to the smooth shiny leaves of many species. The specific epithet refers to the flowers which resemble dead malaria-carrying *Anopheles* mosquitoes.

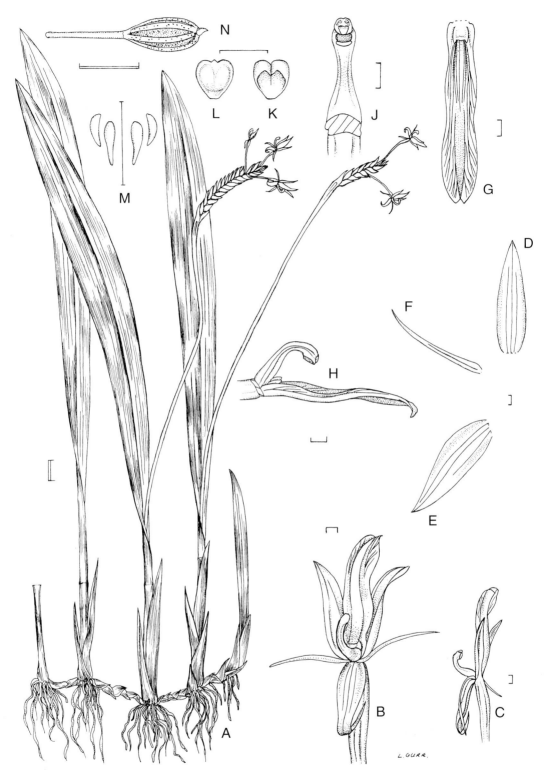

Figure 69. Liparis anopheles J.J. Wood - A: habit. - B: flower, front view. - C: flower, side view. - D: dorsal sepal. - E: lateral sepal. - F: petal. - G: lip, front view. - H: lip and column, side view. - J: column, front view. - K: anther-cap, interior view. - L: anther-cap, back view. - M: pollinia. - N: fruit capsule. A drawn from *Awa & Lee* S. 51113 and B–N from *Wood* 871 (holotype) by Linda Gurr. Scale: single bar = 1 mm; double bar = 1 cm.

70. LIPARIS GRANDIFLORA Ridl.

Liparis grandiflora *Ridl.* in J. Bot.: 333 (1884). Type: Borneo, Sarawak, Mindai-Pramassan, 19 July 1882, *Grabowsky* s.n. (holotype BM).

Epiphyte. **Roots** elongate, smooth, forming a dense mass. **Cataphylls** 2, 2.5–8 cm long, narrowly elliptic, acuminate, sheathing. **Pseudobulbs** (3–)4–7 × 0.8–1(–2.5) cm, cylindrical, closely spaced, tufted, unifoliate. **Leaves** 14–25 × 3–6.5 cm, narrowly elliptic, acute, attenuated and conduplicate below, stiffly coriaceous, main nerves 8–9, prominent, especially on reverse. **Inflorescences** terminal, erect to curving, laxly up to *c.* 15-flowered, flowers borne 1–4 cm apart; peduncle 9 or more cm long, with a 2–2.5 cm acuminate sterile sheath at base; rachis *c.* 20–33 cm long; floral bracts 0.7–1.3 × 0.3–0.4 cm, narrowly elliptic, acute to acuminate. **Flowers** 3 cm long, 1.5–1.8 cm across, sepals, petals and column greenish, or cream flushed pale green, lip apricot with olive-green median ridge, brownish, ochreous, or brick-red to orange. **Pedicel** and **ovary** 1.7–2.5 cm long, slender, straight or gently curved. **Sepals** and **petals** strongly reflexed. **Dorsal sepal** 1.5 × 0.5 cm, oblong-elliptic, obtuse, margins revolute, 5-nerved. **Lateral sepals** 1.6 × 0.5 cm, oblong-elliptic, obtuse, margins gently incurved, 5-nerved. **Petals** 1.5 × 0.11–0.12 cm, linear, obtuse, curved, 1-nerved. **Lip** 2–2.1 cm long, 1.7 cm wide across lobes, *c.* 0.6 cm wide near base, flabellate, bilobulate, cuneate below, curved, margin of lobes crenulate and minutely pubescent, disc with a short, fleshy sulcate basal callus. **Column** 7 mm long, *c.* 2.5 mm wide at base, curved, with small shallow, rounded wings flanking stigma; anther-cap *c.* 1.6 × 1.5 mm, ovate, cucullate, obtuse, sometimes notched in front. Plate 14D & E.

HABITAT AND ECOLOGY: Lowland mixed dipterocarp forest, sometimes on limestone; riverine forest. Alt. lowlands to 400 m. Flowering observed in July, and from September until November.

DISTRIBUTION IN BORNEO: KALIMANTAN TIMUR: Kutai, Long Petak. SABAH: Maliau Basin, Lake Linumunsut. SARAWAK: Belaga District, Sepaku, Upper Belaga River. Marudi District, Mt. Mulu, Mt. Pala.

GENERAL DISTRIBUTION: Endemic to Borneo.

NOTES: This is the largest flowered species of *Liparis* native to Borneo. Belonging to section *Cestichis*, it resembles a larger flowered, though less colourful, version of *L. crenulata* (Blume) Lindl. or *L. latifolia* (Blume) Lindl. (see *Orchids of Borneo*, volume 1: fig. 47).

DERIVATION OF NAME: The specific epithet describes the relatively large flowers for the genus.

Figure 70. Liparis grandiflora Ridl. - A: habit. - B: inflorescence. C: flower, viewed from above. - D: dorsal sepal. - E: lateral sepal. F: petal. - G: lip, front view. - H: close-up of lip margin. - J: column, front view. - K: column, side view. - L: anther-cap, back view. - M: pollinia. A drawn from *Kandau & Ismawi* S. 43766 and B–M from *Leche* 211 by Susanna Stuart-Smith. Scale: single bar = 1 mm; double bar = 1cm.

71. LIPARIS LACERATA Ridl.

Liparis lacerata *Ridl.* in J. Linn. Soc., Bot. 22: 284 (1886). Type: Borneo, Sarawak, Lawas River, *Burbidge* s.n. (holotype BM).

Leptorchis lacerata (Ridl.) O. Kuntze, Rev. Gen. Pl. 2: 671 (1891).

    Terrestrial. Rhizome up to 8 cm or more long, creeping. **Roots** numerous, forming a dense mass, *c.* 1 mm in diameter, smooth. **Cataphylls** 2–3, 3–8 cm long, ovate-elliptic, acute, papery-textured. **Pseudobulbs** 2–7 × 0.8–1.5(–2) cm, oblong-conical, somewhat laterally flattened, borne 1 cm or less apart on rhizome, bifoliate, pale green, shiny. **Leaf-blade** *c.* 11.5–21 × 1.5–2.8 cm, narrowly elliptic to oblong-elliptic, acute, gradually attenuated below; petiole 2.5–4 cm long, conduplicate. **Inflorescences** erect at first, becoming mostly pendulous, with up to 60 or more flowers, each borne 3–5(–10) mm apart; peduncle 5–10 cm long; rachis *c.* 20–35 cm long; floral bracts 3–6 mm long, narrowly triangular, subulate, acute, orange. **Flowers** *c.* 1 cm in diameter, sepals and petals orange or brownish, lip dark orange centrally, fading to paler orange, with a dark red basal callus, sometimes with a broad red line or 2 thin red central lines, column white. **Pedicel** and **ovary** 6–11 mm long, slender. **Dorsal sepal** 5.6–6 × 2 mm, narrowly ovate, obtuse, margins involute. **Lateral sepals** 4.7–6 × 2 mm, narrowly ovate, cymbiform, obtuse, margins involute. **Petals** 5.6–6 × 1.2 mm, ligulate, obtuse, curled back behind ovary. **Lip** 5–6 × 3–4 mm, curved near base, panduriform, broadest at base and distally, side margins curved, entire, apex distinctly bilobed, lobules with 5–6 long lacerate teeth; disc with a basal callus. **Column** 3.5–4 mm long, slender, arcuate. Plate 15A & B.

HABITAT AND ECOLOGY: Lowland dipterocarp forest; mixed hill dipterocarp forest; riverine forest; kerangas forest on podsolic soils; peat swamp forest. Alt. sea level to 600 m. Flowering observed in January, May, June, September and December, but probably throughout the year.

DISTRIBUTION IN BORNEO: SABAH: Nabawan area; Mt. Kinabalu; Beaufort District, Weston area; Sipitang District, Long Pa Sia area. SARAWAK: Bau District; Lawas District; Sibu District.

GENERAL DISTRIBUTION: China (Hainan), Myanmar (Burma), Peninsular Malaysia, Thailand, Sumatra (including Mentawai Islands), Borneo and New Guinea.

NOTES: *Liparis lacerata* is one of several attractive species belonging to section *Coriifoliae*. Species of this section have distinctly articulated leaves which are shed when old and inflorescences with rather thin floral bracts, with the flowers often all opening together and facing in all directions.

DERIVATION OF NAME: The specific epithet is derived from the Latin *lacerus*, lacerate, slashed into narrow, taper-pointed divisions, in reference to the lip.

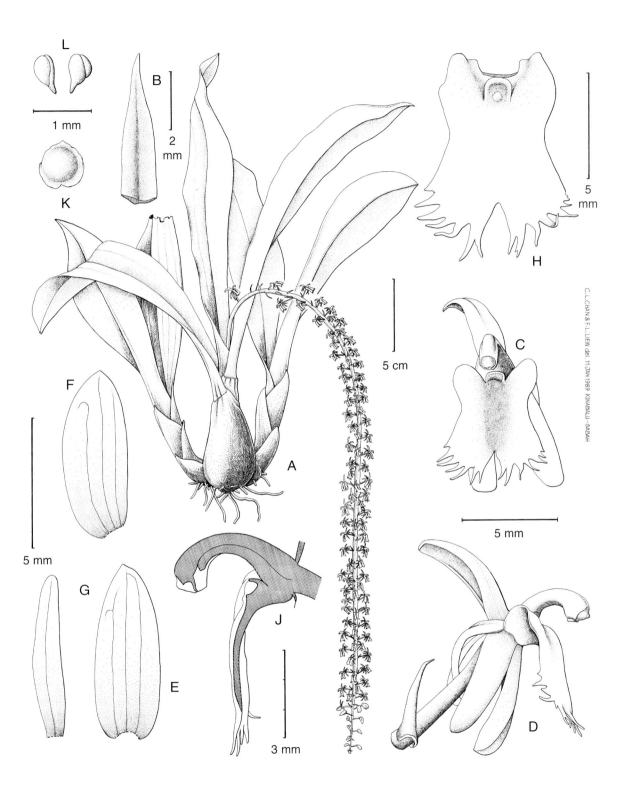

Figure 71. Liparis lacerata Ridl. - A: habit. - B: floral bract. - C: flower, front view. - D: floral bract and flower, side view. - E: dorsal sepal. - F: lateral sepal. - G: petal. - H: lip, front view. - J: lip and column, longitudinal section. - K: anther-cap, back view. - L: pollinia. All drawn from material cultivated at *Tenom Orchid Centre* by C.L. Chan and Liew Fui Ling.

72. NABALUIA EXALTATA de Vogel

Nabaluia exaltata *de Vogel* in Blumea 30: 202, pl. 2a–c (1984). Type: Borneo, Sarawak, Lawas District, Mt. Murud, *Burtt & Martin* B 5259 (holotype E, isotypes K, SAR).

Epiphyte or **lithophyte. Rhizome** short, creeping. **Roots** to 3 mm in diameter. **Cataphylls** of young shoot 8, 10–21 cm long, stiffly herbaceous, margins papery. **Pseudobulbs** (6–)10–18 × 1–3 cm, cylindrical, bifoliate. **Leaf-blade** (8.5–)15–37 × 2.2–4.7 cm, narrowly elliptic, sometimes linear, acute, stiffly coriaceous, main nerves 7–9; petiole (2.5–)6–24 cm long, canaliculate. **Inflorescences** synanthous with newly emerging leaves, 25– to 41–flowered, flowers borne 4–10 mm apart; peduncle 12–20 cm long, sturdy; rachis 14–30 cm long; floral bracts *c.* 2.6–2.7 × 0.15 cm, ovate, truncate. **Flowers** sweetly scented, sepals yellowish green to greenish tan, petals yellowish green to chocolate tan, lip white, mid-lobe with ochre yellow patch, lower half of mid-lobe and side lobes chocolate-brown, column chocolate-brown. **Pedicel** and **ovary** 2–2.4 cm long. **Dorsal sepal** (10–)11.5–12 × 6 mm, obovate, obtuse to acute, somewhat hooded at apex. **Lateral sepals** (10–)11–12 × 5–5.5 mm, oblong to ovate, obtuse to acute, somewhat hooded at apex. **Petals** (10–)11–11.5 × 1.5–2 mm, ligulate, obtuse. **Lip** (9–)10–12 mm long; hypochile 2-lobed, 2–2.5 mm long, 1.5–1.8 mm high, 2–2.7(–3.5) mm wide, lobules 3.5–4 × 0.8 mm, acute; callus projecting over hypochile sac with free, ± erect, flattened, quadrangular to ± ligulate arms *c.* 1–1.5 × 1–1.5 mm which project upwards behind side lobes of hypochile; keels 5, 3 obscure, 2 thick and swollen; epichile (7–)8–9.5 × (3–)3.7–4 mm, deeply retuse, apiculate, recurved. **Column** (7–)7.5–8 mm long, side lobes of hood distinct, 1–1.3 × *c.* 0.7 mm, rounded; anther-cap 1.8–2 × *c.* 1.8 mm, acute; pollinia 4.

HABITAT AND ECOLOGY: Upper montane mossy ericaceous forest; sandstone boulders; recorded as being associated with *Nepenthes edwardsiana* and *N. lowii* on Mt. Trus Madi. Alt. 2000 to 2490 m. Flowering observed in June, August, October and November.

DISTRIBUTION IN BORNEO: SABAH: Mt. Trus Madi. SARAWAK: Gunung Mulu National Park; Mt. Murud.

GENERAL DISTRIBUTION: Endemic to Borneo.

NOTES: Three species are currently recognised in the genus *Nabaluia*, which was proposed by Oakes Ames in 1920. It shares the saccate hypochile of *Pholidota*, but is distinct in having several unique characters. Groups of stellately arranged hairs are present on the inside on the perianth and long slender side lobes project in front of the hypochile between which is a peculiar horseshoe-shaped callus.

DERIVATION OF NAME: The generic name refers to Kinabalu. The specific epithet is derived from the Latin *exaltatus*, raised high, in reference to the erect arms of the horseshoe-shaped callus between the side lobes of the hypochile.

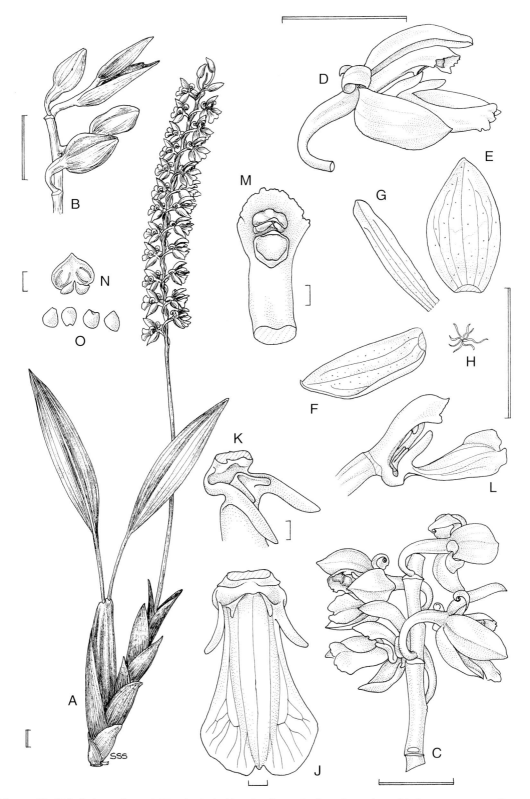

Figure 72. Nabaluia exaltata deVogel - A: habit. - B: flower buds. - C: portion of inflorescence. - D: flower, side view. - E: dorsal sepal. - F: lateral sepal. - G: petal. - H: stellate hair from dorsal sepal. - J: lip, front view. - K: close-up of lip base showing keel. - L: ovary, lip and column, side view. - M: column, front view. - N: anther-cap, interior view. - O: pollinia. All drawn from *Wood* 903 by Susanna Stuart-Smith. Scale: single bar = 1mm; double bar = 1cm.

73. OBERONIA PATENTIFOLIA Ames & C. Schweinf.

Oberonia patentifolia *Ames & C. Schweinf.,* Orchidaceae 6: 83, pl. 90, fig. 1 (1920). Type: Borneo, Sabah, Mt. Kinabalu, Lobong Cave, November 1915, *J. Clemens* 104 (holotype AMES).

Epiphyte. Roots up to *c.* 1 mm in diameter, numerous, fibrous, flexuose, mostly unbranched, furrowed. **Stems** 12–20(–25) cm long, caespitose, flexuose or sinuate, entirely concealed by leaf sheaths. **Leaves** not jointed at base, lower margins 2.2–7.5(–8) cm long, upper margin 3.5–7 cm long, 0.7–1.2(–1.4) cm wide below, progressively smaller towards base and apex of stem, distichous, equitant, imbricate, ensiform, obtuse or minutely truncate to acute, spreading to ascending. **Inflorescences** terminal, densely many-flowered, raceme *c.* 4–5 mm wide, curving, flowers verticillate, slightly laxer toward apex of raceme; peduncle 0.5–1.4 cm long, striate, glabrous below, pubescent above, bearing an irregular whorl of narrowly elliptic, acute bracts 2–5 mm long; rachis 12–18(–20) cm long, striate, densely pubescent; floral bracts 1.5–2.5(–3) mm long, triangular to narrowly elliptic, acute to acuminate, concave, fimbriate, sparsely pubescent on exterior. **Flowers** minute, brownish yellow, buff, pale yellow, lemon-yellow, pale brown, tan brown or white. **Pedicel** and **ovary** *c.* 1.4 mm long, cylindrical, densely pubescent. **Sepals** and **petals** reflexed, sepals pubescent on reverse. **Dorsal sepal** *c.* 1.25 × *c.* 1–1.1 mm, ovate, obtuse, concave. **Lateral sepals** *c.* 1.25 × *c.* 1.1 mm, ovate, obtuse and mucronate, concave. **Petals** *c.* 1 × 0.5–0.6 mm, ovate-oblong, obtuse to acute. **Lip** *c.* 1.4 mm long, 1.1 mm wide at base and near apex, incurved, pandurate, constricted in middle, apically divided into 2 broad, semi-orbicular, obtuse lobules which sometimes slightly overlap, base of lip provided with 2 spreading, oblong, obtuse auricles. **Column** minute; foot absent; pollinia 4. Plate 15D.

HABITAT AND ECOLOGY: Hill forest; lower montane forest, sometimes on ultramafic substrate. Alt. 600 to 1800 m. Flowering observed from October until December.

DISTRIBUTION IN BORNEO: SABAH: Mt. Kinabalu; Beaufort area; Penampang District.

GENERAL DISTRIBUTION: Endemic to Borneo.

NOTES: Over 300 species of *Oberonia* are widely distributed from tropical Africa to the Pacific islands, the largest number occurring in mainland Asia. Seidenfaden (1968) produced a very useful account of the mainland Asian species. Those from the islands of S.E. Asia, including Borneo, however, remain poorly understood and difficult to determine. Some thirty species have been recorded from Borneo (Wood & Cribb, 1994).

Many species of *Oberonia* are pendent in habit, pseudobulbs are absent, and the leaves, which are often fleshy, may be terete or bilaterally flattened and, in one group, are articulated to the sheathing base. The minute, non-resupinate flowers, which are often crystalline in appearance and rarely more than 2 mm long, may be green, yellow, brownish or red. They are more or less arranged in whorls and face in all directions. Curiously, those at the base of the inflorescence are the last to open. The petals often have an irregular, erose margin and the lip, although variable in shape, frequently has small basal "ears" or auricles and a bilobed apex.

The plant figured here closely resembles the illustration accompanying the original description by Ames and Schweinfurth. However, the flowers of material examined for the figure provided

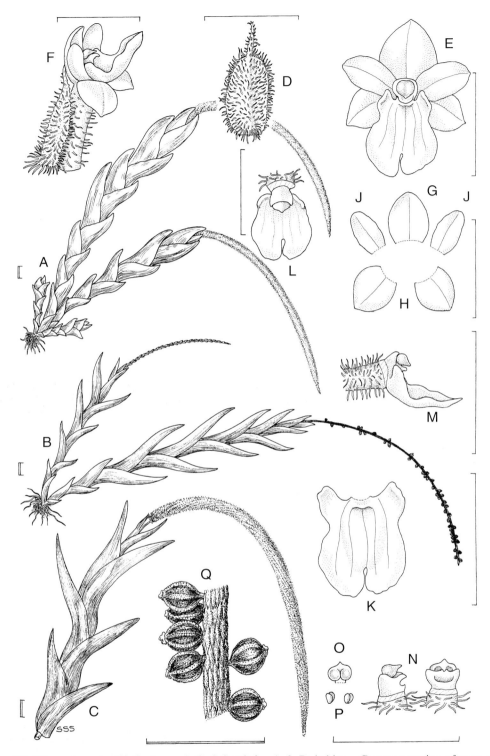

Figure 73. Oberonia patentifolia Ames & C. Schweinf. - A & B: habits. - C: upper portion of stem with inflorescence. - D: floral bract. - E: flower, front view. - F: flower with floral bract attached, side view. - G: dorsal sepal. - H: lateral sepals. - J: petals. - K: lip, front view. - L: upper portion of ovary, lip and column, viewed from above. - M: pedicel with ovary, lip and column, side view. - N: column with anther-cap, side and front views. - O: anther-cap, back view. - P: pollinia. - Q: portion of infructescence. A drawn from *Madani* SAN 89490, B & Q from *Krispinus* SAN 131357, and C–P from *Bacon* 201 by Susanna Stuart-Smith. Scale: single bar = 1 mm; double bar = 1 cm.

here (*Bacon* 201 & *Madani* SAN 89490) have a lip with a narrower sinus between the lobules. In addition, the *Madani* collection has somewhat less divergent, obtuse leaves.

Oberonia patentifolia would appear to be closely related to *O. hispidula* Ames, a species from the Philippines. In *O. hispidula*, however, the leaves are distinctly falcate-incurved, the floral bracts are aristate, and the flowers even smaller and greenish in colour.

DERIVATION OF NAME: The generic name refers to *Oberon*, the mythical prince of the fairies and the little king of the dryads who rode about on the branches of the trees, hiding his many-formed countenance among the leaves. Lindley (1830), in fanciful mood, wrote "so our little herbs, not less changeable in form, lurk in the forests of India and ride triumphantly in their leafy chariot." The specific epithet is derived from the Latin *patens*, spreading, and *folius*, leaved.

74. OCTARRHENA PARVULA Thwaites

Octarrhena parvula *Thwaites*, Enum. Pl. Zeyl.: 305 (1861). Type: Sri Lanka, Central Province, on forest trees, at an elevation of 900 to 1200 m, *Thwaites*, Ceylon Plants 3072 (holotype K).

Phreatia nana J.D. Hook., Fl. Brit. Ind., 5: 811 (1890). Type: Peninsular Malaysia, Perak, *Scortechini* 1432 (syntype K); Perak, Larut, Sept. 1881, *King's Collector* 2428 (syntype K).
P. parvula (Thwaites) Benth. ex J.D. Hook., Fl. Brit. Ind., 5: 811 (1890).
Octarrhena nana (J.D. Hook.) Schltr., Feddes Repert. 9: 217 (1911).
O. amesiana Schltr., Feddes Repert. 9: 439 (1911). Type: Philippines, Mindanao, Mt. Apo, Oct. 1904, *Copeland* 1427 (holotype B, destroyed).

Tufted **epiphyte**. **Roots** numerous, fibrous. **Stems** 3–8(–10) cm long, without pseudobulbs, leafy towards apex, enclosed in brown persistent leaf sheaths below, often curved. **Leaves** distichous; blade 1–2(–2.7) × 0.2–0.3 cm, spreading obliquely, linear to linear-oblong, acute, laterally compressed or terete, straight or somewhat falcate, coriaceous, jointed to basal sheath; sheath 5–8 mm long. **Inflorescence** a lax, few- to 15-flowered, straight or curved axillary raceme; peduncle 0.3–1 cm long, filiform; non-floriferous bracts 1–3, 1–2 mm long, acute; rachis 1.5–3.5 cm long; floral bracts 1.8 × 2 mm, triangular-ovate, acute to acuminate, translucent, partially sheathing pedicel and ovary. **Flowers** opening from the top or middle of the raceme, minute, 2.5 mm across, pale green or yellowish green with a whitish column, ageing to yellowish. **Pedicel** and **ovary** 1.5–1.6 mm long. **Sepals** and **petals** often concave, spreading. **Dorsal sepal** 1.6–1.7 × 1.2 mm, oblong-ovate, acute or subacute. **Lateral sepals** 1.3–1.6 × 1.4 mm, obliquely oblong-ovate, acute or obtuse. **Petals** 1 × 0.6 mm, oblong, obtuse, concave. **Lip** 1 × 0.6 mm, oblong-elliptic, entire, obtuse, concave, sometimes with 2 shallow basal depressions. **Column** 0.6 × 0.6 mm; foot absent; anther-cap ovate; pollinia 8, in 4 pairs.

HABITAT AND ECOLOGY: Hill forest; oak-laurel forest; lower montane Fagaceae, *Agathis*, dipterocarp forest; open kerangas with *Eugenia, Ficus* and *Tristania*; very open, dry forest on sandstone ridges. Alt. 900 to 1800 m. Flowering observed in March, May, July and December.

DISTRIBUTION IN BORNEO: SABAH: Mt. Kinabalu; Sipitang District. SARAWAK: Lawas District, Ba Kelalan. Probably overlooked elsewhere.

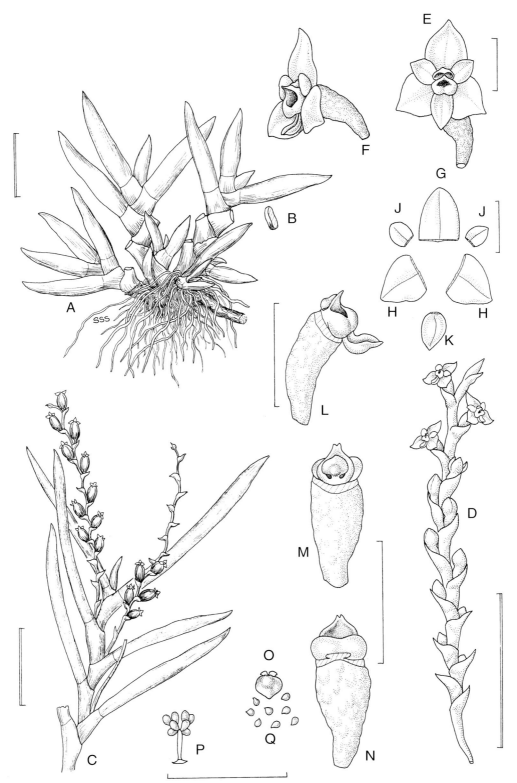

Figure 74. Octarrhena parvula Thwaites - A: habit. B: transverse section through leaf. - C: flowering shoot. D: inflorescence. - E: flower, front view. - F: flower, side view. - G: dorsal sepal. - H: lateral sepals. - J: petals. - K: lip. - L: ovary, lip and column, side view. - M: ovary and column, front view. - N: ovary and column showing anther-cap, back view. - O: anther-cap, back view. - P: pollinarium. - Q: pollinia. A, B, D–O & Q drawn from *Beaman* 10596, C from *Vermeulen & Duistermaat* 1119 and P from after *J.J. Smith* by Susanna Stuart-Smith. Scale: single bar = 1mm; double bar = 1cm.

GENERAL DISTRIBUTION: Sri Lanka, Peninsular Malaysia, Sumatra, Java, Borneo and the Philippines.

NOTES: *Octarrhena* in a genus containing about twenty species distributed from Sri Lanka to the Pacific islands. The majority occur as montane endemics in New Guinea, some of which, such as *O. lorentzii* J.J. Sm., having an unusual habit and attractive orange flowers. *Octarrhena* is distinguished from the closely related *Phreatia* by the laterally compressed or terete leaves, free sepals and footless column. A second species, *O. angraecoides* (Schltr.) Schltr. (syn. *O.condensata* (Ridl.) Holttum), is recorded from Borneo. This differs in having broader leaves, and a densely many-flowered inflorescence of yellowish-orange flowers with a sharply uncinate lip.

DERIVATION OF NAME: The generic name is derived from the Greek *okta*, eightfold, and *arrhen*, male, stamen, in reference to the eight pollinia. The specific epithet is derived from the Latin *parvus*, little, small, and refers to the diminutive habit of this species.

75. PENNILABIUM STRUTHIO Carr

Pennilabium struthio *Carr* in Gard. Bull. Straits Settlem. 5: 151, pl. 4, fig. 4 (1930). Type: Peninsular Malaysia, Pahang, Kuala Teku, *c.* 150 m, Aug. 1928, *Carr* 174 (holotype K).

Epiphyte. **Stem** abbreviated, 1–3 cm long, rooting below, 5- to 8-leaved. **Leaves** spreading; blade 4–9.5 × 0.9–2 cm, linear-lanceolate, falcate, unequally bilobed, longer lobe obtuse, shorter one obtuse or acute and tooth-like, attenuated and slightly twisted at base, rather fleshy, carinate beneath; sheath 3–5 mm long, tubular, recurved towards apex. **Inflorescences** emerging from base of sheaths behind blade, porrect, with up to 20 flowers borne usually one at a time in succession; peduncle 1–1.5 cm long, terete, laterally compressed, bearing a few distant bract-like sheaths; rachis 2.5–5.5 cm long, 2–2.5 mm thick, carinate, strongly laterally compressed, green; floral bracts *c.* 0.8–1 mm long, broadly triangular, acute, concave, carinate, half embracing rachis. **Flowers** lasting one day, non-resupinate, *c.* 1.7 × 1 cm, sepals and petals semi-transparently ochreous or orange-yellow, spotted dark red, particularly centrally, spots numerous or very few, lip side lobes pure white, semi-transparent, mid-lobe white, spur yellowish or orange-yellow, lamellate callus transparently yellowish with a dark red or purple margin, column yellowish with a pale red basal flush. **Pedicel** and **ovary** 0.5–1 cm long, triquetrous, sparsely ramentaceous. **Sepals** and **petals** spreading. **Dorsal sepal** 7.5–10 × 4.8–5.1 mm, elliptic, shortly apiculate, apiculus conical, acute, very shortly clawed at base, concave, sparsely papillose, 3- to 5-nerved. **Lateral sepals** 8–10 × 5 mm, adnate to base of column, obliquely oblong, shortly apiculate, apiculus conical, acute, distal margins sometimes minutely erose, concave, sparsely papillose, recurved, 5-nerved. **Petals** 7–8.5 × 4.8–5 mm, elliptic-obovate, obtuse, shortly clawed at base, erose, concave, incurved, 3-nerved. **Lip** adnate to base of column, 3-lobed; side lobes 7–10 × 3.5–4.5 mm, falcate, linear-cuneate, reflexed from base, anterior margins often contiguous in middle, apex truncate, laciniate into numerous cylindrical shortly hairy appendages 1–1.5 mm long; mid-lobe *c.* 3.5 mm long, *c.* 1.5 mm wide, *c.* 1.8 mm high, fleshy, strongly laterally flattened, linear in basal half, abruptly and obtusely angularly incurved above middle, narrowly obtuse, minutely papillose, porrect in side view; disc with 2 erect, bidentate, triangular, parallel lamellae

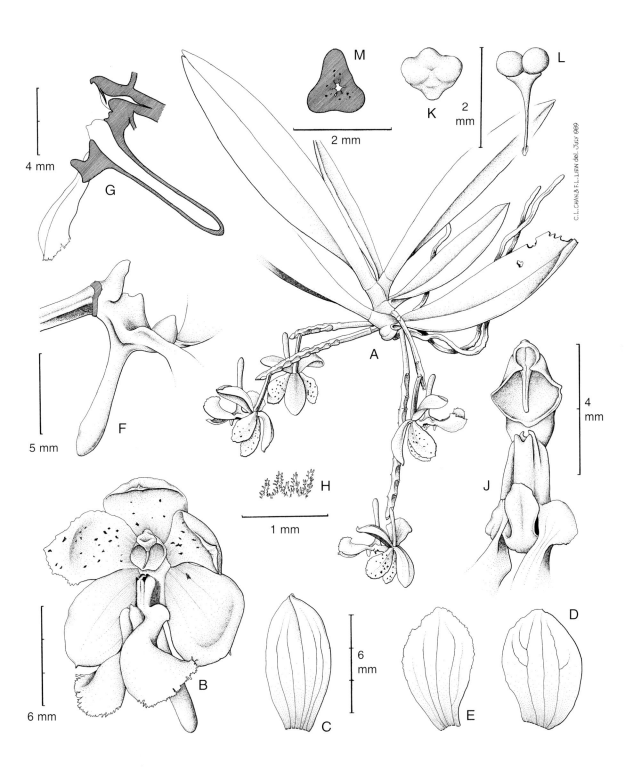

Figure 75. Pennilabium struthio Carr - A: habit. - B: flower, front view. - C: dorsal sepal. - D: lateral sepal. - E: petal. - F: pedicel with ovary, lower portion of lip showing spur, and column, side view. - G: lip and column, longitudinal section. - H: close-up of lip margin showing hairy appendages. - J: lower portion of lip and column, anther-cap removed, front view. - K: anther-cap, back view. - L: pollinarium. - M: ovary, transverse secton. All drawn from *Lamb* AL 382/85 by C.L. Chan.

above spur entrance, whole structure 1.8–2 × 2 mm; spur 8.1–10 mm long, narrowly cylindrical, dilate below the narrowly obtuse apex, straight, porrect, almost parallel with pedicel and ovary. **Column** 2.3–3 mm long, dilate in middle; foot absent; stigmatic cavity very large, suborbicular; rostellum *c*. 1.7 mm long, long-subulate, incurved; anther-cap cucullate; pollinia 2, entire. Plate 15C.

HABITAT AND ECOLOGY: Hill forest; riverine forest. Alt. 200 to 500 m. Flowering observed in February, April, October and November.

DISTRIBUTION IN BORNEO: SABAH: Pensiangan District; Tenom District. SARAWAK: Lawas District, Tebunan Hill, Upper Trusan.

GENERAL DISTRIBUTION: Peninsular Malaysia and Borneo.

NOTES: *Pennilabium* is a small genus of between ten and twelve species distributed from India through Thailand and Malaysia to Indonesia and the Philippines. They have the habit of *Pteroceras*, but the flowers have a lip which is adnate to a footless column and entire instead of sulcate pollinia. A second species, *P. angraecoides* (Schltr.) J.J. Sm., is also found in Borneo. This has a lip with less ragged, serrulate side lobes, a much smaller, triangular, tooth-like mid-lobe and lacking the prominent keels on the disc. It is unclear whether the Sumatran *P. lampongense* J.J. Sm. occurs in Borneo and citations in recent literature refer to *P. struthio*.

The erect, lamellate keels at the entrance to the spur are transparent and may serve as guides for the maxillae of visiting moths.

DERIVATION OF NAME: The generic name is derived from the Latin *penni-*, feathered, and *labium*, lip, in reference to the erose or fimbriate margins of the lip side lobes in some species. The specific epithet refers to the lip side lobes which, according to Carr, "show a passable resemblance to the tail feathers of an ostrich."

76. PERISTYLUS HALLIERI J.J. Sm.

Peristylus hallieri *J.J. Sm.* in Bull. Dép. Agric. Indes Néerl., 22: 1 (1909). Types: Borneo, Kalimantan Barat, Soeka Lanting (Sukalanting), *Hallier* 37 (syntypes BO, L); Soengai Kelasar, *Hallier* 1542 (syntypes BO, L).

An erect, tuberous **terrestrial**. **Stem** 2–3.5 cm long, 7- to 12-leaved. **Leaves** grouped at the base, 4–10(–14) × 0.8–1.3(–2.2) cm, narrowly elliptic, ensiform, acuminate, mucronate, narrowed to a sheathing base, main nerves 3, the median quite prominent. **Inflorescences** laxly many-flowered; peduncle 4–13 cm long; non-floriferous bracts several, 1–1.5(–5) cm long, spreading, sheathing, leafy, narrowly-elliptic, acuminate; rachis (4.5–)8–13(–17) cm long; floral bracts 0.5–1(–2) cm long, narrowly ovate-elliptic, acuminate to long acuminate. **Flowers** with dark green sepals, pale green petals, lip and spur. **Pedicel** and **ovary** 0.6–1 cm long, twisted. **Sepals** and **petals** somewhat connivent, forming a hood. **Dorsal sepal** 3–3.5 × 1.5 mm, oblong-ovate, obtuse, concave. **Lateral sepals** 3–3.5 × 1–1.2 mm, obliquely oblong-elliptic, obtuse, concave. **Petals** 3.5

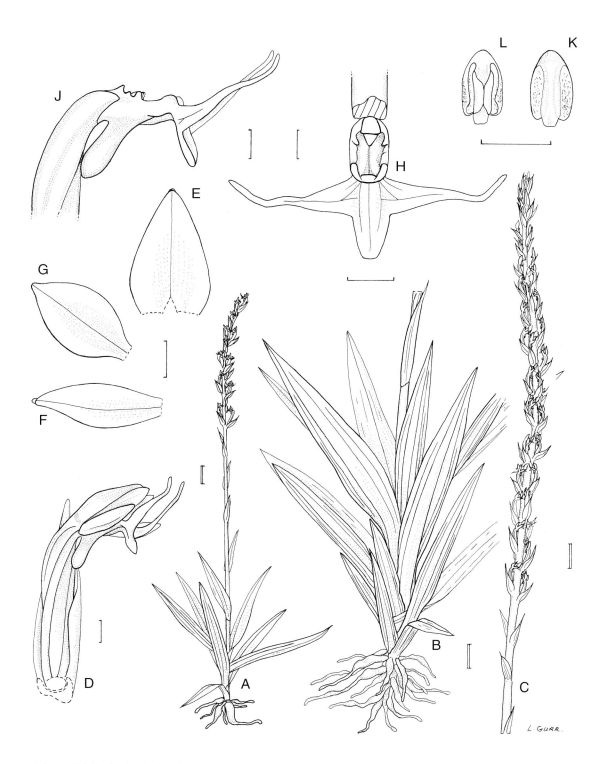

Figure 76. Peristylus hallieri J.J. Sm. - A & B: habits. - C: inflorescence. - D: flower with floral bract, side view. - E: dorsal sepal. - F: lateral sepal. - G: petal. - H: lip and column, front view. J: ovary, lip and column, side view. - K: anther-cap, back view. - L: anther-cap, interior view. A drawn from *Vermeulen & Lamb 310* and *Wood 836,* B drawn from *Vermeulen & Lamb 310* and *Wood 655,* and C–L from *Wood 734* by Linda Gurr. Scale: single bar = 1 mm; double bar = 1 cm.

× 1 mm, obliquely ovate-triangular or oblong, sometimes slightly falcate, obtuse. **Lip** 3-lobed, spurred, adnate to base of column and stigma forming a short tube, 1.5–1.8 mm wide at base; disc with a transverse semicircular basal callus; side lobes 0.5–0.8(–1.1) cm long, filiform, obtuse, spreading; mid-lobe 1–1.2 mm long, triangular or linear, obtuse; spur 2–3 mm long, subclavate to cylindrical, decurved. **Column** 1.3–1.5 mm long; stigma adnate to lip; anther-cap ovate, obtuse; pollinia 2. Plate 16A & B.

HABITAT AND ECOLOGY: Lowland mixed dipterocarp forest; open secondary roadside vegetation; rough mossy ground; landslips; rocky and sandy areas beside rivers; sometimes among mossy rocks in midstream and able to withstand inundation; in shady or exposed situations. Alt. (50–)400 to 1600 m. Flowering observed in February, May, June, and September–December.

DISTRIBUTION IN BORNEO: BRUNEI: Temburong District. KALIMANTAN BARAT: Sukalanting area; Kelasar River. SABAH: Mt. Kinabalu; Crocker Range, Mt. Alab, Kimanis road, Sinsuron road; Sipitang District. SARAWAK: Mt. Matang; Belaga District, Ulu Koyan.

GENERAL DISTRIBUTION: Endemic to Borneo.

NOTES: *Peristylus* is closely related to the very large and cosmopolitan genus *Habenaria* but is less diverse, with around 135 species distributed in Asia, eastwards to New Guinea, Australia and the Pacific islands. In *Habenaria* the two stigmas per flower are each attached to an elongated structure called a stigmatophore. These extend from the column and are thereby free from the lower part of the lip (hypochile). Stigmatophores are absent in *Peristylus*, the two stigmas being convex, cushion-like and either fused with or lying against the hypochile. The genus is represented by a dozen species in Borneo.

DERIVATION OF NAME: The specific epithet honours J.G. Hallier (1868-1932) who collected the type material.

77. PHALAENOPSIS COCHLEARIS Holttum

Phalaenopsis cochlearis *Holttum* in Orchid Rev. 72: 408 (1964). Type: Borneo, Sarawak, locality unknown, *Robert Kho* s.n., cult. Royal Botanic Gardens, Kew (holotype not located).

Polychilos cochlearis (Holttum) Shim in Malayan Nat. J. 36: 22 (1982).

Epiphyte. **Roots** laterally compressed, flexuous, fleshy, glabrous. **Stem** abbreviated, completely covered by persistent leaf sheaths. **Leaves** 2–4, 13–22 × 5–8 cm, oblong-obovate to oblong-elliptic, acute to somewhat obtuse, rather thin-textured, nerves raised on upper surface, recurved. **Inflorescences** suberect, branching; peduncle up to 50 cm long, branches including rachis 4–9 cm long; rachis fleshy, somewhat fractiflex; floral bracts 3–5 mm long, ovate-cucullate, acute. **Flowers** varying in colour from white to pale greenish yellow to yellow with 2 light brownish or orange-brown bars at base of sepals and petals, side lobes of lip white or cream with

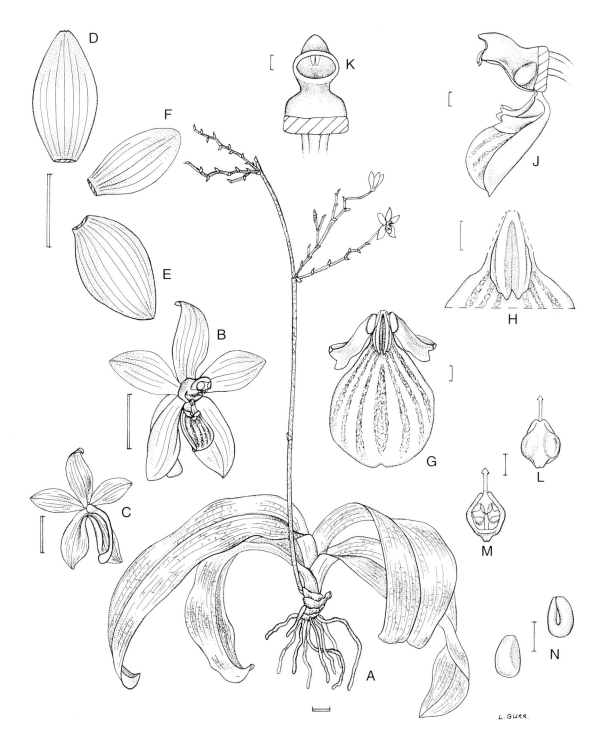

Figure 77. Phalaenopsis cochlearis Holttum - A: habit. - B: flower, oblique view. - C: flower, back view. - D: dorsal sepal. - E: lateral sepal. - F: petal. - G: lip, front view. - H: callus at base of lip. - J: column and lip, side view. - K: column, front view. - L: anther-cap, back view. - M: anther-cap, interior view. - N: pollinia. A drawn from miscellaneous collections, B–N from *Sheridan* s.n. by Linda Gurr. Scale: single bar = 1 mm; double bar = 1 cm.

3 broken cinnabar to orange-brown streaks on forward facing surfaces, mid-lobe primrose-yellow with several reddish brown to orange-brown stripes on each side of median keel. **Pedicel** and **ovary** 1.5–2 cm long. **Sepals** and **petals** minutely papillose at base, spreading. **Dorsal sepal** 1.5–2 × 0.6–0.9 cm, narrowly elliptic, obtuse and dorsally submucronate. **Lateral sepals** 1.6–2.2 × 0.6–0.9 cm, obliquely ovate, acute, dorsally carinate, margins revolute. **Petals** 1.2–1.8 × 0.4–0.7 cm, narrowly elliptic, acute, dorsally somewhat carinate. **Lip** 0.9–1.2 cm long, 0.8–1 cm wide across side lobes, fleshy; side lobes 3–3.5 mm long, oblong-linear, with a central horizontal groove, apex notched; mid-lobe 7–8 mm wide, suborbicular, strongly concave, cochleate, rounded or emarginate; disc between side lobes with a pair of parallel fleshy lamellate calli, in front of which is a bidentate fleshy callus. **Column** 0.6–0.7 cm long, cylindric, fleshy, constricted in middle, minutely papillose on ventral surface at base; anther-cap 2 × 2 mm, cucullate; pollinia 2, cleft. Plate 16C.

HABITAT AND ECOLOGY: Forest on limestone. Alt. 400 to 600 m. Flowering observed in October in cultivation.

DISTRIBUTION IN BORNEO: SARAWAK: Locality unknown.

GENERAL DISTRIBUTION: Endemic to Borneo.

NOTES: The earliest known specimen of *P. cochlearis* was collected during the 1800s and is deposited in the Reichenbach Herbarium (no. 21058) in Vienna. Almost a century passed before material was collected again and formally described as new to science. The species is now widely available in the horticultural trade. *Phalaenopsis cochlearis* belongs to section *Fuscatae* which is also represented in Borneo by *P. fuscata* Rchb.f. (see *Orchids of Borneo*, volume 1: fig. 72). Sweet (1980) cites the holotype as being at Kew but a search could not locate it.

DERIVATION OF NAME: The generic name is derived from the Greek *phalaina*, a moth, and *opsis*, appearance, referring to the delicate moth-like flowers, particularly of the larger white-flowered species. The specific epithet *cochlearis* is a Latin word for spoon-like and refers to the concave mid-lobe of the lip.

78. PHOLIDOTA CLEMENSII Ames

Pholidota clemensii *Ames* in Orchidaceae 6: 66 (1920). Type: Borneo, Sabah, Mt. Kinabalu, Marai Parai Spur, December 1915, *J. Clemens* 390 (holotype AMES).

? *Pholidota dentiloba* J.J. Sm. in Bull. Jard. Bot. Buitenzorg, ser. 3, 11: 107 (1931). Type: Borneo, Kalimantan Tengah, Bukit Raja, *Winkler* 994 (holotype HBG, probably lost).

Erect **epiphyte** or **terrestrial**. **Rhizome** stout. **Roots** 2–3 mm in diameter. **Cataphylls** (5–)9(–13), 0.5–12.5 cm long, acute. **Pseudobulbs** 3–7 × 1 cm, slender, somewhat fusiform, unifoliate. **Leaves** (12.5–)24–40(–44) × (2.2–)3.7–6.5(–7.5) cm, narrowly elliptic to linear-elliptic,

Figure 78. Pholidota clemensii Ames - A: habit. - B: inflorescence. - C: leaf blade, abaxial view. - D: leaf blade, adaxial view. - E: flower, side view. - F: dorsal sepal. - G: lateral sepal. - H: petal. - J: lip, front view. - K: lip and column, side view. - L: column, front view. - M: anther-cap, interior view. - N: anther-cap, back view. A drawn from *Carr* C. 3563, B from *Carr* C. 3563 and *Wood* 873, C from *Beaman* 9046, D from *Chew & Corner* RSNB 4483, and E–N from *Wood* 873 by Linda Gurr. Scale: single bar = 1 mm; double bar = 1 cm

acute, plicate, stiffly herbaceous to rather coriaceous, main nerves 5–7, very prominent on lower surface. **Inflorescences** proteranthous, rarely synanthous, (12-)18- to 25-flowered; peduncle up to 6 cm long, 2.5–4 mm in diameter, erect, elliptic in cross section, possibly elongating after anthesis; non-floriferous bracts absent or solitary; rachis 5–10.5 cm long, internodes 4–7 mm long, with scattered sparse hairs, erect; floral bracts 15–22 × 6.3–9 mm, ligulate, obtuse to acute, stiffly herbaceous, caducous, many-nerved, salmon-pink. **Flowers** rather fleshy, white, occasionally pinkish, pedicel and ovary brownish. **Pedicel** and **ovary** 0.8–1 cm long, narrowly clavate, pedicel twisted. **Dorsal sepal** 9–11.5 × 5–7 mm, ovate, obtuse to acute, margins slightly rolled inwards, (5–)–7-nerved. **Lateral sepals** 9–12 × 5–8 mm, irregularly ovate, deeply concave, saccate at base, acute, apiculate, clasped in a longitudinal fold of the hypochile, 5- to 6-nerved. **Petals** 7.5–10 × 3.5–5 mm, elliptic to oblong, acute, recurved distally, 5- to 7-nerved. **Lip** not extending from lateral sepals; hypochile 3–4 × 2(–3) mm, 3(–4) mm high, slightly laterally compressed, lateral lobes lacking; epichile 6.3–9 × 3–3.7 mm, ligulate, usually widest at middle, truncate, retuse, rounded or acute, sometimes apiculate, curved, margins recurved, distal margins folded, central area with a longitudinal depression separating into two longitudinal, low, rounded, elevated ridges, 5-nerved. **Column** 1.3–3.5 mm long, apical hood 3-lobed, central lobe retuse, laterals triangular; stigmatic cavity *c.* 2 × 1.3 mm; rostellum lobes narrowly triangular, curved; anther-cap 1–1.2 × 1.2–1.5 mm, irregularly orbicular, retuse; pollinia 4. Plate 16D & E.

HABITAT AND ECOLOGY: Lower montane forest with *Agathis borneensis, Lithocarpus,* etc.; open kerangas forest; oak-laurel forest; riparian forest; mossy forest on limestone. Alt. (800–)1300 to 2350 m. Flowering observed from November until July.

DISTRIBUTION IN BORNEO: KALIMANTAN TENGAH: Bukit Raja. SABAH: Mt. Kinabalu; Mt. Trus Madi. SARAWAK: Kapit District, Hose Mountains; Limbang District, Mt. Batu Lawi; Marudi District, Gunung Mulu National Park.

GENERAL DISTRIBUTION: Endemic to Borneo.

NOTES: De Vogel (1988) tentatively placed *P. dentiloba* J.J. Sm., which has slightly larger flowers with a lip epichile having small triangular side lobes, into synonymy under *P. clemensii.* Although these differences appear insufficient to maintain *P. dentiloba* as distinct, he conceded that more material from the type locality is required to ascertain if the lobed epichile is a constant character.

DERIVATION OF NAME: The specific epithet honours the collector of the type, the Reverend Joseph Clemens (1862–1936), who, with his wife Mary Strong Clemens, spent about six weeks on Mt. Kinabalu from 28 October until 12 December, 1915. They spent a further two years making extensive collections on the mountain between 1931 and 1933.

79. PHOLIDOTA SCHWEINFURTHIANA L.O. Williams

Pholidota schweinfurthiana *L.O. Williams* in Bot. Mus. Leafl. 6: 60 (1938). Type: Borneo, Sarawak, Marudi District, Mt. Temabok, Upper Baram Valley, 5 November 1920, *Moulton* SFN 6678 (holotype AMES, isotypes K, SING).

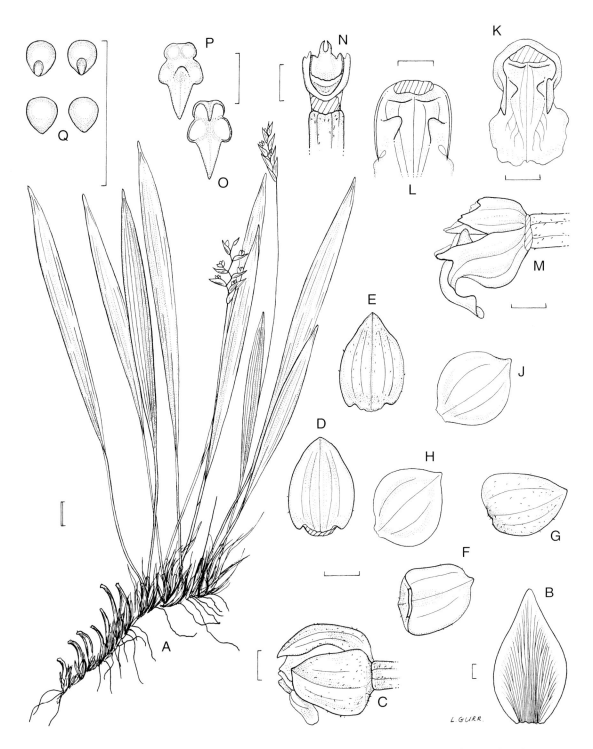

Figure 79. Pholidota schweinfurthiana L.O. Williams. - A: habit. - B: floral bract. - C: flower, side view. - D: dorsal sepal, adaxial surface. - E: dorsal sepal, abaxial surface. - F: lateral sepal, adaxial surface. - G: lateral sepal, abaxial surface. - H: petal, adaxial surface. - J: petal, adaxial surface. - K: lip, front view. - L: lip hypochile, front view. - M: portion of pedicel with ovary, lip and column, side view. - N: column, front view. - O: anther-cap, interior view. - P: anther-cap, back view. - Q: pollinia. All drawn from *Wood 647* by Linda Gurr. Scale: single bar = 1 mm; double bar = 1 cm.

Erect **epiphyte**. **Rhizome** up to 10 cm long, slender, covered in fibrous sheaths. **Roots** to 10 cm or more long, 0.2–0.7(–1) mm in diameter. **Cataphylls** 5–6, 0.5–4.5 cm long, ovate, acute, persistent, often becoming fibrous. **Pseudobulbs** 1.2–2 × 0.3–0.4 cm, cylindrical, slender, ascending, mostly hidden by persistent cataphylls, unifoliate. **Leaf-blade** 6.5–15 × 0.7–0.9 cm, linear to narrowly oblanceolate, usually widest above the middle, acute to acuminate, 24- to 36-nerved, main nerves 5; petiole 3–10 cm long, 0.5–1 mm wide, sulcate. **Inflorescences** synanthous, 5- to 8-flowered; peduncle 7–14 cm long, 0.2–0.4 mm in diameter, wiry, terete to slightly angular; non-floriferous bracts 1–2, ovate, acute to acuminate, imbricate; rachis 2–2.7 cm long, fractiflex, internodes 3–4 mm long, pendulous; floral bracts 6–9 × 4.5–6 mm, ovate, acute to acuminate, spreading, persistent, minutely hairy, palest green flushed red, dirty pink or salmon-pink. **Flowers** non-resupinate, pedicel and ovary dirty pale pink, sepals whitish to dirty pale pink, petals whitish to dirty cream, lip whitish, junction of hypochile and epichile bright yellow, front of column and anther-cap orange. **Pedicel** and **ovary** 2–5.2 mm long, narrowly clavate, angular. **Dorsal sepal** 2.5–3 × 1.5–2.5 mm, broadly ovate, acute, forming a hood with lateral sepals, 3-nerved. **Lateral sepals** 2.5–4 × 1.8–2.5 mm, broadly ovate, acute, shallowly concave, 3- to 4-nerved. **Petals** 2.5–3 × 2–2.3 mm, irregularly ovate, acute to shortly acuminate, 3-nerved. **Lip** 1.5–2.4 mm long (natural position); hypochile 1.5–2.2 × 1–2.2 mm, thickened at the back, side lobes 0.2–0.5 mm long, triangular to narrowly triangular, acute, margins entire to irregular, disc with 5–7 central nerves, 5 of which are more prominent and slightly swollen at the back, lateral nerves near junction with epichile each elevated into a short high triangular to semi-orbicular rounded keel measuring 0.5–0.6 × 0.5–0.6 mm; epichile 1–1.7 × 1.2–2.5 mm, strongly recurved, irregularly transversely elliptic to orbicular, or almost quadrangular, truncate to slightly retuse, with a mucro in the apical sinus, margins entire to irregular. **Column** 1.5–2 mm long, lateral wings triangular, irregularly toothed; stigmatic cavity irregularly elliptic to ovate; rostellum with a longitudinal slit after removal of pollinarium; anther-cap *c.* 1.5 × 0.4–0.7 mm, narrowly triangular.

HABITAT AND ECOLOGY: Ridge-top lower montane forest with *Agathis borneensis*, small rattans, etc.; low open kerangas forest. Alt. 1300 to 1700 m. Flowering observed from October until December.

DISTRIBUTION IN BORNEO: BRUNEI: Locality unknown. SABAH: Sipitang District, Long Pa Sia area. SARAWAK: Marudi District, Mt. Temabok.

GENERAL DISTRIBUTION: Endemic to Borneo.

NOTES: *Pholidota schweinfurthiana* belongs to section *Acanthoglossum* which is represented by seven species in Borneo. Of these, it is most closely related to *P. pectinata* Ames which can be distinguished from *P. schweinfurthiana* by its multi-flowered inflorescence, entire lip hypochile and lower, broader, swollen keels.

DERIVATION OF NAME: This species is named in honour of Charles Schweinfurth (1890–1970) who collaborated with Oakes Ames to produce *The Orchids of Mount Kinabalu, British North Borneo* (1920).

80. PHOLIDOTA SIGMATOCHILUS (Rolfe) J.J. Sm.

Pholidota sigmatochilus (*Rolfe*) *J.J. Sm.* in Blumea 5: 299 (1943). Type: Borneo, Sabah, Mt. Kinabalu, below Pakapaka, *Gibbs* 4260 (holotype BM).

Sigmatochilus kinabaluensis Rolfe in Gibbs, J. Linn. Soc., Bot. 42: 155, pl. 3 (1914).
Chelonistele kinabaluensis (Rolfe) de Vogel in Blumea 30: 203 (1984).

Epiphyte, rarely **terrestrial**. **Roots** 0.7–1.5 mm in diameter. **Cataphylls** 5–7, 0.5–7 cm, ovate-elliptic, acute. **Pseudobulbs** crowded, unifoliate, 1.7–3 × 0.5–0.8 cm, narrowly cylindrical to narrowly ovate, yellowish-green. **Leaf-blade** 4.5–8(–10) × 0.9–2.2(–2.6) cm, narrowly elliptic to linear, rarely ovate-elliptic, acute, tough and thickly coriaceous, conduplicate below, olive-green, stained red; petiole 1–3.5(–8.5) × 0.1–0.2 cm, sulcate. **Inflorescences** synanthous with the more than halfway emerged young leaf, laxly 5- to 9-flowered; flowers borne 4–10 mm apart; peduncle 3–6.5 cm long, increasing to 11 cm long after anthesis, straight, stiff, elliptic in cross-section; rachis (1.5–)2–5.5 cm long, straight, erect; floral bracts 10–15 × 6–8 mm, ovate, acute to acuminate, sometimes with a few scattered hairs at base inside, deciduous, but often still attached when flowers are open. **Flowers** all turned to one side (secund), not opening widely, glabrous, white to cream, rarely with a green throat, column often apple-green. **Pedicel** and **ovary** 10–16 mm long, narrowly clavate. **Dorsal sepal** 8–13(–14) × 4.5–6 mm, ovate to ovate-oblong, obtuse or acute, deeply concave with ± depressed sides, 5(–7)-nerved, mid-nerve prominent or bearing a rounded keel. **Lateral sepals** 9.5–13(–14) × 4–6.3(–7) mm, ovate-oblong, acute to acuminate, 5-nerved, mid-nerve bearing a low rounded keel. **Petals** 8–11(–13) × 1.7–3(–4) mm, ligulate, somewhat asymmetric, obtuse or acute, 3(–5)-nerved. **Lip** strongly sigmoid, tip extending from between lateral sepals; hypochile 3.5–4.5(–5) × 0.8–2.5 mm, 1.5–2.2 mm high, boat-shaped, laterally sometimes somewhat compressed, side lobes absent, 5- to 7-nerved; disc with 2 low keels 1.3–2 mm long on junction with epichile, clasped by margins of hood, wing-like; epichile 5–6(–7) × 2–3(–4) mm, ligulate, acute or obtuse, often narrowed near base, sometimes somewhat swollen distally, distal margins ± folded upwards, 5- to 7-nerved. **Column** 3.5–6 mm long, widest at top when flattened; hood relatively large, truncate to retuse, central portion with a ± irregular to rather distinctly 3-lobed margin, ± clearly separated from side lobes; side lobes rounded, margins turned to front and clasping keels on lip, leaving a narrow opening to anther and stigma; anther-cap 1–1.3 × 0.8–1.3 mm, irregularly cordate, almost entirely hidden behind rostellum. Plate 17A & B.

HABITAT AND ECOLOGY: Upper montane forest, often on ultramafic substrate, growing in well-drained, open sites, particularly among thick moss cushions on the branches of shrubs. Alt. 1800 to 3500 m. Flowering observed throughout the year.

DISTRIBUTION IN BORNEO: SABAH: Mt. Kinabalu and Mt. Tembuyuken.

GENERAL DISTRIBUTION: Endemic to Borneo.

NOTES: Rolfe (1914) created the genus *Sigmatochilus* based upon material collected on Mt. Kinabalu by Lilian Gibbs. The primary character he used when he described *Sigmatochilus* was the "strongly sigmatoid labellum." J.J. Smith (1943) subsequently transferred *S. kinabaluensis* to

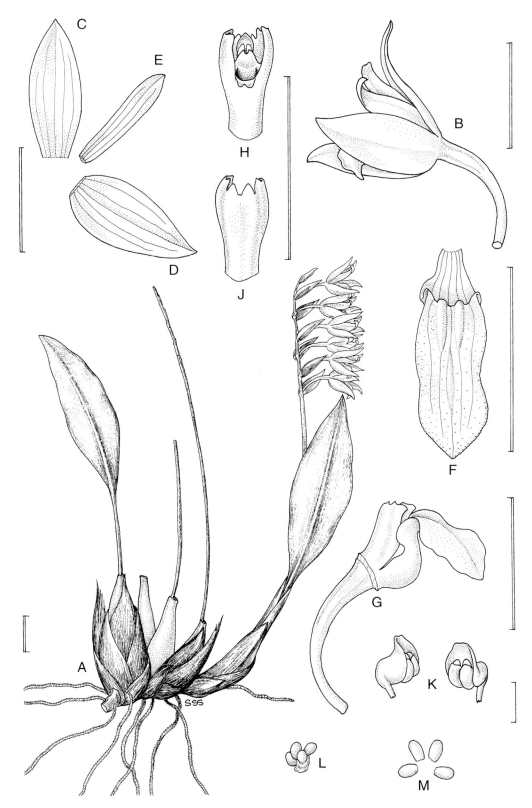

Figure 80. Pholidota sigmatochilus (Rolfe) J.J. Sm. - A: habit. - B: flower, side view. - C: dorsal sepal - D: lateral sepal. - E: petal. - F: lip, front view. - G: pedicel with ovary, lip and column, side view. - H: column, front view. - J: column, back view. - K: anther-cap, side and oblique interior views. - L: pollinarium. - M: pollinia. All drawn from *Fuchs* 21072 by Susanna Stuart-Smith. Scale: single bar = 1 mm; double bar = 1 cm.

Pholidota applying the epithet *P. sigmatochilus*, as the name *P. kinabaluensis* had already been applied to another taxon by Oakes Ames. *Pholidota kinabaluensis* Ames has subsequently been transferred to *Entomophobia* by de Vogel (see Fig. 58). Smith (1943) noted a similarity of the lip of *P. sigmatochilus* to *P. camelostalix* Rchb.f. and *P. clemensii* (see Fig. 78) and decided the taxon should belong to *Pholidota*. Most recently, de Vogel (1986) placed *P. sigmatochilus* into *Chelonistele*. While the leaves are coriaceous and conduplicate, recalling those of *Chelonistele, P. sigmatochilus* lacks the characteristic side lobes on the lip and narrow, recurved petals of that genus.

Barkman and Wood (1997) reinstated *Chelonistele sigmatochilus* within *Pholidota*, including it in section *Acanthoglossum* on account of its rostellum which has a longitudinal slit. Within section *Acanthoglossum*, it is clearly allied to *P. clemensii*, sharing a similarly shaped lip, perianth, floral bracts, and stout, erect inflorescence.

There remains the possibility, given that it shares some morphological features in common with *Chelonistele*, that it may be the product of a former hybridisation event between species of *Pholidota* and *Chelonistele*. Given this hypothesis, one of the assumed parents would likely have been a species similar to *P. clemensii* judging by the high similarity of floral morphology.

DERIVATION OF NAME: The specific epithet is derived from the Latin *sigmoideus*, ie. curved like the letter S, and refers to the shape of the lip.

81. PHOLIDOTA VENTRICOSA (Blume) Rchb.f.

Pholidota ventricosa (*Blume*) *Rchb.f.* in Bonplandia 5: 43 (1857). Type: Java, Tjapus River, Salak, *Blume* s.n. (holotype L).

Chelonanthera ventricosa Blume, Bijdr.: 383 (1825).
Coelogyne ventricosa (Blume) Rchb.f. in Ann. Bot. Syst. (Walp. Ann.) 6: 237 (1861).
Pholidota sesquitorta Kraenzl. in Xenia Orch. 3: 114, t. 266 I, f. 1–7 (1893). Types: ? *Micholitz*, cult. Eisgrub and ? *Micholitz*, cult. Berlin (syntypes probably lost, lectotype plate in Xenia Orch. 3, t. 266, f. 1–7, 1893 chosen by de Vogel).
P. grandis Ridl. in J. Straits Branch Roy. Asiat. Soc. 49: 32 (1908). Type: Peninsular Malaysia, Semangkok Pass, 1905, *Ridley* s.n. (holotype SING).
P. sororia Schltr. in Feddes Repert. Beih. 1: 108 (1911). Type: Papua New Guinea, Torricelli Mountains, *Schlechter* 20277 (holotype B lost, lectotype plate in Fedde Repert. Beih. 21, t. 42, no. 144, 1923–1928 chosen by de Vogel).
P. sororia Schltr. var. *djamuensis* Schltr. in Feddes Repert. Beih. 1: 109 (1911). Type: Papua New Guinea, Kani Mountains, Djamu, *Schlechter* 16612 (holotype B lost, lectotype BO chosen by de Vogel).

Epiphyte, sometimes **terrestrial. Rhizome** creeping. **Roots** up to more than 16 cm long, 1.5–3 mm in diameter. **Cataphylls** 8–14, 1–21 cm long, ovate, acute, imbricate, herbaceous, many-nerved, disintegrating into persistent fibres. **Pseudobulbs** 4–14 × 1–2 cm, cylindrical, with several sharp ridges, borne close together turned to one side of the rhizome, bifoliate. **Leaf-blade**

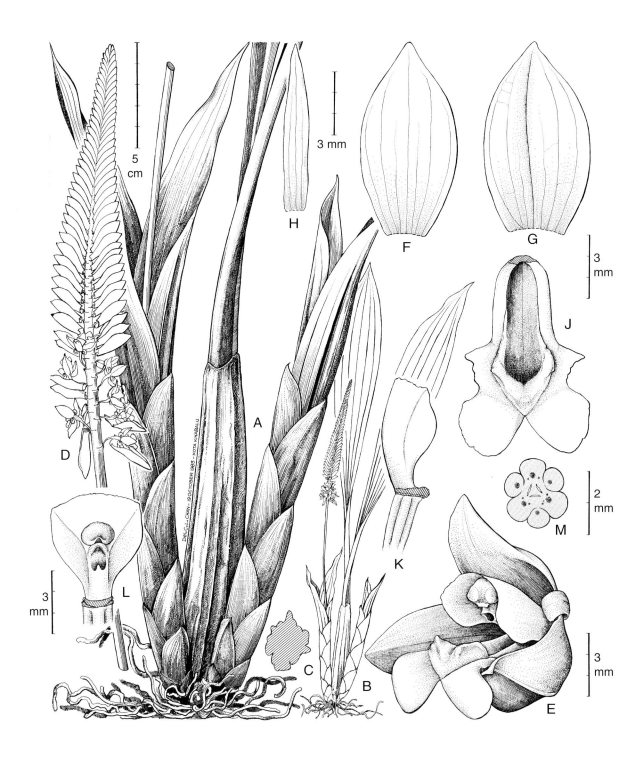

Figure 81. Pholidota ventricosa (Blume) Rchb.f. - A & B: habits. - C: pseudobulb, transverse section. - D: inflorescence. - E: flower, oblique view. - F: dorsal sepal. - G: lateral sepal. - H: petal. - J: lip, front view. - K: pedicel with ovary and column, side view. - L: column, front view. - M: ovary, transverse section. All drawn from *Phillipps & Lamb* s.n. by C.L. Chan.

21–62 × (2–)3.5–7.5(–9.5) cm, narrowly elliptic to linear-elliptic, acute, stiffly rather coriaceous, main nerves 5, very prominent below; petiole (4–)6–10(–24) cm long, (2.5–)4–8 mm in diameter, conduplicate. **Inflorescences** synanthous, appearing with the slightly to almost half emerged leaves, 30–80(–93)-flowered, flowers opening at base first; peduncle (4–)19–25(–30) cm long, erect, straight, irregularly longitudinally grooved; rachis 10–27 cm long, internodes 2–6 mm long, lowermost to 2.5 cm long, erect, straight, stiff, angular, longitudinally concave just above insertion of flower; floral bracts (1.4–)1.8–3 × (0.8–)1–1.2 cm, oblong-ovate, acute, sometimes obtuse, caducous at anthesis, many-nerved. **Flowers** opening quite widely, usually resupinate, mostly turned to one side, white, often flushed yellowish green or pale green, lip with or without a purple spot, often fragrant. **Pedicel** and **ovary** 7–14 mm long, slender, twisted, ovary 3-ribbed. **Dorsal sepal** 6–11 × (1.2–)3–5.5 mm, ovate to ovate-oblong, obtuse to acute, a few scattered minute hairs at base, 5–7-nerved. **Lateral sepals** 6–11 × 3–5.5 mm, elliptic to oblong, acute, a few scattered minute hairs at base, 5(–7)-nerved. **Petals** 6–10 × (0.7–)1–4 mm, oblong to ligulate or linear, truncate, obtuse or acute, rolled backwards, slightly hairy at base, 3-nerved. **Lip** sigmoid; hypochile 4–7 mm long, 2–4 mm wide at back, (3.4–)5–9 mm wide across lobes, 1.2–2 mm high, boat-shaped, narrowed at base, margins recurved, on either side at the front with a horizontal to reflexed, triangular to rounded lobe 1–3 × 1–2.5 mm, front margin of lobes irregularly undulate to serrate, rim along junction of lateral lobes and saccate portion usually fleshy; epichile 2–2.5 × 4–9 mm, recurved, bilobulate, lobes 2–4 × 2–5 mm, semi-orbicular, margins irregular. **Column** 3.5–5 × 3.5–5.5 mm, semi-orbicular, broadly spathulate to cuneate, apical hood large, wide, thin, broadly rounded, irregular; stigmatic cavity elliptic; rostellum rectangular to rounded; anther-cap 1–1.7 × 1.2–1.7 mm, transversely elliptic to orbicular, apex recurved, truncate. Plate 17C & D, 18A & B.

HABITAT AND ECOLOGY: Lower montane forest; open kerangas forest; sometimes on limestone. Alt. 500 to 1700 m. Flowering observed throughout the year, but particularly October and November.

DISTRIBUTION IN BORNEO: KALIMANTAN BARAT, KALIMANTAN TENGAH & KALIMANTAN TIMUR: miscellaneous localities. SABAH: Mt. Kinabalu; Crocker Range; Sipitang District, Long Pa Sia/Long Miau area. SARAWAK: Mt. Dulit, Ulu Tinjar; Gunung Mulu National Park; Limbang District, Mt. Pagon Periuk.

GENERAL DISTRIBUTION: Peninsular Malaysia, Sumatra, Java, Borneo, Sulawesi, the Philippines and New Guinea.

NOTE: Female bees of the genus *Trigonia* and large female wasps (*Vespa multimaculata*) have been observed visiting the flowers and effecting pollination.

DERIVATION OF NAME: The specific epithet is derived from the Latin *ventricosus*, swollen, especially on one side, and refers to the boat-shaped lip hypochile.

82. PODOCHILUS MARSUPIALIS Schuit.

Podochilus marsupialis *Schuit.* in Blumea 43 (2): 489, fig. 1 (1998). Type: Borneo, Sarawak, Sri Aman District, Mt. Silantek, *Schuiteman, Mulder & Vogel* LC 933227 (holotype L, isotypes K, SAR).

Epiphyte. **Roots** shortly hairy. **Stems** 4–18(–24) cm long, tufted, spreading in all directions, rooting at base, or sometimes when in contact with substratum, usually simple, sometimes branching where adventitious roots are produced, densely many-leaved. **Leaves** 6–11 × 1.5–2 mm, narrowly obliquely triangular-subfalcate, acute to acuminate, bilaterally flattened, not articulate, imbricate at base. **Inflorescences** apical, sometimes lateral, 1- to 3-flowered, 3–4 mm long, sometimes rooting at the base; peduncle 2 mm long; rachis 1 mm long; floral bract *c.* 1.3 mm long, broadly ovate, abruptly acuminate. **Flowers** opening only slightly, white, sepals and petals magenta at apex, lip apex deeper magenta. **Pedicel** and **ovary** *c.* 2.4 mm long, glabrous, 6-ribbed. **Dorsal sepal** 2.8 × 1.9 mm, triangular-ovate, obtuse. **Lateral sepals** 3.1 × 1.1–1.2 mm, connate in basal half, strongly oblique, free portion triangular-ovate, obtuse, median adnate to lateral side of column, mid-nerve carinate on reverse. **Mentum** 1.4 mm long, subglobose, obscurely bilobulate. **Petals** 2 × 0.5 mm, linear-oblong, slightly oblique, obtuse. **Lip** 2.8 × 1.4 mm, strongly concave, narrowly obovate, obscurely 3-lobed in upper third, apex cucullate, entire, basal appendage 0.4 mm long, protruding backwards, fleshy. **Column** 1.7 mm long, laterally adnate to lateral sepals for 1 mm; stelidia absent; rostellum 3-toothed, central tooth much shorter than laterals; foot *c.* 1 × 0.4 mm, incurved; anther-cap 1.2 × 0.5 mm, cucullate, apex elongate, bifid; pollinia 4, in two pairs, each pair enveloped in a funnel-shaped sheath 1 mm long; viscidia 2, narrowly lanceolate-oblong. Plate 18C.

HABITAT AND ECOLOGY: Hill dipterocarp forest, growing gregariously on large tree trunks or rotting logs. Alt. 130 to 800 m. Flowering observed in February and July.

DISTRIBUTION IN BORNEO: KALIMANTAN BARAT: Serawai, Merah River. SABAH: Taman Tawau. SARAWAK: Julau District, Lanjak Entimau Protected Forest. Kapit District: Balleh/ Mengiong Rivers, Teneong. Sri Aman District, Mt. Silantek.

GENERAL DISTRIBUTION: Endemic to Borneo.

NOTES: *Podochilus* is a graceful, modestly flowered genus of about 60 species distributed from India and Sri Lanka to China, south and east through Indonesia and New Guinea to the Pacific islands. Sixteen species have been recorded from Borneo. Various species are frequent trunk epiphytes, which, with their diminutive habit and distichous foliage, often resemble leafy liverworts or certain mosses rather than orchids. The recently described *P. marsupialis* belongs to section *Sarganella* which contains four other species, all originating from Sri Lanka and, or, mainland Asia. The vegetative habit of *P. marsupialis* recalls the neotropical genus *Lockhartia*, while the tiny flowers are unique in having the lateral sepals laterally adnate to the column.

DERIVATION OF NAME: The generic name is derived from the Greek *podos*, foot, and *cheilos*, lip, alluding to the introrse basal appendage on the lip. The specific epithet alludes to the pouch-like structure formed by part of the column.

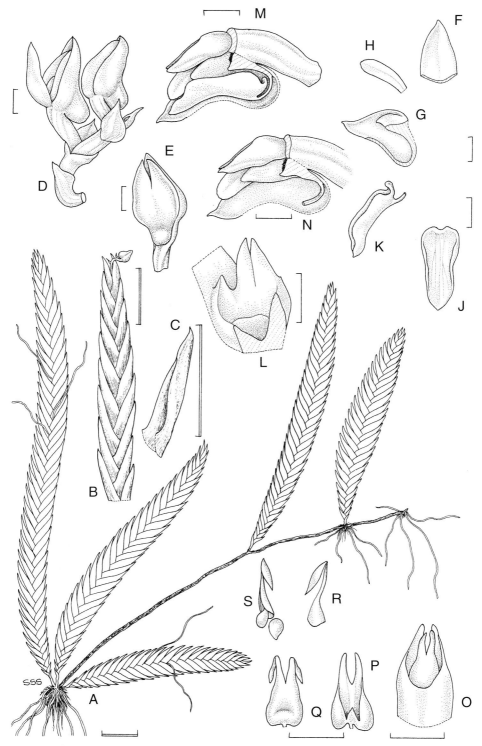

Figure 82. Podochilus marsupialis Schuit. - A: habit. - B: flowering shoot. - C: leaf. - D: inflorescence. - E: flower, viewed from below. - F: dorsal sepal. - G: lateral sepal. - H: petal. - J: lip, front view. - K: lip, side view. - L: view showing adnation of median portion of lateral sepal with column. - M: pedicel with ovary, lateral sepal, petal, lip and column, side view. - N: pedicel with ovary, lateral sepal, petal and column, lip removed, side view. O: column, front view. - P: anther-cap, interior view. - Q: anther-cap, back view. - R: sheath-like stipes and viscidium. - S: stipes and viscidium with two of the four pollinia. A–C drawn from *Church et al. 2095,* D & E from *Leiden cult.* (*Schuiteman et al.*) 59936 and F–S from *Lim* 1.26 by Susanna Stuart-Smith. Scale: single bar = 1 mm; double bar = 1 cm.

83. POMATOCALPA KUNSTLERI (Hook.f.) J.J. Sm.

Pomatocalpa kunstleri (*Hook.f.*) *J.J. Sm.* in Natuurk. Tijdschr. Ned.-Indië 72: 104 (1912). Type: Peninsular Malaysia, Perak, drawing by *Kunstler* (holotype CAL, copy in K).

Cleisostoma kunstleri Hook.f., Icon. Plant. 24: t. 2335 (1894).
Saccolabium pubescens Ridl. in J. Linn. Soc., Bot. 31: 295 (1896). Type: Borneo, Sarawak, Kuching, *Haviland* 2333 (holotype SING, isotype K).
Pomatocalpa merrillii Schltr. in Feddes Repert. Beih. 1: 988 (1913). Type: Philippines, Polillo, *McGregor* 10444 (holotype B, destroyed, isotype AMES).

Epiphyte or **lithophyte**. **Stems** to *c.* 20 cm long, *c.* 6.5 mm wide near base, internodes 1–2 cm long, somewhat flattened, oval in cross-section, producing long roots from nodes, leafy. **Leaf-blade** 10–30 × 2.5–5.5 cm, ligulate, slightly to strongly obtusely unequally bilobed, mid-rib producing a mucro in the sinus, coriaceous; sheath 1–2 cm long. **Inflorescences** mostly 25–40 cm long, erect, borne from nodes near base of stem, the largest often branched many times and paniculate, very finely hairy, branches semi-pendulous; peduncle 10–16 cm long, terete; non-floriferous bracts 3, 3–6 mm long, obtuse; rachis *c.* 8 to around 20 cm long, branches 2–8 cm long; floral bracts *c.* 1 mm long, triangular, acute. **Flowers** *c.* 7 mm in diameter, widely opening, non-resupinate, appearing resupinate on pendulous branches, pale lilac-pink, often spotted darker pink, side lobes of lip white, spur greenish pink. **Pedicel** and **ovary** 2 mm long, narrow, shortly hairy. **Dorsal sepal** 3.75–4 × 1.3–1.5 mm, oblong, obtuse. **Lateral sepals** 3 × 1.3–1.5 mm, obliquely oblong, obtuse. **Petals** 3.5–4 × 1 mm, ligulate, obtuse. **Lip** 3-lobed, *c.* 4 mm long; side lobes *c.* 1.5 × 1 mm, triangular, acute, thick and fleshy; mid-lobe 1 × 1.5 mm, ovate, obtuse, decurved; spur 2.8 × 1.4 mm, narrowly saccate, obtuse, shortly hairy; back wall callus ligulate, bifurcate. **Column** 1.5–2 mm long, obtuse; anther-cap ovate, acute; pollinia 4, appearing as 2 unequal masses. Plate 18D & 19A.

HABITAT AND ECOLOGY: Kerangas forest; sometimes on limestone. Alt. sea level to 400 m. Flowering observed throughout the year.

DISTRIBUTION IN BORNEO: BRUNEI: Locality unknown. KALIMANTAN TIMUR: Lempake, Tanah Merah. SABAH: Dent Peninsula, Tambisan area; Nabawan area. SARAWAK: Bau District, Bau, Jagoi. Kuching District, Kuching; Padawan, Tiang Bekap; Semenggoh Forest Reserve. Lundu District, Lundu.

GENERAL DISTRIBUTION: Peninsular Malaysia, Thailand, Sumatra, Mentawai Islands (Siberut Island), Java, Borneo and the Philippines.

NOTES: *Pomatocalpa* is a genus of between 35 and 40 species distributed from India throughout tropical Asia to New Guinea, Australia and the Pacific islands. Six species are found in Borneo, of which two are thought to be endemic.

DERIVATION OF NAME: The generic name is derived from the Greek *poma*, drinking cup, and *kalpe*, a pitcher, and refers to the shape of the spurred lip. The specific epithet honours J. Kunstler, whose drawing of a specimen from Perak, is the type.

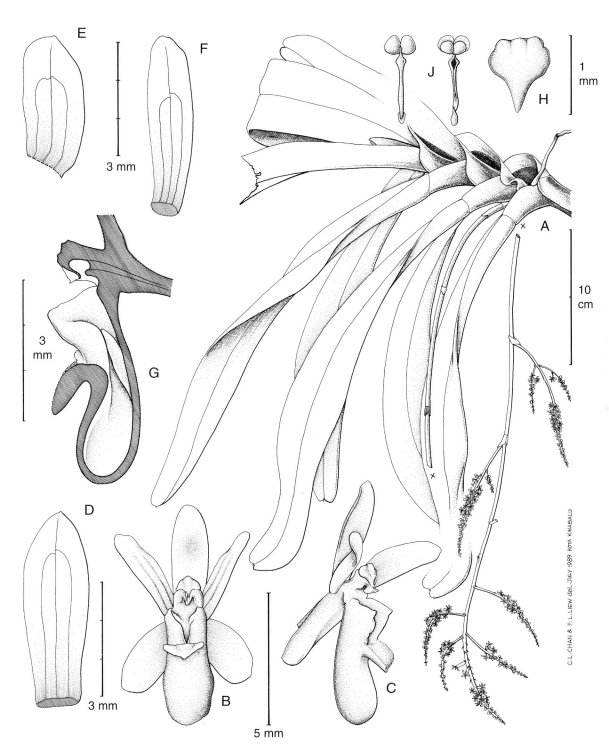

Figure 83. Pomatocalpa kunstleri (Hook.f.) J.J. Sm. - A: habit. - B: flower, front view. - C: flower, side view. - D: dorsal sepal. - E: lateral sepal. - F: petal. - G: lip and column, longitudinal section. - H: anther-cap, back view. - J: pollinarium, front and back views. All drawn from *Lamb* AL 1153/89 by C.L. Chan and Liew Fui Ling.

84. PORPHYRODESME SARCANTHOIDES (J.J. Sm.) U.W. Mahyar

Porphyrodesme sarcanthoides (*J.J. Sm.*) *U.W. Mahyar* in Reinwardtia 10(4): 418 (1988). Type: Sumatra, Bengkulu, Bukit Kaba, Suban Ayam, 1200 m, 1916, *Jacobson* s.n., cult. Bogor (holotype BO).

Renanthera sarcanthoides J.J. Sm. in Bull. Jard. Bot. Buitenzorg, ser. 2, 25: 94 (1917).

Epiphyte. **Stems** up to 85(–250) cm long, internodes 3–3.5 cm long, 0.4–0.5 cm in diameter, terete, pendulous, sparsely branched, rooting at base only, apex curving upward. **Leaf-blade** 11–15(–18) × 0.4–0.9 cm, linear, slightly twisted, narrowed towards a subacute to rather obtuse, apiculate apex, channelled at an obtuse angle, thickly fleshy, shiny dark green; sheath tubular, transversely rugulose. **Inflorescences** 8–10 cm long, simply paniculate, usually shorter than leaves, spreading, many flowered, flowers borne *c.* 2 mm apart; peduncle 0.6–1 cm long, transversely elliptic, with 2–3 triangular, carinate basal sheaths 3–5.5 mm long; rachis 5.5–8 cm long, angular, red, with 3–4 spreading branches 7–*c.* 8.5(–11) cm long; floral bracts *c.* 1.6 mm long, broadly triangular, acuminate, concave, adpressed to pedicel. **Flowers** *c.* 3.5–4.4 mm in diameter, widely opening, pedicel and ovary dark red, sepals and petals bright waxy scarlet-red with paler margins, lip brownish with red blotches, flushed yellow on mid-lobe, column pale ochre-brown, marked red, anther-cap red, apex yellow. **Pedicel** and **ovary** 3–5 mm long, narrow. **Sepals** and **petals** spreading, recurved. **Dorsal sepal** 1.5–2 × 0.9–1.2 mm, almost triangular to oblong, obtuse, concave. **Lateral sepals** 2 × 1–1.3 mm, obliquely ovate, obtuse or subobtuse, concave, mid-rib dorsally thickened. **Petals** 1.1–1.8 × 0.6–0.7 mm, obliquely oblong, narrowed near apex, obtuse and apiculate, apical margins erose. **Lip** 3-lobed, 3 mm long including spur, *c.* 2.25 mm across side lobes; side lobes 2 mm long, almost rectangular but narrowed towards base, thickened inside near base; mid-lobe 1–1.5 × *c.* 1 mm, quadrangular, obtuse; spur *c.* 2 mm long, 0.75 mm wide below, 1.4 mm wide near apex, placed at an acute angle to ovary, oblong-conical, laterally flattened below, apex almost globose. **Column** 1 mm long, incurved, truncate at apex, foot absent; rostellum large; anther-cap cucullate, semiglobose; pollinia 2, sulcate, on a slender stipes. Plate 19B & C.

HABITAT AND ECOLOGY: Lowland mixed dipterocarp forest; hill forest on limestone; reported as growing on the undersides of the lowest crown-layer branches. Alt. *c.* 100 to 400 m. Flowering observed in October and December.

DISTRIBUTION IN BORNEO: KALIMANTAN BARAT: Sintang, Posang River. SABAH: Batu Ponggol; Danum Valley; Sapulut, Labang Basin.

GENERAL DISTRIBUTION: Sumatra, Sulawesi and Borneo.

NOTES: *Porphyrodesme* is a genus of at least two species distributed in New Guinea (*P. papuana* Schltr.) and Sumatra, Sulawesi and Borneo (*P. sarcanthoides* (J.J. Sm.) U.W. Mahyar). A third species from Sarawak, *P. hewittii* (Ames) Garay, is a small plant with a very short stem bearing flat, oblong-elliptic leaves, its appearance being far removed from *Porphyrodesme*. I agree with the view of Ames (1921) that it is probably best placed within *Thrixspermum*. Garay (1972) also

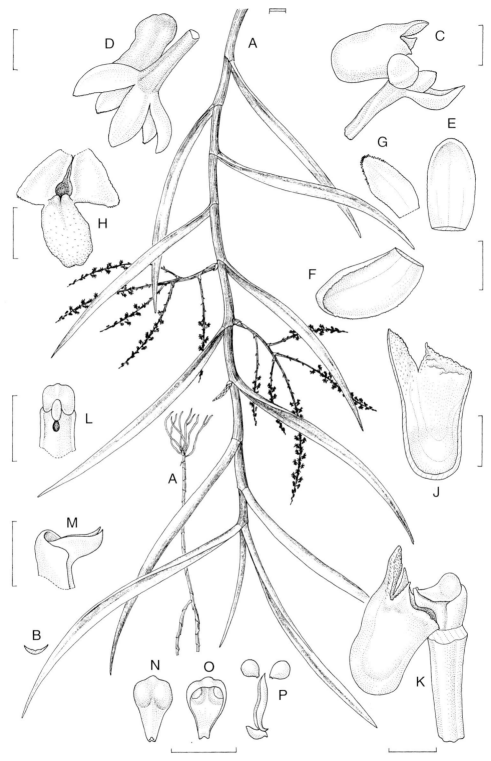

Figure 84. Porphyrodesme sarcanthoides (J.J.Sm.) U.W. Mahyar - A: habit. - B: transverse section through leaf. - C: flower, side view. - D: flower, oblique back view. - E: dorsal sepal. - F: lateral sepal. - G: petal. - H: lip, front view. - J: lip, longitudinal section. - K: pedicel with ovary, lip and column, side view. - L: column, front view. - M: column, anther-cap removed, side view. - N: anther-cap, back view. - O: anther-cap, interior view. - P: pollinarium. A–G and J–P drawn from *Lamb* AL 1200/89, and H from *Suali & Sumbing* SAN 101310 by Susanna Stuart-Smith. Scale: single bar = 1 mm; double bar = 1 cm.

says that *Renanthera moluccana* Blume from Maluku may also possibly belong within *Porphyrodesme*. However, he also transferred *R. elongata* (Blume) Lindl., but studies by Seidenfaden (1988) have shown this to have four rather than two pollinia.

O'Byrne (1998) comments that it is rather surprising that *P. sarcanthoides* should have escaped detection in the Danum Valley for so long, since dislodged specimens can readily be found lying on the ground on the main trails near the Field Centre.

DERIVATION OF NAME: The generic name is derived from the Greek *porphyra*, purple, and *desme*, a bundle, referring to the abbreviated, reddish inflorescence with its scarlet flowers. The specific epithet refers to the general appearance of this species to certain *Sarcanthus* (= *Cleisostoma*).

85. SARCOGLYPHIS POTAMOPHILA (Schltr.) Garay & W. Kittr.

Sarcoglyphis potamophila (*Schltr.*) *Garay & W. Kittr.* in Bot. Mus. Leafl. 30 (3): 58 (1985). Type: Borneo, Kalimantan Timur, Kutai, Long Sele, Long Wahan, flowered August 1901, *Schlechter* 13518 (holotype B, destroyed).

Sarcanthus potamophilus Schltr. in Feddes Repert. 3: 279 (1907).
Cleisostoma potamophilum (Schltr.) Garay in Bot. Mus. Leafl. 23 (4): 173 (1972).

Epiphyte. Roots filiform, glabrous. **Stem** 10–15 cm long, unbranched, erect or porrect. **Leaf-blade** 7–14 × 0.7–2 cm, ligulate to linear-ligulate, unequally bilobed, lobes varying from acute with a deep V-shaped notch (in Sabahan material examined), or bluntly rounded (in Sarawak material examined), coriaceous, tough; sheath 1 × 0.5 cm. **Inflorescences** arising from leaf axils low on stem, horizontal to descending, laxly 7- to 18-flowered; peduncle and rachis 11–18 cm long, dull olive-green, spotted maroon or brown; peduncle quadrangular in cross-section, slender, glabrous, bearing 2–3 small sheaths; rachis terete; floral bracts 1–2 mm long, ovate-deltoid, acuminate, glabrous to minutely papillose. **Flowers** *c.* 1.3 cm in diameter, long-lasting, sepals and petals dull pea-green with two faded reddish brown streaks, lateral sepals may also have a hint of dull reddish brown around base and margins, lip and spur delicate rose-lilac, turning deep pinkish purple on side lobes, column and anther-cap white. **Pedicel** and **ovary** 10 mm long, slender. **Dorsal sepal** 5 × 3.5 mm, ovate to elliptic, acute, concave, glabrous to minutely papillose. **Lateral sepals** 5 × 3.5 mm, oblong, obtuse, slightly oblique, glabrous. **Petals** 5 × 2 mm, obliquely linear-ligulate, obtuse to acute. **Lip** including spur 8 mm long, 3-lobed; side lobes ovate-elliptic, obtuse, erect, upper margins incurved; mid-lobe narrowly elliptic, obtuse, porrect; spur 5 mm long, clavate-cylindric, obtuse, slightly incurved, with a longitudinal septum running its length, and a large trapezoid callus which projects down from the upper wall, almost blocking its entrance. **Column** 2–3 mm long, fleshy; rostellum conspicuous, raised, fleshy, laterally compressed, projecting well beyond column apex; anther-cap 2 mm long, triangular, acuminate; stipes 2 mm long, slender, located in a longitudinal furrow along edge of rostellum; viscidium small, ovate; pollinia 4, appearing as two unequal masses. Plate 19D.

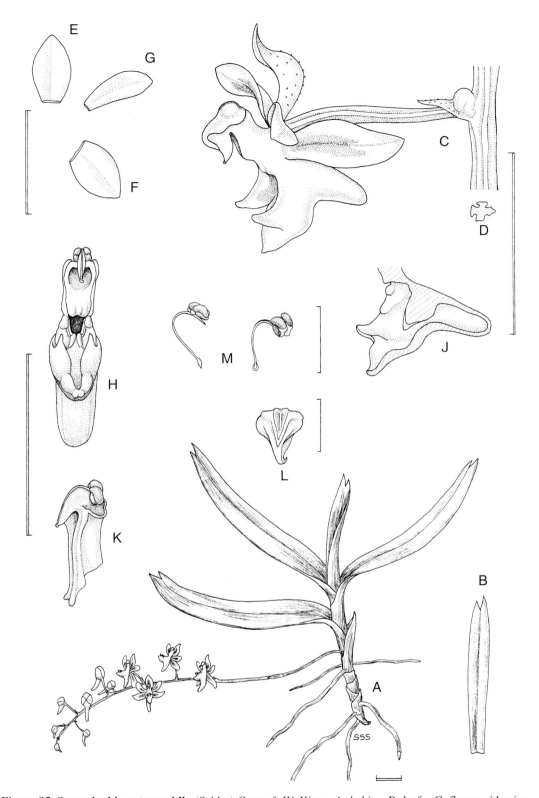

Figure 85. Sarcoglyphis potamophila (Schltr.) Garay & W. Kittr. - A: habit. - B: leaf. - C: flower, side view. - D: transverse section through rachis. - E: dorsal sepal. - F: lateral sepal. - G: petal. - H: lip and column, anther-cap removed, front view. - J: lip, longitudinal section. - K: column with pollinarium, anther-cap removed, side view. - L: anther-cap, front view. - M: two views of pollinarium. A–M drawn after *O' Byrne* in *Malayan Orchid Review* 32 : 52 & 69 (1998) by Susanna Stuart-Smith. Scale: single bar = 1 mm; double bar = 1 cm.

HABITAT AND ECOLOGY: Lowland and hill dipterocarp forest, on branches of emergent trees, often along river banks; kerangas forest. Alt. *c.* 150 m. Flowering observed in February, August, October and November.

DISTRIBUTION IN BORNEO: KALIMANTAN TIMUR: Kutai, Long Sele, Long Wahan. SABAH: Danum Valley; Nabawan area. SARAWAK: Kuching District, Kuching River area.

GENERAL DISTRIBUTION: Endemic to Borneo.

NOTES: *Sarcoglyphis* is a small genus of ten species established by Garay in 1972 and distributed from Myanmar (Burma) to southern China (Yunnan), south to Sumatra, Java and Borneo. Two species, viz. *S. fimbriatus* (Ridl.) Garay and *S. potamophila* (Schltr.) Garay & W. Kittr., are known from Borneo. Although resembling *Cleisostoma*, it is at once distinguished by the short, stout column with a high, raised fleshy, laterally compressed rostellum. Along the edge of the rostellum runs a longitudinal furrow in which the pollinarium is situated.

DERIVATION OF NAME: The generic name is derived from the Greek *sarc*, fleshy, and *glyphon*, carving, in reference to the rostellum which sits on top of the clinandrium like a carved ornament. The specific epithet is derived from the Greek for river-loving, in reference to the preferred habitat of this species.

86. SCHOENORCHIS BUDDLEIFLORA (Schltr. & J.J. Sm.) J.J. Sm.

Schoenorchis buddleiflora (*Schltr. & J.J. Sm.*) *J.J. Sm.* in Nat. Tijdschr. Ned. Ind. 72: 100 (1912). Type: Sumatra, Padang, *Schlechter* s.n. (holotype BO).

Saccolabium buddleiflorum Schltr. & J.J. Sm. in Bull. Dép. Agric. Indes Néerl. 15: 25 (1908).

Epiphyte. **Stems** up to *c.* 40 cm long, internodes 1.2–1.7 cm long, *c.* 1.6 mm in diameter, subterete, branching, longitudinally ribbed, transversely rugulose, pendent, green. **Leaf-blade** 6–12(–15) × 0.15–0.2 cm, subterete, acute, sulcate above, transversely rugose, fleshy, gently curved, shiny green, spotted and stained dark violet, becoming reddish to purplish in exposed sites; sheath tubular, subterete, longitudinally ribbed, slightly longer than internodes. **Inflorescences** emerging *c.* midway along sheath, a many-flowered decurved raceme; flowers borne 3–5 mm apart; peduncle 1.6–2.5 cm long, slender, broader distally, shiny dark green, spotted dark violet; non-floriferous bracts 2 or 3, 0.6–1 mm long, ovate-triangular, acute; rachis 4.5–8 cm long, thickened, *c.* 2 mm in diameter, sulcate, shiny green, spotted dark violet; floral bracts 0.6–0.7 mm long, triangular, acute. **Flowers** 5–6 mm across, white, tipped pale rose-pink in bud, turning pale lilac-pink, spur whitish, anther-cap purple. **Pedicel** and **ovary** 8–9 mm long, slender, cylindrical, gently curved, purple-green. **Sepals** and **petals** recurved distally, 1-nerved. **Dorsal sepal** 3–4 × 1.3–1.8 mm, oblong, obtuse, dorsally carinate. **Lateral sepals** 4.3–6 × 1.6–2 mm, obliquely oblong, falcate, somewhat acute, dorsally carinate. **Petals** 2–3.9 × 1–1.6 mm, oblong-obovate, subfalcate, obtuse, narrowed at base, somewhat thickened along mid nerve on reverse. **Lip** 6 mm

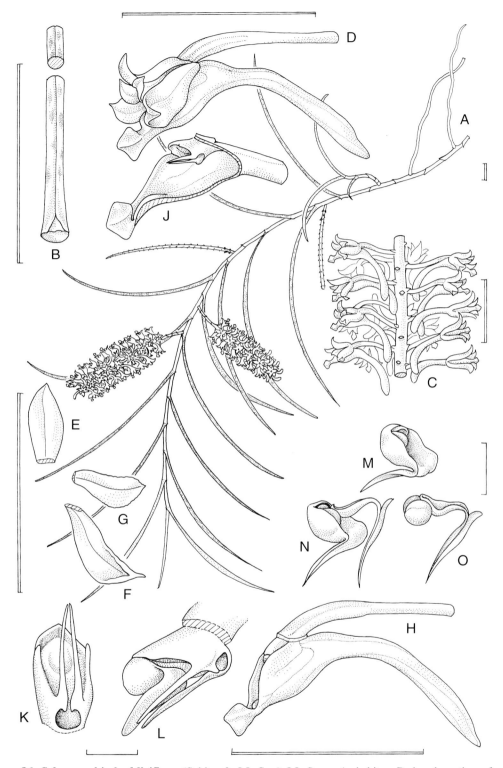

Figure 86. Schoenorchis buddleiflora (Schltr. & J.J. Sm.) J.J. Sm. - A: habit. - B: basal portion of and transverse section through leaf. - C: portion of inflorescence. - D: flower, side view. - E: dorsal sepal. - F: lateral sepal. - G: petal. - H: pedicel with ovary, lip and column, side view. - J: longitudinal section through portion of lip, column and portion of ovary, oblique view. - K: column, front view. - L: column, side view. - M: anther-cap, side view. - N: anther-cap and pollinarium, side view. - O: pollinarium, side view. All drawn from *Argent* C. 14844 by Susanna Stuart-Smith. Scale: single bar = 1 mm; double bar = 1 cm.

long, 3-lobed; side lobes *c.* 4.8–5 mm long, *c.* 4.5 mm wide when flattened, erect, broadly rounded, forming a concave hypochile which clasps the column; mid-lobe 1.2–1.25 × 1 mm, subcubical, with conical fleshy thickening above and below, constricted at base, porrect; spur 0.9–1 cm long, 1 mm wide at base, 1.9–2 mm wide distally, clavate, obtuse, laterally compressed distally, slightly incurved towards lip. **Column** 0.75–1 × 1 mm, oblong; stelidia minute, tooth-like; rostellum 1.2 mm long, bifurcate into filiform laciniae; anther-cap *c.* 1.3 mm long, cucullate, with a recurved elongate, triangular, acute apical appendage; viscidium linear-lanceolate, peltate; stipes linear-spathulate; pollinia 4, in 2 unequal pairs. Plate 20A.

HABITAT AND ECOLOGY: Lower montane mossy forest on sandstone substrate; open forest of small trees on ultramafic substrate. Alt. 1100 to 2200 m. Flowering observed in December (August in cultivation).

DISTRIBUTION IN BORNEO: SABAH: Mt. Kinabalu; Ulu Segama; Crocker Range; Sook District. SARAWAK: Mt. Penrissen.

GENERAL DISTRIBUTION: Sumatra and Borneo.

NOTES: Until now, it had been assumed that *S. buddleiflora* was conspecific with *S. juncifolia* Reinw. ex Blume, a species from Sumatra and Java. Interest in these two species arose during studies at Kew of material received from Borneo. Material of an unknown *Schoenorchis* in cultivation at the Royal Botanic Garden, Edinburgh since 1985 and originating from Mt. Kinabalu, was received on loan for identification. This was thought to represent an undescribed species. Collections from Mt. Penrissen in Sarawak and Ulu Segama in Southeast Sabah, filed under *S. juncifolia* in the Kew herbarium, were found to be identical with the Edinburgh plant.

A search through the descriptions and illustrations of existing taxa soon revealed an illustration of a flower of *S. buddleiflora* published in 1949, two years after J.J. Smith's death in 1947. It is clear, from re-examination of the original descriptions and published drawings, that the two species are distinct. The material from Borneo is morphologically identical with *S. buddleiflora* and represents a new record for the island. Reports of the occurrence of *S. juncifolia* in Borneo are probably erroneous and most likely refer to *S. buddleiflora*. Both species can be readily distinguished using the key below:

Mid-lobe (epichile) of lip directed outward and forward, with conical thickenings above and below, somewhat cubical, *c.* 1.2–1.25 mm long; spur only very slightly incurved towards lip, slender, slightly widened at base and distally. Inflorescences 6–10 cm long. Leaves generally 6–12(–15) cm long, 1.5–2 mm in diameter .... *S. buddleiflora*
Mid-lobe (epichile) of lip sharply recurved, fleshy, but lacking conical thickenings, oblong, *c.* 2 mm long; spur sharply upcurved from about the middle towards and sometimes touching middle of lip, widely funnel-shaped at base, slender distally. Inflorescences 6–7 cm long. Leaves up to 16 cm long, although often shorter, 3 mm in diameter .... *S. juncifolia*

DERIVATION OF NAME: The generic name is derived from the Greek *schoenus*, reed or rush, and *orchis*, orchid, alluding to the linear, rush-like leaves of many species. The specific epithet is a rather fanciful allusion to the inflorescences of the ornamental shrub genus *Buddleja* (Buddlejaceae).

87. SCHOENORCHIS ENDERTII
(J.J. Sm.) E.A. Christenson & J.J. Wood

Schoenorchis endertii (*J.J. Sm.*) *E.A. Christenson & J.J. Wood* in Lindeyana 5, 2: 101 (1990).

Robiquetia endertii J.J. Sm. in Bull. Jard. Bot. Buitenzorg, ser. 3, 11: 153 (1931). Type: Borneo, Kalimantan Timur, West Kutai (Koetai), Long Temelen, *F.H. Endert* 2852 (holotype BO, isotype L).

Epiphyte. **Stem** 0.5–1 cm long. **Leaf-blade** 3–4, 1.5–3.8 × 0.7–1.9 cm, oblong-elliptic, apex obscurely and obtusely unequally bilobed, rigid; sheath 2–4 mm long, dark green, sometimes flushed purple above, purple beneath. **Inflorescence** axillary, densely many-flowered; peduncle 4–10 mm long; non-floriferous bracts 2 mm long, triangular-ovate, acute; rachis 3–8 mm long, fleshy, angular; floral bracts 1–2 mm long, triangular, acute to acuminate, stiff. **Flowers** 4 mm long, sepals and petals white or palest pink, lip white, spur greenish white, column whitish pink with a lilac-pink blotch on each rostellar lobe, anther-cap pale pink, flushed dark lilac-pink. **Pedicel** and **ovary** 4–5 mm long. **Dorsal sepal** 2 × 1 mm, oblong, obtuse, concave-cucullate. **Lateral sepals** 2 × 1.5 mm, obliquely oblong-ovate, obtuse, thickened at apex. **Petals** 1.8 × 0.8 mm, oblong, obtuse. **Lip** 4 mm long, obscurely 3-lobed; side lobes 0.5 mm long, adnate to base of column, erect, rounded; mid-lobe bipartite, lower half *c.* 1.5 mm wide, oblong, shallowly concave, 3-nerved, with a V-shaped basal callus, upper half *c.* 0.9 mm wide, laterally compressed, oblong, truncate, fleshy, spur 3 mm long, decurved, obtuse. **Column** 1 mm long, with 2 ovate, rounded, fleshy basal rostellar protuberances 0.8 mm long, foot absent; anther-cap 0.8 × 0.8 mm, quadrate, cucullate, apex bicuspidate; stipes spathulate; viscidium ligulate; pollinia 4, in 2 unequal halves. Plate 20B & C.

HABITAT AND ECOLOGY: Hill forest on ultramafic substrate; lower montane forest. Alt. 400 to 600 m. Flowering observed in August.

DISTRIBUTION IN BORNEO: KALIMANTAN TIMUR: Kutai area. SABAH: Mt. Kinabalu.

GENERAL DISTRIBUTION: Endemic to Borneo.

NOTES: Originally described as a *Robiquetia*, this species was recently transferred to *Schoenorchis* on account of the four pollinia in two unequal halves and long, linear viscidium. In *Robiquetia* there are only two, cleft pollinia and the viscidium is small.

DERIVATION OF NAME: The specific epithet honours Frederik Hendrik Endert (1891–?), a forestry officer who made extensive collections in Sumatra, Java, Borneo and Sulawesi. He participated in the Central-East Borneo Expedition of 1925 that explored the Mahakam and other east coast rivers, the Kutai area and the mountains around Mt. Kemal.

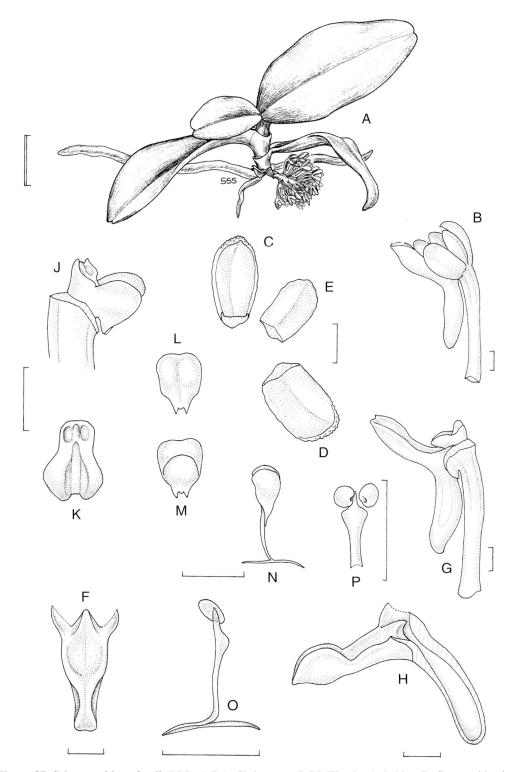

Figure 87. Schoenorchis endertii (J.J.Sm.) E.A. Christenson & J.J. Wood. - A: habit. - B: flower, side view. - C: dorsal sepal. - D: lateral sepal. - E: petal. - F: lip, front view. - G: pedicel with ovary, lip and column, anther-cap detached, side view. - H: lip, longitudinal section. - J: upper portion of ovary and column, anther-cap detached, side view. - K: column, front view. - L: anther-cap, back view. - M: anther-cap, front view. - N: pollinarium with anther-cap attached, side view. - O: pollinarium, side view. - P: stipes and pollinia, front view. All drawn from *Clements* 3392 by Susanna Stuart-Smith. Scale: single bar = 1 mm; double bar = 1 cm

88. SMITINANDIA MICRANTHA (Lindl.) Holttum

Smitinandia micrantha (*Lindl.*) *Holttum* in Gard. Bull. Singapore 25: 106 (1969). Types: Nepal, *Wallich* 7300a (syntype K-WALL); Bangladesh, Sylhet, Chota Nagpur, *De Silva* in *Wallich* 7300b (syntype K-WALL).

Saccolabium micranthum Lindl., Gen. Sp. Orch. Pl.: 220 (1833).
Gastrochilus parviflorus Kuntze, Rev. Gen. Pl. 2: 661 (1891).
Saccolabium fissum Ridl. in J. Linn. Soc., Bot. 32: 361 (1896). Type: Peninsular Malaysia, Langkawi, *Curtis* s.n. (holotype K).
Cleisostoma micranthum (Lindl.) King & Pantl. in Ann. Bot. Gard. Calcutta 8: 234, pl. 312 (1898).
Uncifera albiflora Guill. in Bull. Soc. Bot. France 77: 333 (1930). Type: Cambodia, Angkor Wat, *Thorel* 2073 (holotype P).
Cleisostoma poilanei Gagnep. in Bull. Soc. Bot. France 79: 34 (1932). Type: Cambodia, Pum-lovea and Pum-rong, Kompong-speu Province, *Poilane* 1735 (holotype P).
Cleisostoma petitiana Guill. in Bull. Mus. Nat. Hist. Paris, ser. 2, 17, 5: 434 (1945). Type: Vietnam, Annam, Darlac, *Petit* s.n. (holotype P).
Ascocentrum micranthum (Lindl.) Holttum in Gard. Bull. Sing. 11: 275 (1947).
Pomatocalpa poilanei (Gagnep.) T. Tang & F.T. Wang in Acta Phytotax. Sin. 1, 1: 100 (1951).
Cleisostoma tixieri Guill. in Bull. Mus. Hist. Nat. Paris, ser. 2, 32, 4: 369 (1960). Type: Vietnam, Annam, Dalat, *Tixier* 30/60 (holotype P).

Epiphyte. **Stems** *c.* 10–30 cm long, internodes 1–1.6 cm long, up to *c.* 4 mm wide, branching, becoming pendulous, producing many elongate, smooth roots at nodes. **Leaf-blade** 6–8 × 1.2–1.5 cm, ligulate, unequally obtusely bilobed, lobes 3–4 mm long, fleshy; sheath 1–1.6 cm long, slightly alate. **Inflorescences** borne from near or at base of leaf sheath opposite blade, *c.* 6–15 cm long, densely few- to many-flowered, flowers borne 1–3 mm apart; peduncle 1.5–3 cm long; non-floriferous bracts *c.* 3, 3–5 mm long, ovate, obtuse to acute; rachis 2–10 cm long, rather fleshy; floral bracts 1–2 mm long, triangular-ovate, acute to acuminate. **Flowers** *c.* 5 mm in diameter, rather fleshy, light translucent cream, lip side lobes purple distally, mid-lobe with a large purple patch, column off-white, anther-cap dark purple with 2 yellow spots. **Pedicel** and **ovary** 4 mm long. **Sepals** and **petals** spreading. **Dorsal sepal** 2.5 × 1.8 mm, ovate, apex obtuse to subacute, slightly thickened and sometimes slightly erose. **Lateral sepals** 3 × 2 mm, obliquely ovate, apex obtuse, slightly thickened and cucullate. **Petals** 2.3 × 1 mm, narrowly oblong, ligulate, obtuse, erose. **Lip** 3-lobed; side lobes 1 mm, triangular, rounded, pointing forwards; mid-lobe 3 × 3 mm, oblong, rounded, erose, decurved, disc with a transverse basal thickening at spur entrance; spur 2 × 1.5 mm, ± parallel with ovary, saccate, obtuse, without inner ornamentation. **Column** 1 mm long, with a swollen area either side of rostellum; foot absent; rostellum small; anther-cap ovate, cucullate, acute; pollinia 4, the pairs being completely split, the smaller half in each pair detaching from the larger, attached to a spathulate stipes; viscidium ovate. Plate 20D & E.

HABITAT AND ECOLOGY: Riverine forest on ultramafic substrate. Alt. 200 to 300 m. Flowering observed in July.

DISTRIBUTION IN BORNEO: KALIMANTAN SELATAN: Martapura, Pleihari Reserve.

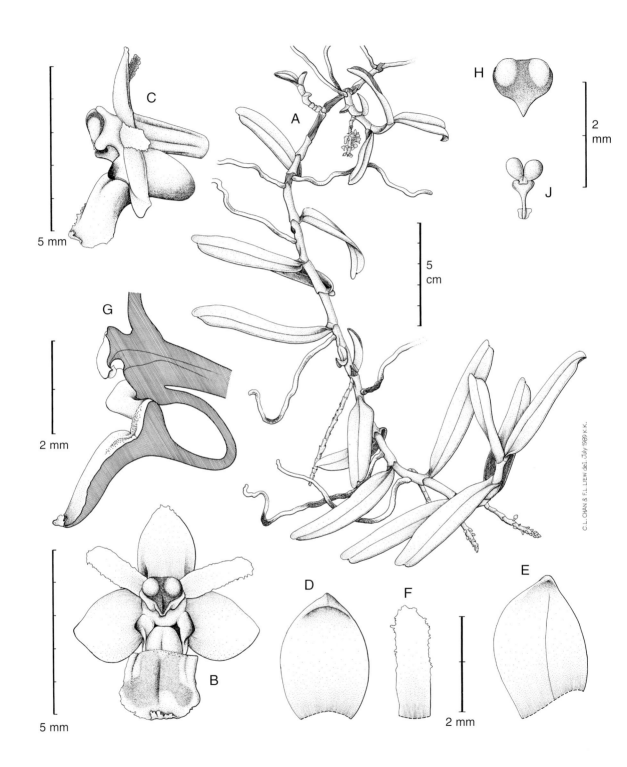

Figure 88. Smitinandia micrantha (Lindl.) Holttum. - A: habit. - B: flower, front view. - C: flower, side view. - D: dorsal sepal. - E: lateral sepal. - F: petal. - G: pedicel with ovary, lip and column, longitudinal section. - H: anther-cap, back view. - J: pollinarium. All drawn from *Lamb* AL 1148/89 by C.L. Chan and Liew Fui Ling.

GENERAL DISTRIBUTION: Bangladesh, India, Nepal, Bhutan, Myanmar (Burma) Thailand, Laos, Cambodia, Vietnam, Peninsular Malaysia and Borneo.

NOTES: A genus of three species, one of which is endemic to Sulawesi. *Smitinandia micrantha* is the most widespread species whose distribution was extended eastward from mainland Asia with the discovery in 1989 by Anthony Lamb of a population in South Kalimantan.

DERIVATION OF NAME: The generic name honours the Thai forest botanist Tem Smitinand (1920–1995). The Latin specific epithet means small-flowered.

89. SPATHOGLOTTIS KIMBALLIANA Hook.f.

Spathoglottis kimballiana *Hook.f.* in Bot. Mag. 121, t. 7443 (1895). Type: Borneo, ex cult. *Sander & Co.* (holotype Bot. Mag. t. 7443).

Terrestrial. Pseudobulbs 1–2 × 1–2 cm, ovoid, enclosed in sheaths. **Cataphylls** 1–2, *c.* 4–16 cm long, acuminate, imbricate. **Leaf-blade** *c.* 27–70 × *c.* (1.5–)2–5 cm, narrowly linear-lanceolate, long acuminate, plicate, main nerves 5–7; petiole *c.* 10–20 cm long, canaliculate. **Inflorescence** emerging from base of pseudobulb, laxly 6- to 12-flowered, flowers borne 0.6–3 cm apart; peduncle *c.* 30–80 cm long, erect; non-floriferous bracts 8–12, 1.5–2.3 cm long, ovate, obtuse to acute, sheathing, uppermost remote, lowermost imbricate; rachis *c.* 5–9 cm long; floral bracts 1–1.5 cm long, broadly ovate-elliptic, obtuse, strongly concave-cymbiform, olive-green, stained red. **Flowers** up to 10, opening 1 or 2 in succession, 5 to 7 cm in diameter, sepals bright yellow, mottled and splashed rose-red on reverse, petals bright yellow, lip similar, with dark red streaks on lower half of side lobes, calli and proximal portion of mid-lobe, column bright yellow. **Pedicel** and **ovary** 2.5–4 cm long, slender. **Dorsal sepal** 2.6–3.3 × 1.3–1.5 cm, oblong-elliptic, obtuse, shortly hairy at base inside. **Lateral sepals** 2.5–2.9 × 1.2–1.5 cm, ovate-elliptic, obtuse, shortly hairy at base inside. **Petals** 2.9–3 × 1.6–1.8 cm, elliptic, obtuse, shortly hairy at base inside. **Lip** 2.4–2.5 cm long; side lobes 1.5 cm long, 1.2 cm wide at apex, *c.* 0.7 cm wide below, broadly oblong-auriculate, falcate, rounded, erect to porrect or incurved; mid-lobe 1.6–1.8 cm long, 0.7–0.8 cm wide across basal auricles, 0.3–0.4 cm wide at centre, 0.6–1.1 cm wide at apex, spathulate, distally narrowly flabellate and obtusely bilobulate, slender below, with two 1 mm long triangular-acute, tooth-like, sometimes falcate, minutely hairy basal auricles; calli 2, 4–5 mm long, fleshy, glabrous, porrect and arranged in a V-shape; disc sparsely shortly hairy between calli and auricles. **Column** 2–2.3 cm long, *c.* 0.6 cm wide near apex, curved; anther-cap *c.* 3.5 × 3 mm, cordate, glabrous; pollinia 8, in 2 groups of 4. Plate 21A.

HABITAT AND ECOLOGY: Open places among vegetation and boulders on riverbanks. Alt. lowlands to 1500 m. Flowering observed in January, and from March until June.

DISTRIBUTION IN BORNEO: SABAH: Labuk and Sugut District, near Telupid; Porog River. Mt. Kinabalu; Ranau area.

GENERAL DISTRIBUTION: Borneo and the Philippines.

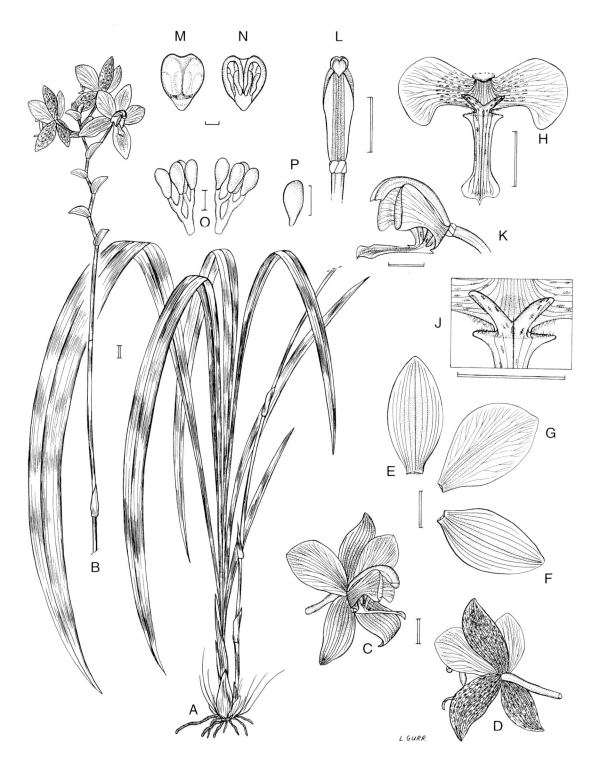

Figure 89. Spathoglottis kimballiana Hook.f. - A: habit. - B: inflorescence. - C: flower, oblique view. - D: flower, back view. - E: dorsal sepal. - F: lateral sepal. - G: petal. - H: lip flattened, front view. - J: close-up of callus. - K: pedicel with ovary, lip and column, side view. - L: column, front view. - M: anther-cap, back view. - N: anther-cap, interior view. - O: pollinarium. - P: pollinium. A drawn from miscellaneous sources and B–P from *Lamb* SAN 93370 by Linda Gurr. Scale: single bar = 1 cm; double bar = 1 cm.

NOTES: Accurate identification of the yellow-flowered species of *Spathoglottis* is somewhat confusing. This is due, in part, to mislabelling of early illustrations, inaccurate keys, natural variation and possible hybridisation (Green, 2001). However, *S. kimballiana* is distinctive and perhaps the most beautiful of the yellow-flowered species of *Spathoglottis* native to Borneo. It is clearly allied to *S. confusa* J.J. Sm. (syn. *S. aurea* Rchb.f., non Lindl.) with which it has been confused both in the literature and in the herbarium. It may be distinguished, however, by the very broad, axe-shaped or ear-like lip side lobes and rose-red mottling on the reverse of the sepals. The distinctly shaped side lobes remain a dominant character in progeny resulting from hybridisation. It may be more widespread than the few collections examined would suggest, since material in herbaria from elsewhere in Borneo may have been incorrectly determined as *S. aurea* Lindl., *S. confusa* J.J. Sm., *S. microchilina* Kraenzl., or even *S. gracilis* Rolfe ex Hook.f. Please note that plate 15F in *A Checklist of the Orchids of Borneo* by Wood & Cribb (1994) was determined as *S. confusa*, but actually depicts *S. kimballiana*. The yellow flowered plant depicted as *S. kimballiana* in figure 161 on page 239 of *National Parks of Sarawak* by Hazebroek & Abang Kashim bin Abang Morshidi (2001) is *S.aurea*.

A narrow leaved variant, described from the Philippines by Oakes Ames as var. *angustifolia,* has also been reported from around Ranau in the hills to the southeast of Mount Kinabalu. This has smaller growth with narrower, rigid leaves measuring only about 1cm in width. The more rounded flowers have clear yellow sepals lacking the characteristic rose-red flushing on the reverse. Red spots are present on the lip as in var. *kimballiana.*

Other yellow-flowered *Spathoglottis* depicted in this series are *S. gracilis* (volume 1: fig. 88) and *S. microchilina* (volume 3: fig. 97).

DERIVATION OF NAME: The generic name is derived from the Greek *spatha*, broad, and *glotta*, tongue, alluding to the broad mid-lobe of the lip found in some species such as *S. gracilis* and *S. plicata* Blume. The specific epithet honours Mr W.S. Kimball of Rochester, New York State, USA, a nurseryman who is described in the Gardeners' Chronicle for 1895 as famous for his collections of pictures, china, and articles of vertu, and especially for his four hundred species and varieties of *Cypripedium*. "In his great orchid house, which is open daily to the public, five thousand plants of this Order may be seen in bloom at one time."

90. THRIXSPERMUM TORTUM J.J. Sm.

Thrixspermum tortum *J.J. Sm.* in Bull. Jard. Bot. Buitenzorg, ser. 2, 13: 40 (1914). Types: Borneo, Kalimantan Barat, Putus (Poetoes) Sibou, *Nieuwenhuis* s.n. (syntype BO); Sumatra, Djambi, *Grootings* s.n., cult. Bogor (syntype BO).

Scandent **terrestrial,** sometimes an **epiphyte. Stem** elongate, 80 cm to over 1 m long, *c.* 4–8 mm wide, internodes 2–4 cm long, rigid, with a few branches, bearing elongate aerial roots, somewhat twisted, compressed, transversely elliptic in cross-section. **Leaves** arranged in two rows, but becoming quaquaversal through twisting of stem; blade 4–5.7(–8) × 2.9–3.5(–3.8) cm, oblong to ovate-elliptic, unequally obtusely bilobulate, with a mucro in the sinus, thick, somewhat fleshy, mid-nerve sulcate above, prominent below; sheath 2–4 cm long, compressed, tubular, transversely elliptic in cross-section. **Inflorescences** emerging at various points through back of leaf sheath, much longer than leaves, spreading, often upcurved; peduncle 10–21 cm long, dilated

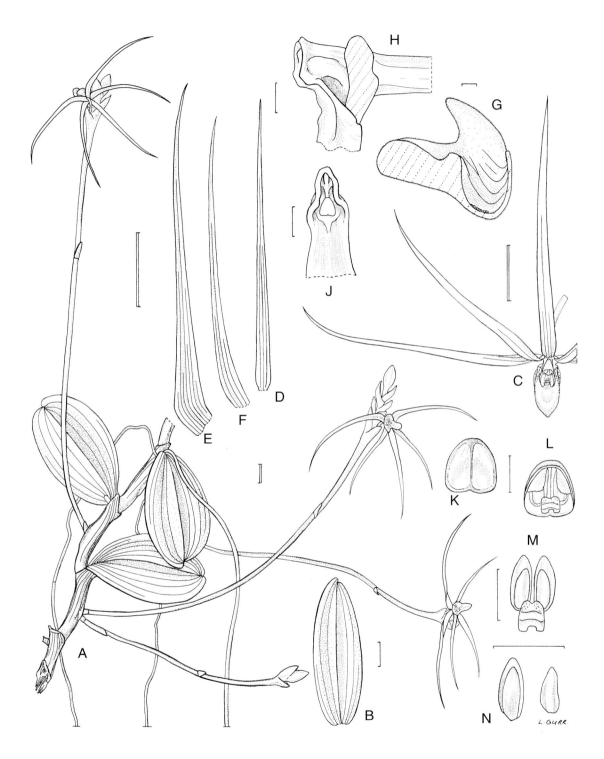

Figure 90. Thrixspermum tortum J.J. Sm. - A: habit. - B: leaf, adaxial surface. - C: flower with lateral sepal and petal cut away, front view. - D: dorsal sepal. - E: lateral sepal. - F: petal. - G: lip, longitudinal section. - H: upper portion of ovary and column, anther-cap detached, side view. - J: column, front view. - K: anther-cap, back view. - L: anther-cap, interior view. - M: pollinarium. - N: pollinia. A drawn from *Wood* 818 and B–N from *Lamb* s.n. by Linda Gurr. Scale: single bar = 1mm; double bar = 1 cm.

to 3–4 mm wide above, transversely elliptic in cross-section; non-floriferous bracts usually 3, 4–8 mm long, ovate, acute, remote; rachis gradually extending from 1.5–5(–8) cm long, thickened, compressed, flexuose, concave above point of flower attachment; floral bracts 6–10 mm long, distichous, triangular, boat-shaped, conduplicate, subfalcate, obtuse to subacute, carinate. **Flowers** produced 1–3 at a time in succession, non-resupinate, ephemeral, fragrant, sepals and petals greenish yellow, sometimes orange distally, basal half of lip yellow to orange-yellow, finely spotted red, upper half pure white, column and anther-cap white. **Pedicel** and **ovary** 0.6–1 cm long, cylindrical. **Sepals** and **petals** spreading. **Dorsal sepal** 4.7–8.8 × 4.1–4.7 cm, narrowly linear, acuminate, shallowly concave. **Lateral sepals** 4.7–8.6 × 0.48–0.6 cm, narrowly linear, falcate-incurved, acuminate, shallowly concave, held at 90 degrees from dorsal sepal, base decurrent with column-foot. **Petals** 4.2–8.2 × 0.3 mm, narrowly linear, acuminate, shallowly concave. **Lip** 1–1.1 cm long, 8 mm high, 3-lobed, immobile, saccate, minutely papillose, sac 4–5 mm long, obtuse, slightly villose on interior walls; side lobes c. 6 × 2.5 mm, erect, triangular, acute, falcate, papillose-ciliate at front; mid-lobe 5–7 × 4–4.6 mm, porrect, conical, fleshy, obtuse, excavate at base. **Column** 1.9 mm long; foot c. 2 mm long; anther-cap cucullate, rounded to quadrangular, somewhat truncate to subretuse; pollinia 4, in 2 unequal masses. Plate 21C.

HABITAT AND ECOLOGY: Lower montane forest; recorded from a steeply sloping roadside cutting, associated with *Rhododendron, Schefflera, Nepenthes fusca, Lycopodium*, ferns, etc., exposed to full sun; rocky, scrubby places. Alt.? 500 to 1500 m. Flowering observed in May, June and November, but probably throughout the year.

DISTRIBUTION IN BORNEO: KALIMANTAN BARAT: Putus Peninsula area. SABAH: Crocker Range, Kimanis road; Mt. Kinabalu.

GENERAL DISTRIBUTION: Sumatra, Java and Borneo.

NOTES: *Thrixspermum* is a little studied and consequently poorly understood monopodial genus of around 100 species (probably less when conspecificity is taken into account) distributed from Sri Lanka and the Himalayan region east to the Pacific islands. The greatest concentration of species occurs in Sumatra, while 24 species are reported from Borneo.

There is much vegetative and floral variation within the genus. *T. amplexicaule* (Blume) Rchb.f., for example, is strictly terrestrial, some like *T. tortum* clamber up through undergrowth, while many others are twig epiphytes found high in the canopy. Some species have broad, rounded sepals and petals while in others these are long and narrow giving the flowers a spidery appearance. The arrangement and shape of the floral bracts is very characteristic. Two sections are generally recognised, viz. *Thrixspermum* (syn. *Orcidice*) in which the flowers are borne in two ranks on a mostly flattened rachis, and *Dendrocolla*, where the flowers face in all directions (quaquaversal). A third section, *Katocolla*, is sometimes recognised to accommodate some species, such as *T. pensile* Schltr., having long, limpid, pendent stems.

It is evident that the complex of species related to the widespread *T. centipeda* Lour. (syn. *T. arachnites* (Blume) Rchb.f.), including *T. longicauda* Ridl., *T. scopa* (Rchb.f. ex Hook.f.) Ridl. and *T. tortum*, etc. is little understood and clearly in need of critical study.

DERIVATION OF NAME: The generic name is derived from the Greek *thrix*, hair, and *sperma*, seed, in reference to the hair-like seeds. The specific epithet is derived from the Latin *tortus*, twisted, and refers to the somewhat twisted main stem.

91. THRIXSPERMUM TRIANGULARE Ames & C. Schweinf.

Thrixspermum triangulare *Ames & C. Schweinf.*, Orchidaceae 6: 217 (1920). Type: Borneo, Sabah, Mt. Kinabalu, 15 November 1915, *J. Clemens* 201 (holotype AMES, isotypes K, SING).

Epiphyte or **lithophyte**. **Roots** elongate, 2–3 mm in diameter, with a few branches, smooth, emerging through leaf sheaths. **Stems** up to 20 cm long, 4–6 mm in diameter, internodes 1–2 cm long, entirely enclosed by leaf sheaths, unbranched. **Leaves** distichous, spreading; blade (4.5–) 9–23 × 1.5–3 cm, linear-lanceolate, linear or ligulate, unequally obliquely or equally obtusely or acutely bilobed at apex, slightly narrowed to a complicate base, coriaceous, rigid and fleshy, many-nerved, mid-nerve sharply carinate beneath; sheath 1–2 cm long, coriaceous, striate-nervose. **Inflorescences** emerging through base of leaf sheaths, usually exceeding leaves, spreading to ascending, flowers opening one at a time in succession; peduncle 9–22 cm long; non-floriferous bracts 3, 3–6 mm long, ovate, obtuse, clasping peduncle, remote; rachis 2.5–16.5(–25) cm long, laterally flattened, fractiflex, becoming arcuate, somewhat fleshy; floral bracts 5–6.5 mm long, distichous, equitant, borne 4–5 mm apart, ovate-triangular, subacute or obtuse, strongly complanate with a broad dorsal keel, clasping, cymbiform. **Flowers** *c.* 3 cm in diameter (when expanded), pale lemon-yellow, side lobes of lip streaked with crimson-red lines, mid-lobe cream flushed crimson-red beneath, column white or pale lemon-yellow, sometimes flushed purple, whole flower fading to orange-red or pinkish purple, unscented. **Pedicel** and **ovary** 6–7 mm long. **Dorsal sepal** 1.4–1.5 × 0.7 cm, oblong-elliptic, obtuse, dorsally toothed, 5- to 6-nerved. **Lateral sepals** 1.4–1.6 × 0.8–0.85 cm, slightly obliquely ovate-oblong, obtuse, dorsally toothed, 5- to 7-nerved. **Petals** 1.3–1.5 × 0.45–0.7 cm, elliptic to oblong, obtuse, 5-nerved. **Lip** 1.2–1.4 cm long, *c.* 1.25 cm wide across side lobes, immobile, somewhat saccate at base, central portion longitudinally thickened and fleshy on both surfaces; side lobes 6–7 × 3.5–4 mm, erect, subquadrate, anterior and lateral margins shallowly concave and forming a sharp angle, thin-textured, dorsal margin convex, sometimes irregular to erose; mid-lobe 5–5.5 mm long, 5–6 mm wide at base, triangular, very thick and fleshy, margins cellular pubescent; disc with a fleshy triangular callus *c.* 3 mm long, 3.5 mm wide at base and 2 mm high, lateral walls at base highest, hairs sometimes present along disc below callus. **Column** 2–3 × *c.* 3.5 mm, stout, fleshy; foot 3–3.5 mm long, winged; anther-cap and pollinia not seen. Plate 21B.

HABITAT AND ECOLOGY: Lower montane forest; upper montane mossy forest, on tree roots, branches, rocks, in ridge scrub, often on ultramafic substrate. Alt. 1200 to 3400 m. Flowering observed throughout the year.

DISTRIBUTION IN BORNEO: SABAH: Mt. Kinabalu and probably elsewhere in the Crocker Range.

GENERAL DISTRIBUTION: Endemic to Borneo.

NOTES: Most Bornean species of *Thrixspermum* are found in the lowlands and extend only to about 1500 m in the mountains. *Thrixspermum triangulare* is unusual in being exclusively montane and has been recorded from the Mt. Kinabalu Summit Trail above Layang Layang at over 3000 m.

DERIVATION OF NAME: The specific epithet refers to the triangular callus on the lip.

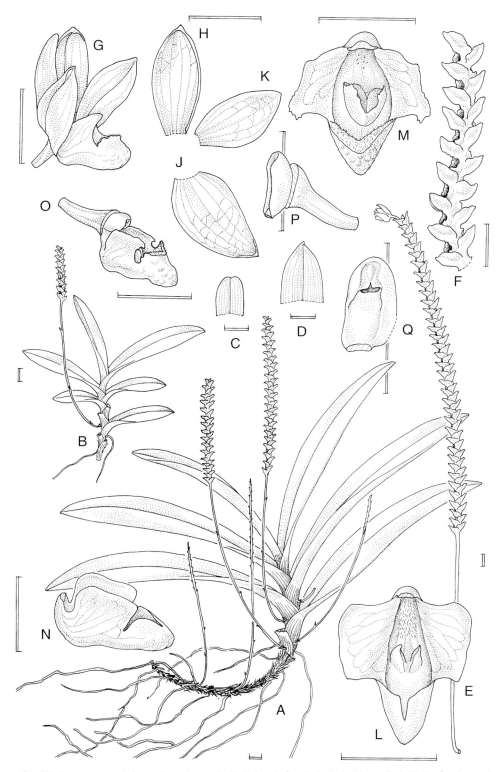

Figure 91. Thrixspermum triangulare Ames & C. Schweinf. - A & B: habits. - C & D: leaf apices, adaxial surface. - E: inflorescence. - F: upper portion of inflorescence showing floral bracts. - G: flower, oblique view. - H: dorsal sepal. - J: lateral sepal. - K: petal. - L & M: lips flattened, front view. - N: lip, side view. - O: pedicel with ovary, lip and column, oblique view. - P: pedicel with ovary and column, side view. - Q: column, oblique view. A & D drawn from *Beaman* 8999, B & C from *Price* 220, E from *Beaman* 9146 and G–K, M, & O–Q from *Gunsalam* 2 by Susanna Stuart-Smith. Scale: single bar = 1 mm; double bar =1 cm.

92. TRICHOGLOTTIS JIEWHOEI
J.J. Wood, A. Lamb & C.L. Chan

Trichoglottis jiewhoei *J.J. Wood, A. Lamb & C.L. Chan* in Sandakania 11: 44, figs. 1–3 (1998). Type: Borneo, Sabah, Crocker Range, Kimanis road, *c.* 1050 m, 1980, flowered in cultivation in Tenom Orchid Centre in April 1985, *Phillipps & Collenette* in *Lamb* AL 341/85 (holotype K).

Epiphyte. Stems up to l m long, 0.5 cm in diameter, semi-pendulous to pendulous, internodes (2.5–)3.3–3.5 cm long, branching, bearing smooth, branching roots from lower nodes. **Leaf-blade** 10.5–17 × 1.5–3 cm, lanceolate to elliptic, obscurely to distinctly unequally acutely bilobed, longest lobe up to 1 cm long, coriaceous; sheath (2.5–)3.3–3.5 cm long. **Inflorescences** emerging from about the middle of the internodes opposite leaf blade, erect, 3-flowered; peduncle 5 mm long, enclosed in 2 or 3 obtuse, sheathing non-floriferous bracts 1–2 mm long, very minutely hairy; rachis 6 mm long, zigzag, very minutely hairy; floral bracts 2–2.1 mm long, ovate, rounded, adpressed to pedicel, very minutely hairy. **Flowers** 3.5–3.8 cm in diameter, with a musky scent, pedicel and ovary yellow, yellowish green or purplish, sepals and petals pale yellow with red or purple blotches and bars, side lobes of lip pale yellow with a purple mark along lower edge near junction with mid-lobe, mid-lobe pale yellow or cream with two central purple lines or bars, scattered smaller purple blotches and numerous long, mostly white hairs, a few purple hairs also present, back wall tongue white-hairy, column purple-red, pinkish to white distally, anther-cap white. **Pedicel** and **ovary** 3.5–4.3 cm long, slender. **Sepals** and **petals** fleshy, minutely papillose, especially on adaxial surface. **Dorsal sepal** 1.9(–2) × 0.3–0.45 cm, ligulate-subspathulate, obtuse. **Lateral sepals** 1.3–1.5 × 0.48–0.5 cm, oblong, obtuse, gently twisted at middle. **Petals** 1.7–1.9 × 0.35–0.38 cm, narrowly spathulate, subfalcate, obtuse. **Lip** 6–8 mm long when flattened; side lobes 2 × 1.5–2 mm, oblong, truncate, erect, slightly divergent, glabrous; mid-lobe 4–5 × 4–5.8 mm, ovate-cordate, obtuse to subacute, decurved, basal portion minutely papillose then becoming villous-hairy, hairs longest along fold along which blade becomes decurved, otherwise glabrous, 3-nerved; spur 2 mm long, saccate, rounded, mouth papillose-hairy; back wall tongue 2 × 2–2.5 mm, narrowly flabellate, truncate, minutely papillose, shortly hairy, with longer hairs along apical margin. **Column** 4–5 × 2.9–3 mm, oblong, glabrous; foot absent; stigmatic cavity broadly ovate; anther-cap 2.5 × 2–2.1 mm, ovate, cucullate, rostrate; viscidium very small; stipes 2 mm long, ligulate; pollinia 4, in 2 unequal pairs. Plate 22A & B.

HABITAT AND ECOLOGY: Hill dipterocarp forest rich in *Castanopsis* and *Lithocarpus* on sandstone ridges. Alt. 1000 to 1400 m. Flowering observed in March and April.

DISTRIBUTION IN BORNEO: SABAH: Crocker Range, Kimanis road; Sinsuron road.

GENERAL DISTRIBUTION: Endemic to Borneo.

NOTES: *Trichoglottis jiewhoei* in a distinctive species distinguished from all others by its erect, 3-flowered inflorescences, gently twisted lateral sepals, narrowly spathulate, subfalcate petals, lip with a narrowly flabellate, truncate back wall callus (tongue), glabrous, oblong, truncate side lobes and a decurved, ovate-cordate mid-lobe bearing long whisker-like hairs.

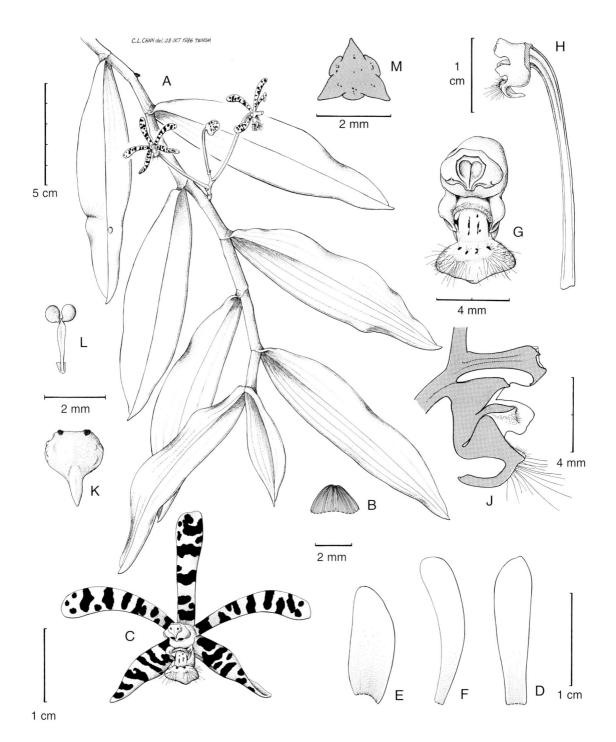

Figure 92. Trichoglottis jiewhoei J.J. Wood, A. Lamb & C.L. Chan. - A: habit. - B: close-up of an obscurely bilobed leaf apex, adaxial surface. - C: flower, front view. - D: dorsal sepal. - E: lateral sepal. - F: petal. - G: lip and column, front view. - H: pedicel with ovary, lip and column, anther-cap detached, side view. - J: ovary, lip and column, longitudinal section. - K: anther-cap, back view. - L: pollinarium. - M: ovary, transverse section. All drawn from *Lamb* AL 341/85 (holotype) by C.L. Chan.

The type was found on the forest floor where it had fallen from the canopy above. It has subsequently flowered well in cultivation at Tenom Orchid Centre (alt. *c.* 180 m), despite its being at a much lower elevation.

DERIVATION OF NAME: The specific epithet honours Mr Tan Jiew Hoe of Singapore in recognition of his services towards the conservation of wild orchids in Malaysia and Singapore.

93. TRICHOGLOTTIS KINABALUENSIS Rolfe

Trichoglottis kinabaluensis *Rolfe* in Gibbs in J. Linn. Soc., Bot. 42: 157 (1914). Type: Borneo, Sabah, Mt. Kinabalu, Gurulau Spur, 1700 m, February 1910, *Gibbs* 3993 (holotype K).

Epiphyte. Stems 20–80 cm long, internodes 2–4 cm long, entirely enclosed by leaf sheaths, rooting at lower nodes. **Leaf-blade** 5–13.3 × 1–2.5 cm, narrowly elliptic to oblong-elliptic, apex acute to acuminate, asymmetrical or sometimes unequally bidentate, often with a marginal constriction 2–3 cm below apex, coriaceous; sheath 2–4 cm long. **Inflorescences** borne from nodes opposite leaf-blades, fasciculate, sessile, with 1–4 flowers open at a time; rachis *c.* 0.5–1 mm long; floral bracts minute, scale-like. **Flowers** *c.* 2–2.5 cm in diameter, pedicel and ovary cream, sepals and petals cream barred pale cinnamon-orange, lip pale creamy yellow, column pale orange to cinnamon-brown, anther-cap creamy yellow. **Pedicel** and **ovary** 1–1.2 cm long, slender. **Sepals** and **petals** spreading. **Dorsal sepal** 8–12 × 3 mm, narrowly oblong-ligulate, obtuse. **Lateral sepals** 8–12 × 2.5–3 mm, narrowly triangular-oblong, obtuse. **Petals** 8–10 × 2 mm, linear-oblong, obtuse, slightly curved. **Lip** 3-lobed, with a small hairy tongue above entrance to spur; side lobes 3–4 × 1.5 mm, linear, obtuse, recurved; mid-lobe clawed, claw 3–4 mm long, narrow, gently curved, expanded into a 5–6 × 5–6 mm ovate, obtuse decurved blade, with a median thickened area on the disc; spur 2–4 mm long, oblong, obtuse. **Column** 4 mm long; foot absent; anther-cap 1.5 × 1.5 mm, cucullate, smooth. Plate 23A & B.

HABITAT AND ECOLOGY: Hill forest; lower montane forest. Alt. 900 to 1700 m. Flowering observed from January until March, May, July and August.

DISTRIBUTION IN BORNEO: SABAH: Mt. Kinabalu; Crocker Range, Sinsuron road.

GENERAL DISTRIBUTION: Endemic to Borneo.

NOTES: *Trichoglottis kinabaluensis* had, until recently, only been collected at moderate elevations on Mt. Kinabalu. Lamb (pers. comm.) has recently found it growing near the Rafflesia Centre along the Sinsuron road.

DERIVATION OF NAME: The specific epithet refers to Mt. Kinabalu, the type locality.

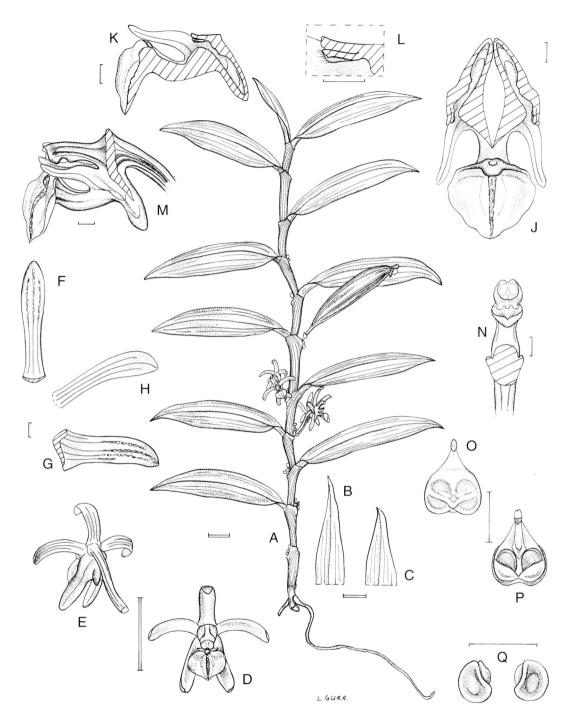

Figure 93. Trichoglottis kinabaluensis Rolfe. - A: habit. - B & C: leaf apices, adaxial surface. - D: flower, front view. - E: flower, back view. - F: dorsal sepal. - G: lateral sepal. - H: petal. - J: lip, spread out, front view. - K: lip, longitudinal section. - L: close-up of hairy tongue above entrance to spur. - M: pedicel with ovary, lip and column, side view. - N: column, front view. - O: anther-cap, back view. - P: anther-cap, interior view. - Q: pollinia. A drawn from miscellaneous collections, B & C from *J. & M.S.Clemens* 26126, and D–Q from *Raman* 5/99 by Linda Gurr. Scale: single bar = 1 mm; double bar = 1 cm.

94. TRICHOGLOTTIS MAGNICALLOSA Ames & C. Schweinf.

Trichoglottis magnicallosa *Ames & C. Schweinf.*, Orchidaceae 6: 221 (1920). Type: Borneo, Sabah, Mt. Kinabalu, July-August 1916, *Haslam* s.n. (holotype AMES, isotypes BO, K).

Epiphyte. Roots emerging about halfway along leaf sheath, elongate, 3–4 mm in diameter, sparsely branched, rigid, tough, longitudinally sulcate. **Stems** 50–100 cm long, 4–5 mm in diameter, internodes 2.5–5 cm long, pendulous, shorter stems suberect to arcuate, sometimes somewhat fractiflex, entirely concealed by leaf sheaths. **Leaves** distichous, horizontally spreading; blade 12–24 × 3–4.5 cm, usually 13–15 × 3.5–4 cm, oblong-elliptic or oblong, apex obliquely, obtusely bilobed, lobules separated by a small mucro, complicate at base, thickly coriaceous, many-nerved, mid-nerve prominently carinate beneath; sheath 2.5–5 cm long, tubular, coriaceous; striate-nervose. **Inflorescences** emerging from middle or lower half of leaf sheaths, sessile, 1- (to 3-) flowered; floral bracts 3–4 mm long, ovate, obtuse. **Flowers** about 4.5 cm in diameter, showy, rigid and fleshy, whitish yellow, lemon-yellow or creamy-white, with orange-brown or purplish-pink blotches or rings, lip yellow. **Pedicel** and **ovary** 2.5–2.7 cm long, distinctly 6-keeled, 3 prominent keels, 3 less prominent keels. **Sepals** and **petals** spreading. **Dorsal sepal** 2.3–2.7 × 1–1.3 cm, oblanceolate, obtuse and broadly rounded, main nerves 6. **Lateral sepals** 2.2–2.4 × (1–)1.5–1.6 cm, broadly elliptic or oblanceolate, obtuse and broadly rounded, main nerves 5. **Petals** 2.2–2.4 × 0.9–1 cm, 2 mm wide at base, cuneate-spathulate, broadly rounded, retuse, main nerves 3, mid-nerve sometimes distinctly carinate beneath. **Lip** adnate to base of column, 1.1–1.2 cm long, 2.8–2.9 mm wide at base, narrowly linear, 3-lobed; side lobes 4–5 × 1 mm, linear, slightly broader above, apex obliquely rounded, fleshy; mid-lobe *c.* 9 mm long, 2.5–2.6 mm wide at middle, *c.* 1.6 mm wide at apex, linear, thick, fleshy, slightly broader above then narrowed to an obscurely emarginate or retuse apex, with an obscure, obtuse tooth on each side near base, densely pubescent, less so or glabrous at apex, disc with a prominent, fleshy, keel-like, pubescent callus 5–6 mm long, almost occupying entire width of mid-lobe and terminating *c.* 4 mm from apex, distal portion of callus 2 mm high; base of lip with an irregularly ovate, pubescent lamella. **Column** 3–4.5 × 4 mm, stout, pubescent; anther-cap 3 × 3 mm, ovate, subacute, densely pubescent. Plate 23C.

HABITAT AND ECOLOGY: Hill forest, sometimes on ultramafic substrate; recorded as epiphytic on Annonaceae. Alt. 500 to 900 m. Flowering observed from June to August and November.

DISTRIBUTION IN BORNEO: SABAH: Mt. Kinabalu; Crocker Range.

GENERAL DISTRIBUTION: Endemic to Borneo.

NOTES: *Trichoglottis magnicallosa* has relatively large flowers for the genus which measure about 4.5 cm in diameter. Although closely related to *T. uexkulliana* J.J. Sm., described from Kalimantan, it can be distinguished by its longer sepals and petals and lip with non falcate-incurved side lobes and a longer, narrower mid-lobe.

DERIVATION OF NAME: The specific epithet is derived from the Latin *magni*, large, and *callosus*, callose, bearing a callus or hardened thickening, referring to the prominent swelling on the mid-lobe of the lip.

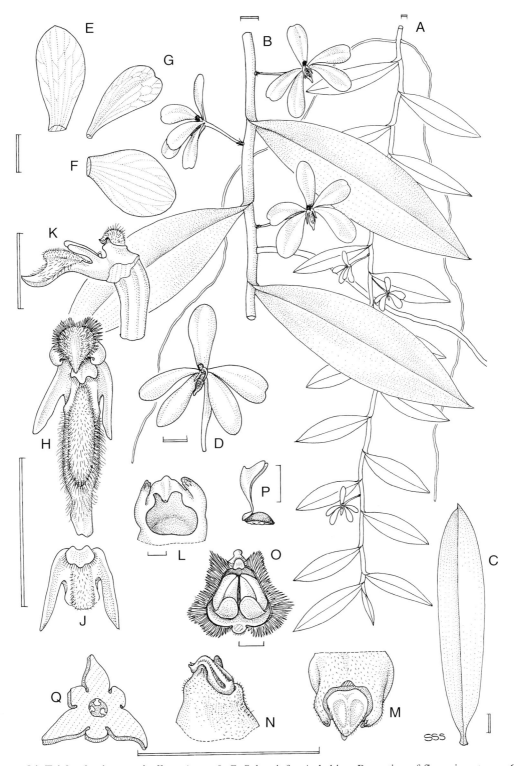

Figure 94. Trichoglottis magnicallosa Ames & C. Schweinf. - A: habit. - B: portion of flowering stem. - C: leaf, adaxial surface. - D: flower, front view. - E: dorsal sepal. - F: lateral sepal. - G: petal. H: lip, column and anther-cap, front view. - J: close up of base of lip. - K: pedicel with ovary, lip and column, side view. - L: column, anther-cap removed, front view. - M: column, anther-cap removed, viewed from above. - N: column, anther-cap removed, side view. O: anther-cap, interior view. - P: stipes and viscidium, side view. Q: ovary, transverse section. A & B drawn from *Lamb* AL 864/87, C from *J. & M.S. Clemens* 27163, and D–Q from *Cribb* 89/51 by Susanna Stuart-Smith. Scale: single bar = 1 mm; double bar = 1 cm.

95. TRICHOGLOTTIS TINEKEAE Schuit.

Trichoglottis tinekeae Schuit. in Blumea 43, 2: 492, fig. 2 (1998). Type: Borneo, Sabah, Sipitang District, east of trail from Long Pa Sia to Long Miau, 1100 m, Dec. 1986, *Vermeulen & Duistermaat* 117 (holotype L, isotype K).

Epiphyte. Roots emerging from base of leaf sheath on lower part of stem, elongate, 1–2 mm in diameter, sparsely branched, glabrous. **Stems** 5–45 cm long, 2.5–3.5 mm in diameter, internodes 1–1.5 cm long, spreading to pendulous, somewhat flexuose, terete, entirely concealed by leaf sheaths. **Leaves** spreading, twisted at base 90° so as to lie in one plane with stem, often suffused purple; blade 4.5–9 × 0.6–1.3 cm, narrowly elliptic to linear, obliquely short-bilobulate, shorter lobule apiculate, longer lobule truncate, mucronate, coriaceous, mid-nerve carinate beneath; sheath 1–1.5 cm long, tubular, finely transversely wrinkled, with an apical subulate appendage 2–6 mm long opposite blade. **Inflorescences** up to 8 from node bases, appearing almost opposite leaf-blade of node below, only one or rarely two flowering at a time, subsessile, 1-flowered; peduncle *c.* 1 mm long. **Flowers** *c.* 1.5 cm in diameter, wide-opening, rigid, fleshy, sepals and petals whitish at base, yellow to the apex, with one or two orange-brown blotches in the basal half, lip white at base, apex of mid-lobe with a large yellow to pale orange-brown blotch, side lobes white or orange-brown, front margin violet, column greenish at base, yellow at apex, or dorsally white and ventrally orange, lateral margins violet, margins of stigma red, anther yellow. **Pedicel** and **ovary** 4–5 mm long, terete, curved, glabrous. **Sepals** and **petals** spreading, 3-nerved. **Dorsal sepal** 8–9 × 2.4–2.5 mm, linear-oblong, obtuse. **Lateral sepals** 7.8–8 × 3.5–3.8 mm, obliquely ovate, obtuse. **Petals** 7–8 × 2 mm, linear-elliptic to linear-oblanceolate, slightly falcate, obtuse. **Lip** rigidly adnate to base of column, 8 × 5.5–6.7 mm, straight, 3-lobed, with an erect basal, rectangular lamella 2.3 × 1.3 mm, having a short median, longitudinal crest on underside in apical half, and very finely and densely pubescent on underside in basal half and on crest; side lobes 2 × 0.5 mm, linear, slightly broader towards base, truncate, porrect, glabrous, connected along their basal half by a transverse septum which, together with the erect basal margins of the lip, forms a pocket-like cavity covered by the basal lamella like a lid; interior of cavity densely pubescent; mid-lobe 7–8 × 5.5–6.7 mm, broadly elliptic, broadly rounded, margins in apical half crenulate, with a basal U-shaped patch of soft, erect, hyaline hairs surrounding a shallow concavity. **Column** 5–6.5 × 2.3 mm, slender, slightly curved, minutely papillose at the sides near the apex; stelidia and foot absent; stigma pyriform; anther-cap 2.3 mm high, galeate, glabrous; pollinia 4, in 2 spherical masses, each pair somewhat unequal. **Fruit** *c.* 2 × 0.7 cm, sharply 3-angular in cross section. Plate 24A & B.

HABITAT AND ECOLOGY: Open, disturbed kerangas forest; primary lower montane forest with *Agathis*; forest on limestone; occasionally growing on mossy limestone cliffs. Alt. 450 to 1700 m. Flowering observed in March, May, June, August and December.

DISTRIBUTION IN BORNEO: SABAH: Sipitang District, Long Pa Sia area; Pensiangan District, Batu Ponggol; also noted by Lamb (pers. comm.) as being on Batu Urun. SARAWAK: Baram District, Kelabit Highlands.

NOTE: This recently described species is distinguished by having appendages on the leaf sheaths and a spurless lip with a large, almost flat, elliptic mid-lobe with a basal tuft of very long hairs.

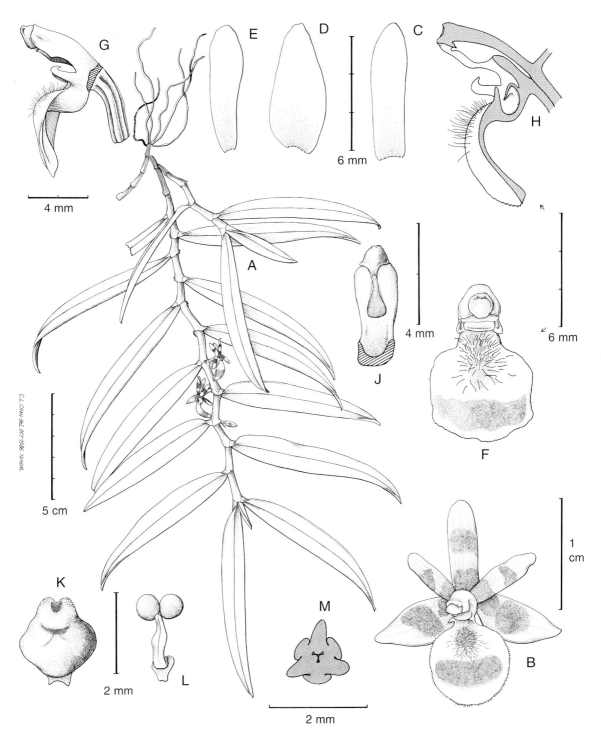

Figure 95. Trichoglottis tinekeae Schuit. - A: habit. - B: flower, front view. - C: dorsal sepal. - D: lateral sepal. - E: petal. - F: lip, front view. - G: pedicel with ovary, lip and column, side view. - H: pedicel with ovary, lip and column, longitudinal section. - J: column and anther-cap, front view. - K: anther-cap, back view. - L: pollinarium. - M: ovary, transverse section. All drawn from *Lamb* AL 330/85 and *Lamb* AL 815/87 by C.L. Chan.

DERIVATION OF NAME: *Trichoglottis tinekeae* is named in honour of Mrs Tineke Roelfsema of Wageningen in The Netherlands, who is an ardent supporter of the study and cultivation of orchids at the Rijksherbarium and Hortus Botanicus of Leiden University.

96. TRICHOTOSIA VESTITA (Lindl.) Kraenzl.

Trichotosia vestita (*Lindl.*) *Kraenzl.* in Engl., Pflanzenr. Orch. Monandr. Dendr. IV, 50, II, B, 21: 151 (1911). Type: Singapore, *Wallich* 2005 (holotype K-LINDL).

Dendrobium vestitum Wall. ex Lindl., Gen. Sp. Orch. Pl.: 82 (1830).
Eria vestita (Lindl.) Lindl. in Bot. Reg. 30, misc. 76, no. 79 (1844).
Pinalia vestita (Lindl.) Kuntze, Rev. Gen. Pl. 2: 679 (1891).

Epiphyte. **Stems** up to 2 m long, 0.7–1 cm in diameter, internodes 2.5–5 cm long, leafy, erect to spreading, often becoming pendulous. **Leaves** densely covered by reddish hairs giving the sheaths a furry appearance and the blades a softly velvety texture; blade (12–)14–20 × 2–4.8 cm, narrowly elliptic, acute to acuminate, tough, coriaceous; sheath 2.5–5 cm long. **Inflorescences** 20–38 cm long, borne from nodes opposite leaf-blades, many-flowered, flowers borne 1–2(–4) cm apart, densely long reddish-hairy all over, pendulous; peduncle 2–6 cm long, to *c.* 5 mm in diameter; non-floriferous bracts 2–3, 0.5–1.5 cm long, ovate, obtuse, lowermost entirely clasping peduncle; rachis *c.* 18–32 cm long, to *c.* 3 mm in diameter, zigzag; floral bracts 0.8–1 × 1.2–1.8 cm, broadly ovate, obtuse, rounded, concave, persistent, pale green, tipped purple. **Flowers** 2–3 cm long, not opening widely, sepals cream to pale yellow, covered in long reddish or golden hairs outside, petals pinkish white to yellow, stained pink or purple at apex, lip almost white, mid-lobe pale pink with red or purple margin, callus apex yellow, column and anther-cap blackish purple. **Pedicel** and **ovary** 3–6 mm long, densely reddish hairy, appearing like fur. **Dorsal sepal** 2–2.5 × 0.6–0.7 cm, oblong to triangular, acute to acuminate, concave. **Lateral sepals** 2–2.5 × 0.8–1 cm, obliquely triangular, acute to acuminate. **Mentum** 1–1.2 × 0.4–0.5 cm, conical, obtuse. **Petals** 1.8–2 × 0.3–0.4 cm, ligulate, obtuse, glabrous. **Lip** *c.* 2.5 cm long, 0.3–0.4 cm wide at base, 1.3–1.4 cm wide across side lobes, narrowly clawed, minutely papillose; side lobes 4 mm high, free apical portion 3 mm long, oblong, obtuse, erect; mid-lobe 7 × 8–9 mm, subquadrate to broadly oblong, corners rounded, ± truncate, with a small knob in the sinus, gently undulate; disc with 2 low keels between side lobes commencing above base of claw and terminating *c.* midway, with a central narrowly oblong-elliptic, 3-ridged, papillose-hairy callus commencing at *c.* middle of disc and terminating below apex of mid-lobe. **Column** 1.2–1.3 cm long; foot 1–1.2 cm long; anther-cap 4 × *c.* 1.5 mm, narrowly oblong; pollinia 8. Plate 24C.

HABITAT AND ECOLOGY: Kerangas forest; riverine forest; hill forest on sandstone; mixed dipterocarp forest. Alt. sea level to 500 m. Flowering observed throughout the year.

DISTRIBUTION IN BORNEO: BRUNEI: Belait District. KALIMANTAN SELATAN: Banjarmassin. SABAH: Nabawan area; Sandakan District, Wanyang River; Telupid to Ranau road; Lamag District, Namatoi River; Keningau District; Pingas Pingas River; Pinangah Forest Reserve.

Figure 96. Trichotosia vestita (Lindl.) Kraenzl. - A & B: habits. C: floral bract and flower, side view. - D: dorsal sepal. - E: lateral sepal. - F: petal. - G: lip, front view. - H: pedicel with ovary, lip and column, side view. - J: pedicel with ovary and column, side view. - K: column, front view. - L: anther-cap, front view. - M: anther-cap, side view. N: pollinia. A & B drawn from *Madani* SAN 88841 and C–N from *Carr* C.111 by Susanna Stuart-Smith. Scale: single bar = 1 mm; double bar = 1 cm.

Kabili-Sepilok Forest Reserve. SARAWAK: Belaga District, Sekapang Panjang waterfalls; Kapit District, Hose Mountains, Simpurai River; Kuching District, Kuching; Mt. Santubong; Marudi District, Lio Matoh; Miri District, Niah; Samarahan District, Samarahan; Simunjan District, Sadong.

GENERAL DISTRIBUTION: Peninsular Malaysia, Sumatra and Borneo.

NOTES: There are about 50 species of *Trichotosia* distributed from mainland Asia, through S.E. Asia to New Guinea and the Pacific islands. Established by Blume in 1825, *Trichotosia* has been variously treated as a subgenus or section of *Eria* by many authors until Kraenzlin (1911) found reasons to re-establish it at generic level. However, Kraenzlin's original transcription was too broad and included taxa which we today consider to be within *Eria* section *Cylindrolobus*. Consequently, subsequent authors, such as Holttum, chose to disregard Kraenzlin's proposal.

Holttum (1957) may be correct when he states that "*Trichotosia*... have no more claim to... ... separation than some of the other secitons." Nevertheless, most recent workers shave upheld Kraenzlin's view for reasons of practicality rather than scientific evidence, since it officers a convenient way of reducing the size of the large, unwieldy and polyphyletic *Eria*. Certainly species of *Trichotosia* have a distinctive aspect and are, with a little practice, readily distinguishable from sections of *Eria*, including *Cylindrolobus*.

Twenty-seven named species have been recorded in Borneo, of which fifteen are thought to be endemic. Accurate determination of material from Borneo and neighbouring islands, however, remains problematic and it is clear that the genus is in need of a modern monographic treatment. The endemic *T. brevipedunculata* (Ames & C. Schweinf.) J.J. Wood was figured in *Orchids of Borneo*, volume 3: fig. 99.

The species depicted as *T. ferox* Blume by Wood *et al.*, (1993, plate 81C) is in fact *T. vestita*.

DERIVATION OF NAME: The generic name is derived from the Greek *trichotos*, hairy, referring to the hairy leaves, inflorescences and flowers of the majority of species. The specific epithet is derived from the Latin *vestitus*, clothed, also referring to the hairy covering.

97. VANILLA HAVILANDII Rolfe

Vanilla havilandii *Rolfe* in Kew Bull. 1918: 236 (1918). Types: Borneo, Sarawak, Kuching, *Haviland* s.n. (syntype K); Matang, edge of tea estate, January 1915, *Ridley* s.n. (syntype K).

Climber. **Stem** up to 13 m or more long, stout, upper internodes 4–10 cm long, rooting at nodes. **Leaf-blade** (7–)9–17(–19) × (1.5–)2.5–5 cm, narrowly elliptic to elliptic, gradually to abruptly attenuated to a long-acuminate apex, rather coriaceous, fleshy, bluish-green, semi-gloss; petiole 0.8–1 cm long, sulcate, fleshy. **Inflorescences** produced on apical, pendulous portion of stems, arising from leaf axils, successively many-flowered, one or two flowers open at a time, borne 2–3 mm apart; peduncle absent, or at most 2 mm long; rachis 3–4 cm long, sometimes with an additional branch; floral bracts 3–5 × 3 mm, suborbicular to ovate-oblong, obtuse, concave.

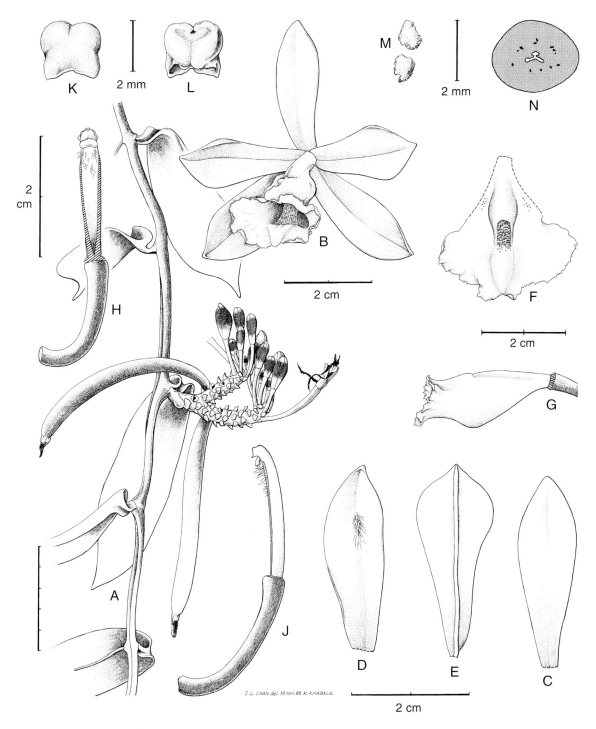

Figure 97. Vanilla havilandii Rolfe. - A: habit. - B: flower, oblique view. - C: dorsal sepal. - D: lateral sepal. - E: petal. - F: lip, spread out, front view. - G: ovary and lip, side view. - H: pedicel with ovary and column, front view. - J: pedicel with ovary and column, side view. - K: anther-cap, back view. - L: anther-cap, front view. - M: pollinia. - N: ovary, transverse section. All drawn from material cultivated at *Tenom Orchid Centre* by C.L. Chan.

Flowers *c.* 3 cm long, *c.* 4–4.5 cm across, having a musty smell, sepals and petals pale green, whitish at base and distally, lip white, column white with mauve spotting and streaks along the sides. **Pedicel** and **ovary** 2–5 cm long, cylindrical, curved. **Sepals** and **petals** spreading, fleshy. **Dorsal sepal** 3 × 1 cm, oblong-elliptic, with a tuft of laciniate processes on inner surface. **Lateral sepals** 2.8–3 × 1 cm, oblong-elliptic, obtuse, with or without a smaller tuft of laciniate processes. **Petals** 3 × *c.* 2.3–2.4 cm, oblong-elliptic to spathulate, obtuse, carinate on reverse, glabrous. **Lip** 2.5–3 cm long, *c.* 3 cm wide at broadest point, subentire, adnate to column below forming a short tube, free portion broadly rounded and unevenly lobulate, crenulate-erose, shallowly emarginate; disc thickened, with a central crest of dense, deflexed, lacerate-fimbriate lamellae. **Column** 2 cm long, very narrowly winged, straight or slightly curved, sparsely hairy below rostellum; anther-cap *c.* 2 × 2 mm, quadrate, cucullate; pollinia granular. **Fruit** 11–15 × *c.* 1.5 cm, subcylindrical, pendulous. Plate 24D.

HABITAT AND ECOLOGY: Lowland and hill dipterocarp forest; secondary forest; swamp forest; sometimes on limestone. Alt. lowlands to *c.* 500 m. Flowering observed in June, July and September.

DISTRIBUTION IN BORNEO: SABAH: Tenom District, near Kallang Waterfall, Mendalom Forest Reserve; Mt. Kinabalu, near Nalumad. SARAWAK: Bau District, Bau Hills; Kuching District, Kuching, Santubong; Lundu District, Datu Protected Forest Reserve; Miri District, Baram.

GENERAL DISTRIBUTION: Endemic to Borneo.

NOTES: Although relatively few in number, identification of South East Asian *Vanilla* can be problematic. Flowering in the wild is often erratic, while collections in herbaria frequently bare fruit only, or are more often sterile. When the once fleshy flowers are present, they are often greatly shrunken, poorly preserved and consequently difficult to examine. For this reason type material is often less than perfect. Ideally, as with all orchids, the best results are obtained from fresh flowers placed into liquid preservative.

Both Ridley (1896) and Rolfe (1896) had earlier referred the subject of this figure, if rather reluctantly, to the widespread *V. albida* Blume. Rolfe, however, later considered it to be distinct enough on account of the longer inflorescences with more numerous, crowded bracts, and named it *V. havilandii*. In addition, it appears to lack the patch of thread-like processes present at the apex of the lip of *V. albida*.

The plant figured here originates from Tenom District in Sabah and seems to fit the circumscription of *V. havilandii*.

DERIVATION OF NAME: The generic name is derived from the Spanish *vainilla*, a little pod, in reference to the characteristic fruits. The specific epithet honours George D. Haviland (1857–1901), a surgeon and naturalist, who was Director of the Raffles Museum in Singapore in 1891–93. Subsequently he became the Sarawak medical officer and Curator of the Sarawak Museum around 1892.

98. VANILLA KINABALUENSIS Carr

Vanilla kinabaluensis *Carr* in Gard. Bull. Straits Settlem. 8: 176 (1935). Types: Borneo, Sabah, Mt. Kinabalu, Kadamaian River, near Minitinduk Gorge, *Carr* 3157 (syntype SING, isosyntype K); Mt. Kinabalu, Dallas, J. & M.S. *Clemens* 26300 (syntype BM, isosyntype SING) and J. & M.S. *Clemens* 26725 (syntype BM, isosyntypes K, SING).

Climber. Stem 8 m or more long, stout, up to 2 cm in diameter, internodes up to 10 cm or more long, rooting at nodes. **Leaf-blade** 32–35 (–42) × 6.5–10.75(–11) cm, oblong or oblong-elliptic, shortly acuminate, obtuse or subtruncate, thick, coriaceous, flaccid, margins rather thickened; sheath 1.5–2 cm long, sulcate, slightly twisted. **Inflorescences** arising from leaf axils, elongate, bearing many flowers in succession; peduncle 0.5–1 cm long; rachis 10–57 cm long, 1–1.2 cm in diameter, fleshy; floral bracts 4–6 × 4 mm, broadly ovate, subacute, fleshy, spreading. **Flowers** 6–8 cm in diameter, faintly scented, sepals and petals pale yellow or lemon-yellow, lip white, sometimes yellow, nerves sometimes reddish purple, with a pale yellow suffused red subapical blotch, lamellae white, apical papillae dark crimson to purple, column creamy white. **Pedicel** and **ovary** 3.5–4.5 cm long, stout. **Dorsal sepal** 6–6.2 × 1.7–2 cm, oblanceolate, obtuse, margins incurved, 9-nerved. **Lateral sepals** 5.3–5.6 × 2–2.15 cm, oblong-oblanceolate or obovate, falcate, subacute, margins incurved, dorsally carinate, 9-nerved, outer nerves branched above base. **Petals** 6–6.1 × 2.2–2.4 cm, narrowly obovate, somewhat sigmoid, obtuse, thickened inside along median line, dorsally carinate with a low, rounded, rather flattened keel, 9-nerved. **Lip** shallowly 3-lobed towards apex, 4.8 cm long, *c.* 3.85–4 cm wide across side lobes when flattened, cuneate, margins adnate to sides of column for entire length forming a tube; disc provided with a broad median, shortly 3-ribbed subapical keel and terminating below apex of the adnate part in a brush-like bundle of incurved cuneate lamellae, the whole bundle *c.* 4.2–4.8 × 6–8.5 mm, with rounded denticulate apex and several smaller similar lamellae in front, the lateral of which on either side being longer and tooth-like; side lobes *c.* 1.3–1.6 × 1.6–1.8 cm, semi-orbicular, margins strongly undulate; mid-lobe 0.5–0.7 × 1.4 cm, shortly and very broadly ovate, obtuse, margins strongly undulate, provided in middle with a large dense mass of branched papillae which is linked to the bundle of lamellae below by a short, broad, 5-ribbed keel. **Column** 4.25–4.4 cm long, *c.* 3.5 mm wide, its margins up to stigma adnate to blade of lip, straight, slightly curved at apex, with a few papillae on ventral surface; rostellum large, quadrate, apex covering stigma; anther-cap *c.* 3.5 × 3.3–3.4 mm, oblong, shortly retuse above, slightly recurved above middle. Plate 24E.

HABITAT AND ECOLOGY: Hill dipterocarp forest on sandstone and shale. Alt. 700 to 900 m. Flowering observed from February to April, June to September, November; probably throughout the year.

DISTRIBUTION IN BORNEO: SABAH: Mt. Kinabalu; Mt. Trus Madi; Tenom District, Paling Paling Hills; Crocker Range, above Sinsuron; Kimanis road.

GENERAL DISTRIBUTION: Peninsular Malaysia and Borneo.

NOTES: Eight species of *Vanilla* have been recorded from Borneo, three of which, viz. *V. abundiflora* J.J. Sm., *V. borneensis* Rolfe and *V. havilandii* Rolfe, are thought to be endemic. Carr

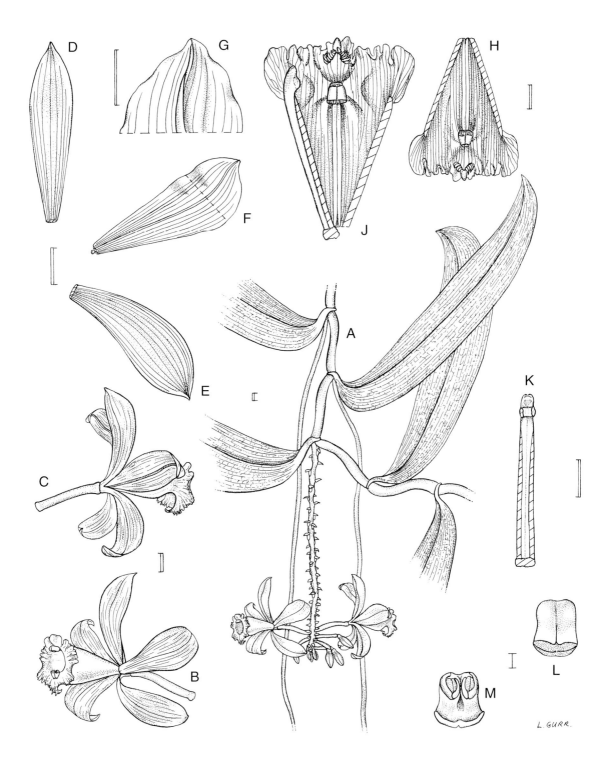

Figure 98. Vanilla kinabaluensis Carr. - A: habit. - B: flower, viewed from above. - C: flower, side view. - D: dorsal sepal. - E: lateral sepal. - F: petal. - G: apex of petal, abaxial surface, showing carinate mid-nerve. - H: lip, detached from column and spread out, front view. - J: column with lip partly detached and spread out, front view. - K: column, front view. - L: anther-cap, back view. - M: anther-cap, interior view. A & B drawn from miscellaneous specimens, C after sketches by *J.J.Wood & A. Lamb*, D–K from *Lamb* AL 1230/90 & *Wood* 909, and L–M from *Wood* 909 by Linda Gurr. Scale: single bar = 1 mm; double bar = 1 cm.

compared *V. kinabaluensis* with *V. abundiflora*, described from Kalimantan. *Vanilla abundiflora*, however, has smaller flowers with a less deeply three-lobed lip lacking a median keel, and distinctive branched appendages on the front surface of the column.

DERIVATION OF NAME: The specific epithet refers to Mt. Kinabalu, the type locality.

99. VENTRICULARIA BORNEENSIS J.J. Wood

Ventricularia borneensis *J.J. Wood* in Orchid Rev. 106 (1221): 175, figs. 97 & 98 (1998). Type: Borneo, Sabah, Sipitang District, Ulu Long Pa Sia, 8 km north west of Long Pa Sia, above Maga River, *c.* 1350 m, 24 October 1985, *Wood* 664 (holotype K).

Pendulous **epiphyte**. **Stem** up to *c.* 60 cm long, slender, simple, rooting from lower nodes, internodes 1–1.8 cm long. **Leaf-blade** 5.5–8 × 0.6–0.7 cm, linear-lanceolate, acute, somewhat fleshy-textured, coriaceous, dark olive-green; sheath 1.2–1.8 cm long, prominently nerved, rugulose (in dried material). **Inflorescences** 2-flowered, emerging from base of leaf sheath opposite blade, abbreviated; peduncle 2 mm long, fleshy; rachis 1–1.5 mm long; floral bracts 0.8–1 mm long, ovate, obtuse, scale-like, adpressed to base of pedicel. **Flowers** resupinate, pale lemon-yellow, lip with creamy white hairs, column palest lemon-yellow, creamy white distally, anther-cap white. **Pedicel** and **ovary** 5–6 mm long, narrowly clavate, gently curved. **Sepals** and **petals** spreading to reflexed, fleshy, adaxial surface minutely papillose, shortly hirsute proximally. **Dorsal sepal** 4–4.2 × 2 mm, oblong, obtuse to subacute, 3-nerved. **Lateral sepals** 4.5 mm long, 3 mm wide at base, 2.1 mm wide at middle, triangular-ovate, obtuse to subacute, with an obtuse basal auricle, 2-nerved. **Petals** 4.1 × 1.9 mm, oblong, subacute, 3-nerved. **Lip** immobile, 3-lobed, slightly thickened at base; side lobes *c.* 0.5 × 0.8 mm, oblong, rounded, erect, villous-hirsute on exterior and interior; mid-lobe 2.8–3 × 1 mm wide at base, narrowly triangular-lanceolate, subacute, glabrous, porrect to deflexed, apex sometimes slightly upcurved; spur 3.5–3.7 × 3–3.1 mm, gibbous, constricted towards entrance, dilated, saccate, obtuse, with villous hairs along interior walls. **Column** 1.1–1 × 1.8–1.9 mm, oblong, shortly hirsute; foot absent; rostellum elevate, carinate, acute; anther-cap *c.* 1.7 × 1.5 mm, cucullate, apiculate; viscidium minute; stipes *c.* 1 mm long, twice as long as diameter of pollinia, spathulate; pollinia 4, in 2 very unequal pairs. Plate 25A & B.

HABITAT AND ECOLOGY: Lowland forest; kerangas forest; lower montane *Lithocarpus/Castanopsis* ridge forest. Alt. 200 – *c.* 1350 m. Flowering observed in October.

DISTRIBUTION IN BORNEO: SABAH: Sipitang District, Maga River area and Long Pa Sia; Tawau District, Luasong. SARAWAK: Kelabit Highlands, Bario area.

GENERAL DISTRIBUTION: Endemic to Borneo.

NOTES: Vegetatively, *Ventricularia* recalls certain *Trichoglottis,* particularly *T. lanceolaria* Blume, but the floral morphology is quite different. In particular, the usually hairy tongue at the base of the lip which is characteristic of *Trichoglottis,* is absent in *Ventricularia.* The elevate,

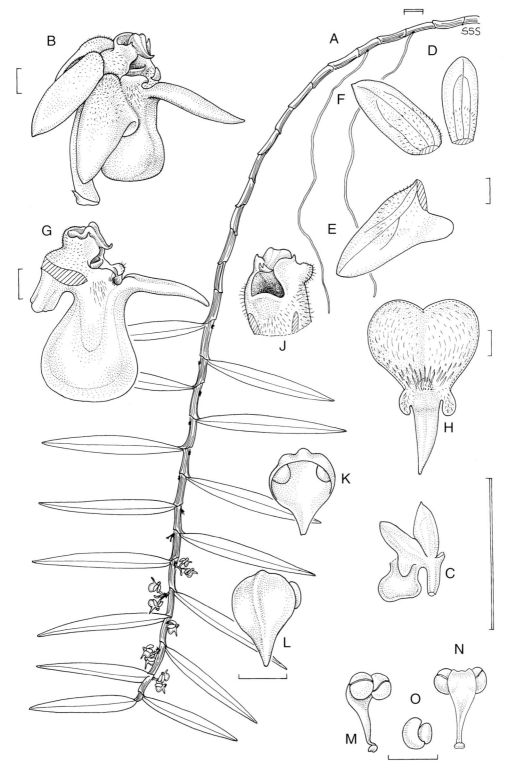

Figure 99. Ventricularia borneensis J.J. Wood. - A: habit. - B: flower, side view. - C: pedicel with ovary, dorsal sepal, lateral sepal and spur, side view. - D: dorsal sepal. - E: lateral sepal. - F: petal. - G: pedicel with ovary, lip and column, side view. - H: lip from above showing transverse section through spur. - J: column, anther-cap removed, oblique view. - K: anther-cap, interior view. - L: anther-cap, back view. - M: pollinarium, oblique view. - N: pollinarium, back view. - O: two pollinia. All drawn from *Wood* 664 (holotype) by Susanna Stuart-Smith. Scale: single bar = 1 mm; double bar = 1 cm.

keeled rostellum is similar to that of *Sarcoglyphis*, while the tiny viscidium and spathulate stipes, which is distally broadened below the pollinia, recall those of *Malleola*.

The recently described *V. borneensis* extends the range of the genus eastward. Seidenfaden (1988) records the only other species, *V. tenuicaulis* (Hook.f.) Garay, as being quite common in Peninsular Malaysia and southern Thailand where it extends as far as about latitude 16° north.

DERIVATION OF NAME: The generic name is derived from the Latin *ventriculus*, meaning a belly, and refers to the inflated, belly-shaped spur of the lip.

100. VRYDAGZYNEA GRANDIS Ames & C. Schweinf.

Vrydagzynea grandis *Ames & C. Schweinf.*, Orchidaceae 6: 16 (1920). Type: Borneo, Sabah, Mt. Kinabalu, Kiau, *J. Clemens* 340 (holotype AMES, isotype K).

Terrestrial. **Rhizome** elongate, *c.* 5 mm in diameter, nodes prominent, internodes 1.6–3 cm long. **Roots** few, borne from nodes, fibrous, densely brown-lanuginose, simple or with a few branches. **Stem** 19–28 cm long, fleshy. **Leaves** 5 or 6 grouped mostly about middle of stem; blade 6.5–10.5(–11) × (2.5–)4–5 cm, ovate to broadly elliptic, sometimes oblique, acute, cuneate at base, chartaceous, main nerves mostly 5, lowermost blades often distant and smaller, uppermost reduced and bract-like; petiole sheathing, winged, 1.5–3.5 cm long, infundibuliform, adpressed. **Inflorescence** densely many-flowered, *c.* 2 cm wide; peduncle 2–10 cm long; rachis 3–6.5 cm long; floral bracts 8–9 × 2.6–2.8 mm, linear-oblong, narrowed distally, obtuse to acute, 1-nerved, longitudinally sulcate. **Flowers** not opening widely, resupinate, ovary white tipped green, sepals green tipped white, petals whitish, lip white. **Pedicel** and **ovary** 1 cm long (pedicel 2 mm long). **Sepals** and **petals** connivent. **Dorsal sepal** *c.* 5.4–6 × 2.1–2.5 mm, narrowly elliptic, narrowed to a very broadly rounded, fleshy, thickened apex, 1-nerved, forming a hood with the petals. **Lateral sepals** 5.5–6 × 2–2.5 mm, similarly shaped. **Petals** *c.* 5 × 2 mm, ligulate to narrowly elliptic, rather abruptly narrowed to a broadly rounded apex. **Lip** spurred; blade 3–3.8 × 2.4–2.8 mm, deltoid, lingulate, obtuse, lateral margins involute, finely papillose, with 2 short lamelliform basal appendages; spur 5–7 mm long, 2.5–3 mm in diameter at middle, cylindrical, somewhat inflated in front, interior with 2 small ovate-oblong stalked glands, stalks running up towards column. **Column** 2.2 mm long, thick, slightly dilated above, raised into a rostellar thickening at each corner; stigma bilobed; pollina 2. Plate 25C.

HABITAT AND ECOLOGY: Hill and lower montane forest. Alt. 900 to 1500 m. Flowering observed from January until March, and November.

DISTRIBUTION IN BORNEO: SABAH: Mt. Kinabalu.

GENERAL DISTRIBUTION: Endemic to Borneo.

NOTES: *Vrydagzynea* is widely distributed from northern India to the Pacific islands, with 15 species recorded from Borneo. The inflorescence is usually rather short and densely many-

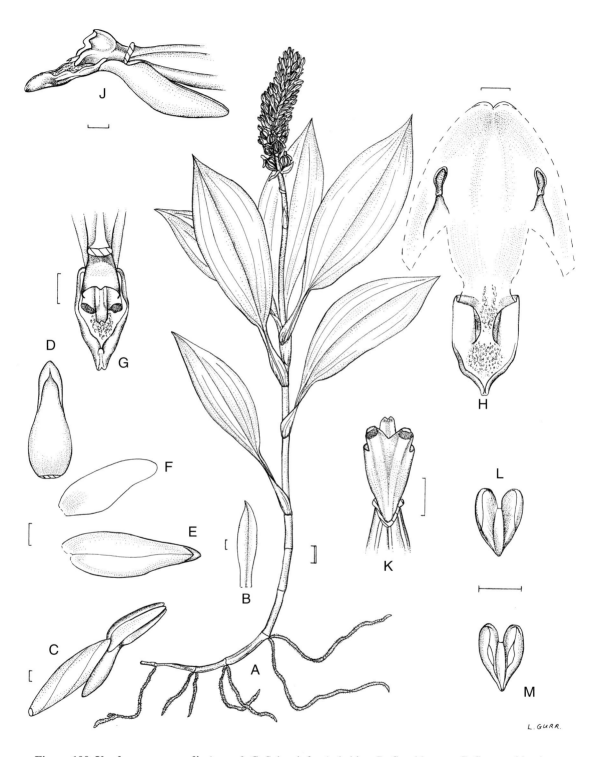

Figure 100. Vrydagzynea grandis Ames & C. Schweinf. - A: habit. - B: floral bract. - C: flower, side view. - D: dorsal sepal. - E: lateral sepal. - F: petal. - G: portion of ovary and lip, front view. - H: lip with spur opened out to show stalked glands. - J: pedicel with ovary, lip and column, side view. - K: column, front view. - L: anther-cap, back view. - M: anther-cap, interior view. A drawn from *Carr* 3195 & *J. & M.S. Clemens* 28344, B from *J. & M.S. Clemens* 28344, and C–M from *Lamb* AL 51/83 by Linda Gurr. Scale: single bar = 1 mm; double bar = 1 cm.

flowered. The small flowers seldom open widely and are probably frequently autogamous, judging from the high rate of capsule development in many species. The spur of the lip projects between the lateral sepals and always contains two stalked glands.

DERIVATION OF NAME: The generic name commemorates the Dutch pharmacologist Theodore Daniel Vrydag Zynen, a contemporary of Blume who, through his writings, brought to public attention many new medicines of vegetable origin. The Latin specific epithet *grandis*, meaning large, refers to the habit of this species.

REFERENCES

Errata and Nomenclatural Changes in Volumes One, Two and Three

Christenson, E.A. (2001). *Phalaenopsis. A Monograph.* Timber Press, Oregon.

Christenson, E.A. & Whitten, M.W. (1995). *Phalaenopsis bellina* (Rchb.f.) Christenson, a segregate from *P. violacea* Witte (Orchidaceae : Aeridinae). *Brittonia* 47: 57–60.

Kaiser, R. (1993). *The scent of orchids—olfactory and chemical investigations.* Givaudan-Roure, Editiones Roche, Basel, Switzerland.

O'Byrne, P. (1998). Three rare black-haired Dendrobiums from Indonesia. *Malayan Orchid Review* 32: 72–76 & 80–83.

Szlachetko, D.L. (1995). *Systema Orchidalium.* W. Szafer Institute of Botany, Polish Academy of Sciences, Kraków.

Vermeulen, J.J. (2002). A Taxonomic Revision of *Bulbophyllum* (Orchidaceae) 2. Sections *Altisceptrum* and *Hirtula. Gard. Bull. Sing.* 54: 1–151.

Wood, J.J. (2001). *Dendrochilum of Borneo.* Natural History Publications (Borneo) in association with The Royal Botanic Gardens, Kew.

Chapter 1—Revised classification system of Bornean orchids

Dressler, R.L. (1990). The major clades of the Orchidaceae—Epidendroideae. *Lindleyana* 5: 117–125.

Dressler, R.L. (1993). *Phylogeny and Classification of the Orchid Family.* Cambridge Univ. Press.

Pridgeon, A.M., Cribb, P.J., Chase, M.W. & Rasmussen, F.N., eds. (1999). *Genera Orchidacearum Volume 1 General Introduction, Apostasioideae, Cypripedioideae.* Oxford University Press.

Chapter 2—Revised Keys to Genera

Garay, L.A., Hamer, F. & Siegerist, E.S. (1994). The genus *Cirrhopetalum* and the genera of the *Bulbophyllum* alliance. *Nord. J. Bot.* 14, 6: 609–646.

Siegerist, E.S. (2001). *Bulbophyllum and Their Allies. A Grower's Guide.* Timber Press, Portland, Oregon, USA.

Chapter 3—Descriptions and Figures

Ames, O. (1921). Orchidaceae, pp. 134–204 in Merrill, E.D. A bibliographic enumeration of Bornean plants. *J. Straits Branch Roy. Asiat. Soc.* Special number.

Ames, O. & Schweinfurth, C. (1920). *The Orchids of Mount Kinabalu, British North Borneo*. Merrymount Press, Boston.

Barkman, T.J. & Wood, J.J. (1997). Reinstatement of *Pholidota sigmatochilus*: an enigmatic species from Mt. Kinabalu, Sabah, Malaysia. *Lindleyana* 12, 3: 153–157.

Burkill, I.H. (1923). A new Malayan orchid. *Dendrobium citrino-castaneum. Gard. Bull. Straits Settlem.* 3: 12.

Carr, C.E. (1932). Some Malayan Orchids III. *Gard. Bull. Straits Settlem.* 7, 1: 1–60.

Carr, C.E. (1935). Some Malayan Orchids V. *Gard. Bull. Straits Settlem.* 8: 69–129.

Comber, J.B. (2001). *Orchids of Sumatra*. Natural History Publications (Borneo) in association with the Royal Botanic Gardens, Kew and Singapore Botanic Gardens.

Garay, L.A. (1972). On the systematics of the monopodial orchids I. *Bot. Mus. Leafl.* 23, 4: 149–212.

Green, E. (2001). Antique's giant *Spathoglottis. Orchid Digest* 65 (2) :88–90.

Hazebroek, H.P. & Abang Kashim bin Abang Morshidi (2001). *National Parks of Sarawak*. Natural History Publications (Borneo), Kota Kinabalu.

Holttum, R.E. (1957). *A Revised Flora of Malaya. An illustrated systematic account of the Malayan flora, including commonly cultivated plants. Vol. 1. Orchids of Malaya*. 2nd ed. Government Printing Office, Singapore.

Holttum, R.E. (1964). *A Revised Flora of Malaya. An illustrated systematic account of the Malayan flora, including commonly cultivated plants. Vol. 1. Orchids of Malaya*. 3rd ed. Botanic Gardens, Singapore.

Kraenzlin, F. (1910). Orchidaceae-Monandrae-Dendrobiinae—Pars 1. Genera n. 275–277. In Engler, A. *Pflanzenr*. iv. 50. II. B. 21: 1–382.

Kraenzlin, F. (1911). Orchidaceae-Monandrae-Dendrobiinae—Pars 2. Genera n. 278–280 (Erieae). In Engler, A. *Pflanzenr*. iv. 50 II. B. 21: 1–182.

Kruizinga, J., Scheindelen, H.J. van, & de Vogel, E.F. (1997). Revision of the genus *Bromheadia* (Orchidaceae). *Orchid Monogr.* 8: 79–118, figs. 29–55, pl. 4b–5b.

Lindley, J. (1830–40). *The Genera and Species of Orchidaceous Plants*. Ridgways, Piccadilly, London.

O'Byrne, P. (1998). New and unusual Aeridinae from Borneo and Sulawesi. *Malayan Orchid Review* 32: 47–54 & 65–70.

O'Byrne, P. (1998). Three rare black-haired Dendrobiums from Indonesia. *Malayan Orchid Review* 32: 72–76 & 80–83.

O'Byrne, P. (2000). The Borneo form of *Dendrobium dearei*. *Malayan Orchid Review* 34 : 59-60.

O'Byrne, P. (2001). *A to Z of South East Asian Orchid Species*. Orchid Society of South East Asia, Singapore.

Ormerod, P. (1995). A reinstatement of *Rhomboda* Lindl. (Orchidaceae subtribe Goodyerinae). *Orchadian* 11 (7): 323–339.

Reichenbach, H.G. (1885). *Dendrobium parthenium*, n. sp. *Gardeners' Chronicle*, ser. 2, 24: 489.

Ridley, H.N. (1896). An enumeration of all Orchideae hitherto recorded from Borneo. *J. Linn. Soc., Bot.* 31: 261–306, 3 pl.

Ridley, H.N. (1907). *Materials for a Flora of the Malayan Peninsula*. Part I: 7–233.

Ridley, H.N. (1908). New or Rare Malayan Plants. Series III. *J. Straits Branch Roy. Asiat. Soc.* 49: 11–52.

Rolfe, R.A. (1894). New Orchids. Decade 8. *Kew Bull.* 1894: 154–159.

Rolfe, R.A.(1896). A Revision of the Genus *Vanilla. J. Linn. Soc., Bot.* 32: 439–478.

Rolfe, R.A. (1914). In Gibbs, L.S. A contribution to the flora and plant formations of Mount Kinabalu and the highlands of British North Borneo. *J. Linn. Soc., Bot.* 42: 1–240, 8 pl.

Seidenfaden, G. (1968). The genus *Oberonia* in mainland Asia. *Dansk. Bot. Arkiv* 25, 3: 1–125.

Seidenfaden, G. (1979). Orchid Genera in Thailand VIII. *Bulbophyllum* Thou. *Dansk. Bot. Arkiv* 33, 3: 1–228.

Seidenfaden, G. (1984). A note on the section *Cymboglossum* Schltr. of *Eria* (Orchidaceae). *Nordic J. Bot.* 4: 39–45.

Seidenfaden, G. (1988). Orchid Genera in Thailand, XIV. Fifty-nine vandoid genera. *Opera Bot.* 95: 1–398.

Seidenfaden, G. & Wood, J.J. (1992). *The Orchids of Peninsular Malaysia and Singapore*. Olsen & Olsen, Fredensborg, Denmark.

Smith, J.J. (1920). Orchidaceae Novae Malayenses IX. *Bull. Jard. Bot. Buitenzorg*, ser. 3, 2: 15–127.

Smith, J.J. (1930). Icones Orchidacearum Malayensium I. *Bull. Jard. Bot. Buitenzorg*, supplement, 2, 1–2, tabulae 1–50.

Smith, J.J. (1934). Icones Orchidacearum Malayensium I. *Bull. Jard. Bot. Buitenzorg, supplement,* 2, 3–4, tabulae 51–100.

Smith, J.J. (1943). Orchidaceae Novae Malayenses XVII. *Blumea* 5: 297–315.

Sweet, H.R. (1980). *The genus Phalaenopsis.* The Orchid Digest Inc.

Turner, H. (1992). A revision of the orchid genera *Ania* Lindley, *Hancockia* Rolfe, *Mischobulbum* Schltr. and *Tainia* Blume. *Orchid Monogr.* 6: 43–100, figs. 25–55, pl. 5b–7d.

de Vogel, E.F. (1986). Revisions in Coelogyninae (Orchidaceae) II. The genera *Bracisepalum, Chelonistele, Entomophobia, Geesinkorchis* and *Nabaluia. Orchid Monogr.* 1: 17–51, figs. 8–27, pl. 2b–4d.

de Vogel, E.F. (1988). Revisions in Coelogyninae (Orchidaceae) III. The genus *Pholidota. Orchid Monogr.* 3: 1–118, 31 figs., 6 pl.

Wood, J.J. (2001). *Dendrochilum of Borneo.* Natural History Publications (Borneo) Kota Kinabalu in association with The Royal Botanic Gardens, Kew.

Wood, J.J., Beaman, R.S. & Beaman, J.H. (1993). *The Plants of Mount Kinabalu 2. Orchids.* The Royal Botanic Gardens, Kew.

Wood, J.J. & Cribb, P.J. (1994). *A Checklist of the Orchids of Borneo.* Royal Botanic Gardens, Kew.

Wood, J.J. & Ormerod, P. (1998). Notes on *Habenaria* Willd. and *Peristylus* Blume. *Orchid Rev.* 106 (1222): 238–239.

IDENTIFICATION LIST

This list is based on a selection of specimens. Each species is arranged according to the descriptive part of the volume. Collections are cited, under each locality, in alphabetical order using the collector's name. A few sight records are included where no collection exists. Herbaria abbreviations are provided in brackets.

1. **Abdominea minimiflora**: SABAH: Kota Belud District, canyon of Penataran River south of Melangkap Tamis, NW side of Mt. Kinabalu, 450–500 m, *Beaman* 8856 (K). Batu Urun, 300–360 m, *Lamb* AL 368/85 (K). Lamag District, Tekala, *Majawat* SAN 90801 (K, SAN).

2. **Agrostophyllum glumaceum**: BRUNEI: Temburong District, Kuala Belalong area, 75 m, *Argent et al.* 9153 (E, K), & Wong WKM 279 (BRUN, K). KALIMANTAN (province unknown): Mt. Ilas Bungaan, 600 m, *Kostermans* s.n. (BO, K, L). SABAH: Mt. Kinabalu, kampung Nalumad, *Andau* 777 (K, Sabah Parks Herbarium, Kinabalu Park); Mt. Kinabalu, Dallas, 900 m, *J. & M.S. Clemens* 27481 (K) & 27745 (K). Tongod District, Ulu Menanam, 150 m, *Dewol & Kodoh* SAN 88745 (K, L, SAN, SAR). Pun Batu, 600 m, *Vermeulen* 447 (K, L, UKMS). SARAWAK: Marudi District, Mt. Api, *Anderson* J.A.R.S. 30765 (AMES, K, L, SAR). Mukah District, Iju Hill, Balingian, 20 m, *Jugah ak. Kudi* S. 23784 (K, L, SAR). Limbang District, Mt. Mulu/Rob. Anderson Camp, *Nielsen* 648 (AAU). Serian District, Tebakang/Majang Hill, 200 m, *de Vogel, Yii & Kessler* 8815 (L). Kapit District, Batu Tiban, 900 m, *Yii et al.* S. 51605 (K, L, SAR).

3. **Ania borneensis**: SABAH: Mt. Kinabalu, Dallas, 900 m, *J. & M.S. Clemens* 30125 (K). Mt. Kinabalu, Kiau, 1200 m, *Lamb* AL 275/84 (K). Tambunan District, Crocker Range, Sinsuron road, *Lamb* AL 321/85 (K).

4. **Ania ponggolensis**: SABAH: Pensiangan District, Batu Ponggol, *c.* 900 m, *C.L. Chan* s.n. (K).

5. **Aphyllorchis montana**: KALIMANTAN BARAT/KALIMANTAN TENGAH border: Mt. Raya, Njarumkop, 700 m, *Elsener* 31 (BO, K). SABAH: Mt. Kinabalu, Eastern Shoulder, 750 m, *Chew et al.* 214 (K). Tenom District, Mandalom Forest Reserve, 750 m, *Lamb* AL 1400/92 (K). SARAWAK: Kuching District, Mt. Matang, *Beccari* 3648 (FI). Lawas District, Ba Kelalan, 1200 m, *Brooke* W. 10533 (L). Belaga District, Dema Hill, *Burtt* B. 11317 (E, SAR). Marudi District, Melinau Paku River, *Hansen* 110 (K). Lundu District, Lundu, *Mjöberg* s.n. (AMES). Bintulu District, Merurong Plateau, Jelalong River, 300 m, *Othman et al.* S. 48805 (K, SAR).

6. **Appendicula bilobulata**: SARAWAK: Known only from the type.

7. **Appendicula calcarata**: SABAH: Mt. Kinabalu, near Bundu Tuhan, 750 m, *Carr* SFN 27882 (SING). Mt. Kinabalu, Dallas, 900 m, *J. & M.S. Clemens* 26520 (BM, K). SARAWAK: Marudi District, Tama Abu Range, 1300 m, *Dayang Awa & Bernard Lee* S. 51186 (K, L, SAR).

8. **Appendicula congesta**: SABAH: Mt. Kinabalu, West Mesilau River, 1600-1700 m, *Beaman* 7459 (K), 7526 (K) & 9035 (K). Mt. Kinabalu, Kilembun River Head, 1400 m, *J. & M.S. Clemens* 32433 (BM, K). Mt. Kinabalu, Muru-tura Ridge, 1500 m, *J. & M.S. Clemens* 34362 (BM). Mt. Lumaku, 1200 m, *J.B. Comber* 118 (K). Mt. Kinabalu, Penibukan, 1800 m, *Gibbs* 4058 (BM).

9. **Appendicula foliosa**: SABAH: Mt. Kinabalu, East Mesilau River between golf course site and Mesilau Cave, 1700–1900 m, *Beaman* 8003 (K). Mt. Kinabalu, Sayap, 800–1000 m,

Beaman 9782 (K). Mt. Kinabalu, Tenompok, 1500 m, *J. & M.S. Clemens* 28736 (BM, K) & 30169 (K). Lahad Datu District, Pangaruan River, Takun, *Saigol* SAN 93105 (K, L, SAN). SARAWAK: Lundu District, Mt. Gading, 200 m, *Purseglove & Shah* P. 4533 (K, L, SAR). Limbang District, Batu Buli, 1600–1800 m, *Vogel, Schuiteman & Roelfsema* LC 980111 (L). Marudi District, Mt. Murud, 1600 m, *Yii* S. 44421 (K, L, SAR).

10. **Appendicula pilosa**: KALIMANTAN TIMUR: Cehan (Tjehan) River, *Nieuwenhuis* s.n. (BO). SARAWAK: Kuching District, Matang Road, *Mjöberg* s.n. (AMES). Sri Aman District, Entalang River, Sekarang, *Ilias Paie* S. 45190 (K, L, SAR). Serian District, Mt. Ampungan, 700 m, *Roefsema, Schuiteman & Vogel* LC 970036 (L). Kapit District, Hose Mountains, Batu Hill, 1200 m, *de Vogel* LC 913920 (L). Lundu District, Lundu, *Yong* 249 (K).

11. **Appendicula rostellata**: KALIMANTAN TENGAH: Liangangang, *Hallier* s.n. (BO). SABAH: Mt. Kinabalu, near Bundu Tuhan, 900 m, *Carr* SFN 27803 (AMES). Lamag District, Namatoi River, 390 m, *Madani* SAN 88849 (K, L, SAN).

12. **Ascidieria longifolia**: BRUNEI: Mt. Retak, Wong, WKM 846 (BRUN, K). SABAH: Tambunan District, Crocker Range, Kota Kinabalu to Tambunan road, 1400 m, *Beaman* 7363 (K) & *Beaman* 10492 (K). Mt. Kinabalu, golf course site, 1700–1800 m, *Beaman* 10674 (K). Mt. Kinabalu, Tenompok, 1500 m, *Carr* SFN 27049 (SING). Mt. Kinabalu, Penataran River, 1700 m, *J. & M.S. Clemens* 34441 (BM, K). Mt. Tembuyuken, 1860 m, *Gibot* SAN 68555 (K, SAN). Mt Kinabalu, Kundasang, 1400 m, *Kidman Cox* 2517 (K). SARAWAK: Marudi District, Mt. Api, 1500 m, *Argent & Jermy* 985 (SAR). Marudi District, Mt. Mulu, *Burtt & Woods* B. 2091 (E). Belaga District, Semawat River, Upper Belaga River, 300 m, *Hansen* 826 (L, SAR). Marudi District, Mt. Murud, summit, 1900–2400 m, *Mjöberg* 26 (AMES) & *Mjöberg* 35 (AMES). Marudi District, Mt. Murud, 1800 m, *Nooteboom & Chai* 1947 (L). Lundu District, Mt. Pueh, 1100 m, *Purseglove & Shah* P. 4719 (K, SING), & 1200 m, *Purseglove & Shah* P. 4740 (AMES, K, L, SAR, SING). Marudi District, Bario/Pa Berang, 1000 m, *Roelfsema, Schuiteman & Vogel* LC 970178 (L). Marudi District, Apad Keruma/Kelabit Highlands, 1500 m, *Yii* S. 55929 (K, L, SAR, SING).

13. **Brachypeza indusiata**: SABAH: Tenom District, Crocker Range, Tenom Valley, 200–300 m, *Lamb* AL 367/85 (K).

14. **Bromheadia borneensis** var. **borneensis**: BRUNEI: Belait District, Ulu Ingei, 170 m, *Boyce* 299 (BRUN, K). Tutong District, Ulu Tutong, 300–400 m, *Johns* 7618 (BRUN, K). KALIMANTAN TIMUR: Mt. Samonggaris, *Amdjah* 980 (BO, K, L). SABAH: Labuk & Sugut District, Tawai Hills, 300–500 m, *Beaman* 10268 (K). Telupid, near Tangkunan Hill, *Gibot* SAN 93991 (K, SAN). Near Nabawan, *c.* 600 m, *Lamb* SAN 93455 (K, SAN). Sandakan District, Kapur River, *Meijer* SAN 22899 (K, SAN). Sandakan District, Mt. Tawai, 700 m, *Vermeulen* 816 (K, L). SARAWAK: Kuching District, Matang, *Beccari* 1675 (FI). Bintulu District, Niah-Jelalong Protected Forest, *Brunig* S. 962 (SAR). Miri District, Lambir Hills, 200 m, *Burtt* B. 11669 (E, SAR). Lundu District, Biawak, *Munting* S. 56380 (L). Marudi District, Mt. Mulu/Tapin River, 40 m, *Nielsen* 278 (K).

14a. **Bromheadia borneensis var. longiflora**: SARAWAK: Mt. Pueh, *Native collector* 5026 (AMES).

15. **Bromheadia divaricata**: SABAH: Mt. Kinabalu, Summit Trail near Pondok Lowi, 2100–2300 m, *Beaman* 11770 (K). Mt. Kinabalu, Summit Trail between Power Station and Kemburongoh, 1800–2100 m, *Beaman* 11942 (K). Mt. Kinabalu, Kemburongoh, 2200 m, *Carr* SFN 27467 (SING). Mt. Kinabalu, Marai Parai Spur, 2100 m, *J. & M.S. Clemens*

32887 (BM).

16. **Bromheadia graminea**: KALIMANTAN BARAT: Serawai, Merah River, 700 m, *Church et al.* 1800 (AMES, BO, K). SARAWAK: Belaga District, Usun Apau Plateau, 900 m, *Lai et al.* S. 68976 (L). Kapit District, Melatai Hill, 900 m, *Yii* S. 48472 (K, SAR). Kuching District, Mt. Penrissen, 800 m, *Schuiteman, Mulder & Vogel* LC 933152 (L).

17. **Bulbophyllum coriaceum**: SABAH: Mt. Kinabalu, Panar Laban, 3500 m, *Beaman* 8303 (K). Mt. Kinabalu, Lumu-Lumu, 1800 m, *Carr* SFN 27833A (K, SING). Mt. Kinabalu, Mesilau Cave, 1800 m, *Chew & Corner* RNSB 4784 (K, SING). Mt. Kinabalu, Minetuhan Spur, 1400 m, *J. & M.S. Clemens* 34194 (BM). Mt. Kinabalu, Pinosuk Plateau, 1500–1700 m, *Vermeulen* 472 (K, L).

18. **Bulbophyllum lemniscatoides**: KALIMANTAN TIMUR: Sangkulirang, *de Vogel et al.* 992 (K).

19. **Calanthe otuhanica**: SABAH: Known only from the type.

20. **Cheirostylis spathulata**: SABAH: Mt. Sidungol, 30–40 m, *Keith* 9331 (K).

21. **Cleisostoma suaveolens**: SARAWAK: Marudi District, Mt. Mulu/Deer Cave, *Lai & Jugah* S. 44278 (SAR). Marudi District, Mt. Mulu, 200 m, *Lamb* s.n. (K). Bau District, Bungo Range, 200–300 m, *de Vogel* LC 27602 (L, SAR).

22. **Cordiglottis fulgens**: SABAH: Nabawan area, cult. *Tenom Orchid Centre* (TOC).

23. **Cordiglottis westenenkii***: KALIMANTAN TIMUR: Mt. Sungai Pendan, 900 m, *de Vogel & Cribb* 9197 (K).

24. **Corybas serpentinus**: SABAH: Mt Tawai, *Lamb* (sight record).

25. **Cymbidium kinabaluense**: Known only from the type.

26. **Cystorchis macrophysa**: SARAWAK: Kuching District, Mt. Matang, 600 m, *Anderson*, J.A.R.S. 25104 (SAR). Marudi District, Melinau Gorge, *Burtt & Woods* B. 2318 (E, K). Marudi District, Mt. Mulu, 500 m, *Hansen* 341 (K). Marudi District, Mt. Mulu, 30 m, *Lamb & Collenette* in *Lamb* AL 8 (K). Marudi District, Mt. Matang, 600 m, *Stevens et al.* 266 (AMES, SAR).

27. **Cystorchis salmoneus**: SABAH: Known only from the type.

28. **Cystorchis saprophytica**: KALIMANTAN BARAT: Tilung Hill, 750 m, *Winkler* 1487 (HBG). SABAH: Mt. Lotong, Maliau Basin, *c.* 1100 m, *Lamb & Argent* SAN 92202 (K, SAN). Sipitang District, Long Pa Sia to Meligan trail, 900 m, *Lamb* in *Wood* 735 (K). SARAWAK: Marudi District, Sekelun Hill, 1400 m, *Beaman* 11507 (UNIMAS). Lawas District, Batu Buli, 1200–1400 m, *Lamb* ALSAR 220/85 (K).

29. **Dendrobium aurantiflammeum**: BRUNEI: Temburong District, Belalong Hill, 850 m, *Prance* 30687 (BRUN, K). SABAH: Crocker Range, Kimanis road, 1400 m, *Dransfield* JD 5528 (K). Crocker Range, Mt. Alab, 900 m, *Lamb* SAN 88575 (SAN). Mt. Kinabalu, Langanan River, *Lohok* 14 (K). SARAWAK: Belaga District, Mt. Dulit, Ulu Koyan, 1100 m, *Richards* in *Synge* S. 481 (K).

30. **Dendrobium bifarium**: KALIMANTAN SELATAN: Banjarmasin, *Motley* s.n. (K). SABAH: Mt. Kinabalu, West Mesilau River, 1600 m, *Beaman* 9036a (K). Mt. Kinabalu, Golf Course Site, 1700 m, *Beaman* 8484 (K) & 1700–1800 m, *Beaman* 10670 (K). Mt. Kinabalu, Marai Parai, 1500 m, *J. & M.S. Clemens* 32310 (BM, K). Meliau River, Labuk, 30 m, *Collenette* 513 (K). Mt. Trus Madi, 1800 m, *J.B. Comber* 133 (K). Mt. Kinabalu, Pig Hill, 1650–1950 m, *Sutton* 18 (K). Mt. Napotong, 1000–1400, *Vermeulen* 463 (K, L). Mt. Kinabalu, Pinosuk Plateau, near Kundassang, 1500 1700 m, *Vermeulen* 490 (K, L). SARAWAK: Sri Aman District, Triso Protected Forest, *Anderson*, J.A.R.S. 12208 (SAR).

Kapit District, Batu Tiban, 1400 m, *Mjöberg* s.n. (AMES). Marudi District, Dulit/Long Kapa, 600 m, *Shackleton* S. 78 (K). Bau District, Buso, *Sinclair* SFN 38481 (E, K, SING).

31. **Dendrobium cinnabarinum**: SABAH: Kinabatangan District, Maliau Basin, 900 m, *Lamb* AL 912/88 (K), & 1000–1100 m, *Lamb* AL 1422/92 (K). Sipitang District, trail from Batu to Angol, south of Meligan, 1400 m, *Wood* 743 (K). Sipitang District, Mt. Lumaku, 1300 m, *Wood* 812 (K). SARAWAK: Marudi District, Mt. Api, 1200 m, *Argent & Jermy* 964 (E, K, SAR). Marudi District, Bario/Pa Umor, 1100 m, *Beaman* 11245 (UNIMAS). Bintulu District, Merurong Plateau, 800 m, *Brunig* S. 975 (SAR) & *Brunig* S. 17475 (SAR). Limbang District, Mt. Buda, 900–1000 m, *Chai* S. 39878 (K, SAR). Marudi District, Mt. Dulit, 1900 m, *Hartley* S. 518 (K), & 1200–1500 m, *Mjöberg* 21 (AMES).

32. **Dendrobium dearei**: SABAH: Dent Peninsula, Tambisan, sea level, *Lamb* AL 1375/91 (K).

33. **Dendrobium derryi**: SABAH: Crocker Range, *Lamb* SAN 91588 (SAN). SARAWAK: Kuching District, Mt Serapi, 700 m, *Schuiteman, Mulder & Vogel* LC 932556 (L). Marudi District, Bario/Pa Berang, 1000 m, *Roelfsema, Schuiteman & Vogel* LC 970154 (L). Kapit District, Hose Mountains, Batu Hill, 1600–1700 m, *de Vogel* LC 913760 (SAR), 1300–1600, *de Vogel* LC 913816 (L), 1200 m, *de Vogel* 913939 (L), & 800 m, *de Vogel* LC 914207 (L).

34. **Dendrobium hallieri**: KALIMANTAN BARAT: Kelam Hill, *Hallier* 2313 (BO, K). KALIMANTAN TENGAH: East of Sampit, sea level, *Alston* 13205 (BM, BO, K).

35. **Dendrobium hendersonii**: KALIMANTAN SELATAN: Banjarmasin, *Motley* 1013 (K). SABAH: Tenom District, Tenom Valley, *Lamb* AL 370/85 (K). Tenom District, Padas Gorge, 200 m, *Lamb* AL 1112/89 (K). SARAWAK: Bau District, Bidi, *Brooks* s.n. (SAR). Kuching District, Kuching, *Hewitt* s.n. (SAR). Limbang District, Buyo Hill, *Moulton* 7 (SAR). Bau District, Bau/Krokong, 200 m, *Schuiteman, Mulder & Vogel* LC 932602 (L). Belaga District, Belaga, *Yong* 177 (K).

36. **Dendrobium kurashigei**: KALIMANTAN TIMUR: Apo Kayan, Mt. Sungai Pendan, 1300–1500 m, *de Vogel & Cribb* 9519 (K) & *de Vogel & Cribb* 9177 (K). SABAH: Mt. Kinabalu, Pinosuk Plateau, 1720 m, *Beaman* 8501 (K). Mt. Kinabalu, Menteki Ridge, 1900–2100 m, *Beaman* 11189 (K). Mt. Kinabalu, West Mesilau River, 1700 m, *Brentnall* 123 (K). Sipitang District, Ulu Long Pa Sia, 1400 m, *Wood* 648 (K). Crocker Range, Mt. Alab, 1800 m, *Wood* 776 (K). Mt. Kinabalu, between East Mesilau and Menteki Rivers, 1650 m, *Wood* 832 (K). SARAWAK: Marudi District, Mt. Mulu, 1200 m, *Burtt & Woods* B. 2195 (E). Limbang District, Batu Buli, 1500 m, *Vogel, Schuiteman & Roelfsema* LC 980059 (L).

37. **Dendrobium lamrianum**: SABAH: Mt. Kinabalu, Kiau View Trail, 1650 m, *Lamb & Phillipps* in *Lamb* AL 161/83 (K).

38. **Dendrobium lawiense**: SARAWAK: Limbang District, Mt. Batu Lawi, *Moulton* 8 (BO) & 9 (BO). Marudi District, Mt. Dulit, Dulit Ridge, 1300 m, *Synge* S. 473 (K, SING).

39. **Dendrobium metachilinum**: SABAH: Sandakan District, Telupid area, 200 m, *Lamb* AL 1126/89 (K).

40. **Dendrobium parthenium**: SABAH: Mt. Kinabalu, Hempuen Hill, 600–900 m, *Lamb* AL 311/85 (K), & *Lamb* SAN 91514 (K, SAN). Mt. Kinabalu, Hempuen Hill, 900 m, *Wood* 837 (K).

41. **Dendrobium radians**: SARAWAK: Marudi District, Mt. Dulit, 1200 m, *Richards* in *Synge* S. 421 (AMES, K, L, SING).

42. **Dendrochilum alpinum**: SABAH: Mt. Kinabalu, above Panar Laban, *Barkman* 4 & *Barkman* 6 (K, Sabah Parks Herbarium, Kinabalu Park). Mt. Kinabalu, Gurulau Spur,

2400–2700 m, *J. & M.S. Clemens* 50669 (BM). Mt. Kinabalu, summit area, 3500 m, *Sato et al.* 1375 (UKMS).

43. **Dendrochilum corrugatum**: SABAH: Mt. Kinabalu, Marai Parai, 1700 m, *Carr* 3128, SFN 27428 (BM, K, SING). Mt. Kinabalu, Penataran Basin, 1700 m, *J. & M.S. Clemens* 40135 (BM). Mt. Kinabalu, Marai Parai, 1500 m, *Collenette* 31 (BM).

44. **Dendrochilum gravenhorstii**: KALIMANTAN BARAT: Upper Kapuas, Talaj River, *Gravenhorst* s.n. (BO, L). SABAH: Sandakan District, Telupid Hap Seng, 100–200 m, *Leopold & Taha* SAN 83505 (K, L, SAN, SAR). Keningau to Sepulot road, 6 km from Nabawan, *c.* 600 m, *Vermeulen & Lamb* 325 (L). SARAWAK: Marudi District, route to Pa Ukat, Bario, 1000 m, *Awa et al.* S. 50416 (AAU, K, L, MEL, SAR, SING).

45. **Dendrochilum jiewhoei**: Known only from the type.

46. **Dendrochilum joclemensii**: SABAH: Mt. Kinabalu, Kemburongoh, 2300 m, *Carr* 3622, SFN 27908 (K, SING), 2100 m, *Carr* 3751 (SING), & 2200 m, *Carr* SFN 36567 (SING). Mt. Kinabalu, Lumu-Lumu, 2100 m, *J. & M.S. Clemens* 27860 (BM). Mt. Kinabalu, Mesilau Cave/Janet's Halt, 2400 m, *Fuchs & Collenette* 21404 (K). Mt. Kinabalu, Summit Trail, 2200 m, *Sato* 2100 (UKMS), & 2300 m, *Sato* 2158 (UKMS).

47. **Dendrochilum kelabitense**: Known only from the type.

48. **Dendrochilum longipes**: KALIMANTAN BARAT: Serawai, summit of Raya Hill (Bukit Raya), 2300 m, *Church et al.* 2610 (AMES, BO, K). KALIMANTAN TENGAH: Raya Hill and upper Katingan (Mendawai) River area, Upper Samba River, *Mogea* 3963 (BO, L). KALIMANTAN TIMUR: Mt. Kemal, summit, 1800 m, *Endert* 3991 (L). Mt. Buduk Rakik, north of Long Bawan, *Kato et al.* B. 11071 (L). SARAWAK: Lawas District, Mt. Murud, 1900 m, *Burtt & Martin* B. 5399 (E, SAR). Marudi District, Mt. Mulu, 1800 m, *Lewis* 345 (K). Marudi District, Mt. Murud, summit, 1900–2400 m, *Mjöberg* 66 (AMES). Limbang District, Batu Lawi, 2000 m, *Nooteboom & Chai* 2278 (K, L, SAR). Marudi District, Mt. Murud, 2300 m, *Yii* S. 44432 (K, L, SAR, SING).

49. **Dendrochilum longirachis**: KALIMANTAN TIMUR: Apo Kayan, north of Long Bawan, *Kato et al.* B. 10110 (BO, L). SABAH: Mt. Kinabalu, Penibukan, 1200 m, *J. & M.S. Clemens* 30601 (BM, BO, E, L). Crocker Range, Sinsuron road, *Lamb* SAN 89674 (SAN). Mt. Kinabalu, Mamut River, 1200 m, *Sato et al.* 1558 (UKMS). Sipitang District, Ulu Long Pa Sia, 1260 m, *Wood* 694 (K). SARAWAK: Marudi District, Mt. Murud, *Nooteboom & Chai* 1890 (L, SAR). Belaga District, Mt. Dulit, Upper Koyan River, 1200 m, *Richards* in *Synge* S. 483 (K, SING).

50. **Dendrochilum simplex**: KALIMANTAN TENGAH: Liangangang, *Hallier* 2646 (BO, K, L). SABAH: Beluran District, Mt. Monkobo, *Aban* SAN 95241 (K, L, SAN). Mt. Kinabalu, Pig Hill, 570–600 m, *Barkman* s.n. (K). Mt. Kinabalu, Penibukan, 1400 m, *J. & M.S. Clemens* 40548 (BM), & 1200 m, *J. & M.S. Clemens* 50105 (BM, K). Penampang District, Tunggul Togudon, *Fidilis* SAN 127819 (K, SAN). Keningau to Sepulot road, 6 km from Nabawan, 400 m, *Vermeulen & Lamb* 324 (K, L). SARAWAK: Marudi District, Bario/Pa Umor, 1100 m, *Beaman, T.E.* 103 (K, MSC, UNIMAS). Limbang District, Batu Lawi, 1500 m, *Awa & Lee* S. 50792 (AAU, K, KEP, L, MEL, SAR, SING). Kapit District, Mengiong/ Entulu Rivers, *Lee* S. 54798 (K, SAR). Marudi District, Mt. Murud, summit, 1900–2400 m, *Mjöberg* 49 (AMES). Kapit District, Hose Mountains, Batu Hill, 1200 m, *de Vogel* LC 913963 (K, L).

51. **Dendrochilum subulibrachium**: KALIMANTAN TIMUR: Long Petak, 800 m, *Endert* 3221 (L). SARAWAK: Limbang District, Batu Lawi, 1600 m, *Awa & Lee* S. 50756 (K, L,

SAR, SING). Marudi District, Mt. Mulu, 1800 m, *Lewis* 349 (K).

52. **Dendrochilum transversum**: SABAH: Mt. Kinabalu, Marai Parai, 2100–2700 m, *J. & M.S. Clemens* 33130 (BM, K), & 2400–3000 m, *J. & M.S. Clemens* 33173 (BM).

53. **Dilochia parviflora**: KALIMANTAN TIMUR: Mt. Kemal, 1800 m, *Endert* 4262 (BO, L). SABAH: Mt. Kinabalu, Tenompok, 1500 m, *Carr* 3246 (SING). SARAWAK: Marudi District, Bario/Marario River, 1000 m, *Anderson, J.A.R.S.* 20078 (K). Marudi District, Mt. Api, 900 m, *Anderson, J.A.R.S.* 30879 (AMES, K, L, SAR). Limbang District, Mt. Buda, 600 m, *Chai* S. 39452 (K, SAR).

54. **Dilochia rigida**: SABAH: Beluran District, Mt. Monkobo, *Aban* SAN 95245 (K, SAN). Crocker Range, Mt. Alab, 1900–1950 m, *Beaman* 8243 (K). Mt. Trus Madi, 2430 m, *Collenette* 649 (K). Mt. Trus Madi, 1950 m, *J.B. Comber* 147 (K). Mt. Kinabalu, Marai Parai, 1500 m, *Carr* 3119, SFN 26550 (SING). Mt. Kinabalu, Kemburongoh, 2400 m, *J. & M.S. Clemens* 29123 (BM). Mt. Kinabalu, Pinosuk Plateau, 1500 m, *Lamb* s.n. (K). Mt. Kinabalu, Mt. Nungkek, 1400 m, *Sands* 3989 (K). Sipitang District, Ulu Long Pa Sia, above Maga River, 1500 m, *Wood* 651 (K). Crocker Range, Mt. Alab, 1800 m, *Wood* 769 (K). SARAWAK: Marudi District, Mt. Murud, summit, 2300–2400 m, *Beaman* 11455 (K, UNIMAS). Marudi District, Mt. Mulu, Camp 3, *Blicher & Rantai* S. 57905 (SAR). Kapit District, Batu Tiban, 1400 m, *Mjöberg* s.n. (AMES). Marudi District, Mt. Dulit, *Synge* S. 534 (K, L). Kapit District, Hose Mountains, Batu Hill, 1200 m, *de Vogel & Lai S. Teck* 9330 (L).

55. **Diplocaulobium brevicolle**: SABAH: Nabawan, 600 m, *Lamb* AL 429/85 (K), & *Lamb* 550/86 (K). SARAWAK: Kuching District, Kuching, *Haviland* s.n. (K). Julau District, Lanjak Entimau Protected Forest, Upper Ensirieng River, 300 m, *Schuiteman, Mulder & Vogel* LC 932706 (L). Serian District, Tebakang/Majang Hill, 200 m, *de Vogel* LC 27476 (L). Lundu District, Upper Sematan River, 50 m, *de Vogel* LC 27486 (L), & *de Vogel* 27487 (L).

56. **Diplocaulobium longicolle**: SARAWAK: Locality unknown, *Hose* s.n. (SING). Marudi District, Dulit/Long Kapa, 300 m, *Richards* 2334 (K).

57. **Dossinia marmorata**: SARAWAK: Miri District, Mt. Subis, *Anderson, J.A.R.S.* 16025 (K, SAR). Bau District, Bau/Jebong Hill, 200 m, *Anderson, J.A.R.S.* 26823 (SAR). Bau District, Bau Hills, *Brooks* 51 (BM). Kuching District, Mt. Siburan, *Burtt & Woods* B. 1927 (E). Bau District, Bidi, *J. & M.S. Clemens* 20716 (K, L). Lundu District, Mt. Gading, *George* S. 43404 (SAR). Kuching District, Braang, *Haviland* 264 (SAR). Bau District, Bau/Jebong Hill, 100 m, *Lehmann* S. 29402 (SAR). Bau District, Bau/Mt. Setiak, 100 m, *Martin* S. 38661 (K, L, SAR, SING). Marudi District, Mt. Mulu/Mt. Pala, *Roelfsema, Schuiteman & Vogel* LC 970522 (L). Kuching District, Padawan/Manok Hill, 300 m, *Wright & Chai* S. 27418 (K, L).

58. **Entomophobia kinabaluensis**: BRUNEI: Temburong District, Bukit Retak, *Wong* WKM 810 (BRUN, K). SABAH: Mt. Kinabalu, Mamut Copper Mine, 1400–1500 m, *Beaman* 10348 (K). Mt. Kinabalu, Mahandei River, 1100 m, *Carr* 3112 (SING). Mt. Kinabalu, Marai Parai, 2100 m, *Carr* 3475, SFN 28059 (SING). Mt. Kinabalu, Penibukan, 1200–1500 m, *J. & M.S. Clemens* 30607 (K). Crocker Range, Kimanis road, 1050 m, *Lamb* AL 859/87 (K). Sipitang District, trail between Long Pa Sia and Long Samado, 1300 m, *Vermeulen & Duistermaat* 991 (K, L). SARAWAK: Marudi District, Mt. Api, 1600 m, *Argent & Jermy* 1007 (E). Marudi District, Bario/Pa Umor, 1100 m, *Beaman* 11260 (K, MSC, UNIMAS). Bintulu District, Merurong Plateau, *Brunig* S. 17468 (SAR). Limbang District, Mt. Buda,

1000 m, *Chai* S. 39886 (K, SAR). Marudi District, Mt. Mulu, *Mohtar et al.* S. 49635 (SAR). Kuching District, Mt. Penrissen, *Paie* 16312 (K, SAR). Kapit District, Hose Mountains, Batu Hill, 1200 m, *de Vogel & Lai S. Teck* 9265 (L).

59. **Epigeneium zebrinum**: SARAWAK: Kuching District, Sijingkat, *Hewitt* s.n. (SAR, SING).

60. **Eria aurantia**: SABAH: Sipitang District, Ulu Long Pa Sia, near Pa Sia River, 1260 m, *Wood* 701 (K).

61. **Eria cymbidifolia var. cymbidifolia**: SABAH: Mt. Kinabalu, Mesilau Cave, 1900–2200 m, *Beaman* 9571 (K). Mt. Kinabalu, golf course site, 1700–1800 m, *Brentnall* 133 (K). Mt. Kinabalu, Kemburongoh/Lumu-Lumu, 1800–2400 m, *J. & M.S. Clemens* 27160 (BM, K). Mt. Kinabalu, Marai Parai Spur, 2000 m, *Collenette* 62 (BM). Mt. Kinabalu, Tinekuk Falls, 2000 m, *J. & M.S. Clemens* 40911 (BM, K). Tambunan District, Mt. Trus Madi, 2200 m, *Sands* 4023 (K).

61a. **Eria cymbidifolia var. pandanifolia**: SABAH: Mt. Kinabalu, Kinateki River Head, 2700 m, *J. & M.S. Clemens* 35183 (BM). Mt. Kinabalu, Gurulau Spur, 2400–2700 m, *J. & M.S. Clemens* 50648 (BM). SARAWAK: Kapit District, Hose Mountains, Upper Mujong River 1000 m, *Paie* S. 19962 (K, SAR).

62. **Eria pseudocymbiformis var. pseudocymbiformis**: SABAH: Mt. Kinabalu, Mamut Copper Mine, 1600–1700 m, *Beaman* 9951 (K). Mt. Kinabalu, Lumu-Lumu, 2100 m, *J. & M.S. Clemens* 29835 (BM). Mt. Kinabalu, Marai Parai/Nungkek, 1400 m, *J. & M.S. Clemens* 32555 (BM). SARAWAK: Mt. Dulit, *Synge* S. 416 (K, L, SING).

62a. **Eria pseudocymbiformis var. hirsuta**: SABAH: Mt. Kinabalu, Pig Hill, 2000–2300 m, *Beaman* 9894 (K). Mt. Kinabalu, Kiau, 900 m, *J. Clemens* 77 (AMES, BM, K, SING). Mt. Kinabalu, Minetuhan, 1800–2400 m, *J. & M.S. Clemens* 33787 (BM). Mt. Kinabalu, Mesilau Camp, *Poore* H. 301 (K). Mt. Kinabalu, Tenompok/Kemburongoh, 2000 m, *Meijer* SAN 20417 (K). Mt. Kinabalu, Park Headquarters, 1700–2000 m, *Vermeulen & Chan* 391 (K). Crocker Range, Kota Kinabalu to Tambunan road, 1350–1500 m, *Beaman* 10484 (K). Tambunan District, Mt. Trus Madi, 1560 m, *Wood* 870 (K).

63. **Eria saccifera**: SABAH: Mt. Kinabalu, Tenompok, 1500 m, *J. & M.S. Clemens* 30153 (K). Crocker Range, Kimanis road, 1290 m, *Dewol & Abas* SAN 89073 (K, L, SAN). Tenom District, above Kallang Waterfall, 1260 m, *Lamb* AL 347/85 (K). Sipitang District, Mt. Lumaku, 1300 m, *Wood* 813 (K). SARAWAK: Kuching District, Mt. Penrissen, 1100–1200 m, *Beaman* 12011 (K). Bau District, Bidi, *Brooks* s.n. (SAR). Kapit District, Hose Mountains, Simpurai River, 1200 m, *Burtt & Martin* B. 5045 (E). Kuching District, Mt. Penrissen, 900–1000 m, *Jacobs* 5088 (L). Marudi District, Bario/Pa Berang, 1000 m, *Roelfsema, Schuiteman & Vogel* LC 970137 (L), & 1000–1200 m, *Roelfsema, Schuiteman & Vogel* LC 970295 (L). Julau District, Lanjak Entimau Protected Forest/Upper Ensirieng River, 200 m, *Schuiteman, Mulder & Vogel* LC 932945 (L). Kapit District, Hose Mountains, Batu Hill, 800 m, *de Vogel* LC 914224 (L), & *de Vogel* LC 914314 (L). Limbang District, Mt. Pagon, 1500–1600 m, *de Vogel* LC 914808 (L), & 1300, *de Vogel* LC 914939 (L). Belaga District, Belaga, 100 m, *Yong* 86 (K).

64. **Geesinkorchis alaticallosa**: SABAH: Nabawan area, 400–500 m, *Lamb* AL 351/85 (K). Nabawan area, 400–500 m, *Wood* 751 (K). SARAWAK: Marudi District, Bario/Pa Umor, 1100 m, *Beaman, T.E.* 121 (K, MSC, UNIMAS), & *Beaman, T.E.* 207 (UNIMAS). Bau District, Bungo Range/Mt. Bungo, 1100 m, *Brunig* S. 7626 (SAR). Miri District, Lambir Hills, 500 m, *Burtt* B. 11648 (E, SAR). Kuching District, Mt. Matang, 900 m, *Haviland* 605 (SAR). Miri District, Niah, *Vermeulen* LC 26553 (L).

65. **Geesinkorchis phaiostele**: SARAWAK: Lundu District, Mt. Rumput, *Anderson, J.W.* 175 (SING), & *Anderson, J.W.* 225 (SING). Kuching District, Mt. Santubong, 800 m, *Beaman* 11575 (K, UNIMAS). Kuching District, Mt. Santubong, *Beccari* 2147 (FI). Lundu District, Mt. Pueh, *Beccari* 2451 (FI). Lundu District, Mt. Berumput, *Burtt & Martin* B. 2782 (E). Simunjan District, Mt. Gaharu, *Burtt & Martin* RBGE 622541 (E). Kuching District, Mt. Santubong, 800 m, *Tan, K.W.S.* 28845 (SAR).

66. **Goodyera condensata**: SABAH: Mt. Kinabalu, Tenompok, 1500 m, *Carr* 3437 (K). Mt. Kinabalu, Kadamaian River, 1800 m, *Carr* SFN 27592 (SING). Mt. Kinabalu, Mesilau River, 21002700 m, *J. & M.S.Clemens* 51733 (BM). Crocker Range, Mt. Alab, Sinsuron falls, 1400m, *Lamb* AL 1551/92 (K).

67. **Habenaria lobbii**: SARAWAK: Bau District, Krian Hill, *Anderson, J.A.R.S.* 25139 (SAR). Bau District, Bau/Kroking, 100 m, *Beaman* 11570 (K, UNIMAS). Bau District, Tiang River, *Brooke, W.* 8896 (L). Bau District, Bau Hills, 300 m, *Brooke, W.* 9903 (L). Kuching District, Padawan/Tiang Bekap, 300 m, *Chew Wee-Lek* CWL 1298 (A, SAR, SING). Bau District, Bau/Mt. Setiak, *Martin* S. 38657 (SAR). Belaga District, Pantu Hill/Tasu Hill, 600 m, *Yii* S. 53580 (L), & *Yii* S. 55892 (K, SAR).

68. **Hetaeria anomala**: SABAH: Mt. Kinabalu, East Mesilau River, Mesilau Cave Trail, 1700–1900 m, *Beaman* 7960 (K). Mt. Kinabalu, Mesilau Cave, 2000–2100 m, *Beaman* 8132 (K). Mt. Kinabalu, West Mesilau River, 1600–1700 m, *Beaman* 8702 (K). Mt. Kinabalu, Mesilau Cave Trail, 1700–1900 m, *Beaman* 9129 (K). Mt. Kinabalu, West Mesilau River, 1600 m, *Collenette* 640 (K). Mt. Kinabalu, Mesilau River, 1800 m, *Lamb* SAN 91536 (K, SAN).

69. **Liparis anopheles**: SABAH: Mt. Trus Madi, 1560 m, *Surat* in *Wood* 871 (K). SARAWAK: Marudi District, Tama Abu Range, 1600 m, *Awa & Lee* S. 51113 (K, L, SAR). Marudi District, Mt. Murud, summit, 1900–2400 m, *Mjöberg* 28 (AMES).

70. **Liparis grandiflora**: SARAWAK: Belaga District, Sepaku, Upper Belaga River, *Kandau & Ismawi* S. 43766 (K, SAR). Marudi District, Melinau River, 50 m, *Kerby* RBGE 773431 (E, K). Marudi District, Mt. Mulu/Mt. Pala, *Roelfsema, Schuiteman & Vogel* LC 970538 (L).

71. **Liparis lacerata**: SABAH: Beaufort District, south of Weston, 20 m, *Beaman* 8061 (K). Nabawan area, 400–500 m, *Lamb* AL 1107/89 (K). Mt. Kinabalu, Lohan River, 60 m, *Lamb* SAN 93351 (K, SAN). SARAWAK: Sibu District, Rankan Panjang, sea level, *Anderson, J.A.R.S.* 9443 (SAR). Kuching District, Bako, 90 m, *Carrick & Enoch* 480 (SING). Lundu District, Lundu, *Foxworthy* 107 (AMES). Lundu District, Sematan, *Foxworthy* 137 (AMES). Kuching District, Selabat, 90 m, *Haviland* s.n. (SAR). Belaga District, Iban River, 300 m, *Lee* S. 45449 (SAR). Marudi District, Melinau River, *Roelfsema, Schuiteman & Vogel* LC 970604 (L). Daro District, Bruit Island, *Sanusi bin Tahir* S. 9212 (SAR). Marudi District, Teraja, on Sarawak-Brunei border, 10 m, *de Vogel* LC 914628 (L).10487 (K).

72. **Nabaluia exaltata**: SABAH: Tambunan District, Mt. Trus Madi, 2490 m, *Wood* 903 (K). SARAWAK: Lawas District, Mt. Murud, Camp 3, 1600 m, *Burtt & Martin* B. 5260 (E, K). Lawas District, Mt. Murud, summit, 1900–2400 m, *Mjöberg* 36 (AMES, K). Marudi District, Mt. Mulu, summit, 2300 m, *Yii & Talib* S. 58564 (K, SAR).

73. **Oberonia patentifolia**: SABAH: Mt. Kinabalu, Dallas/Tenompok, *J. & M.S. Clemens* 26783 (BM, K). Mt. Kinabalu, Tenompok, 1500 m, *J. & M.S. Clemens* 27003 (BM, SING). Mt. Kinabalu, Tinekuk Falls, 1800 m, *J. & M.S. Clemens* 40922 (BM). Penampang District, Tunggul Penampang to Tambunan road, *Krispinus* SAN 13157 (K, SAN). Mt. Kinabalu,

Hempuen Hill, *Madani* SAN 89490 (K, SAN). Beaufort District, Beaufort, *Melegrito* 3241 (K).

74. **Octarrhena parvula**: SABAH: Mt. Kinabalu, Lohan/Mamut Copper Mine, 900 m, *Beaman* 10596 (K). Mt. Kinabalu, golf course site, 1700–1800 m, *Beaman* 10676 (K). Mt. Kinabalu, Minitinduk Gorge, 900 m, *Carr* SFN 2669 (SING). Mt. Kinabalu, Kinunut Valley, 1100 m, *Carr* SFN 27190 (SING). Mt. Kinabalu, Kinunut Valley Head, 1200 m, *Carr* SFN 27195 (SING). Sipitang District, Long Pa Sia to Long Samado trail, 1300 m, *Vermeulen & Duistermaat* 950 (K, L). Sipitang District, Rurun River headwaters, 1700 m, *Vermeulen & Duistermaat* 1080 (K, L). Sipitang District, east of Long Pa Sia to Long Miau trail, 1100 m, *Vermeulen & Duistermaat* 1119 (K, L).

75. **Pennilabium struthio**: SABAH: near Nabawan, 400–500 m, *Lamb* AL 382/85 (K). SARAWAK: Lawas District, Tebunan Hill, Upper Trusan River, 500 m, *Lee* S. 52372 (K, L, SAR).

76. **Peristylus hallieri**: SABAH: Mt. Kinabalu, Bundu Tuhan, 1200 m, *Carr* 3489, SFN 27455 (BM, K, SING). Mt. Kinabalu, Mt. Nungkek, 800 m, *J. & M.S. Clemens* 32775 (BM). Crocker Range, Sinsuron road, 1440 m, *Collenette* 2296 (K). Sipitang District, Siang Mayo River near Mendulong, 400–500 m, *J.B. Comber* 124 (K). Crocker Range, Moyog, 1500 m, *Lamb* AL 499/85 (K). Mt. Kinabalu, Park Headquarters, 1500 m, *Lamb* AL 24/82 (K). Sipitang District, trail from Maligan to Long Pa Sia, 1500 m, *Vermeulen & Duistermaat* 925 (K, L). Sipitang District, Ulu Long Pa Sia, beside Maga River, 1260 m, *Wood* 655 (K). Crocker Range, Sinsuron road, 1500 m, *Wood* 734 (K). Sipitang District, trail from Long Pa Sia to Maligan, near Angol, 800 m, *Lamb* in *Wood* 738 (K). Mt. Kinabalu, between Mesilau East River and Menteki River, 1650 m, *Wood* 836 (K). SARAWAK: Lubok Antu District, Marup, *Beccari* 3317 (FI). Matu District, Igan River, *Beccari* 3929 (FI). Kuching District, Mt. Matang, *Collenette* 828 (K). Belaga District, Mt. Dulit, Upper Koyan River, *Richards* in *Synge* S. 543 (K).

77. **Phalaenopsis cochlearis**: SARAWAK: Locality unknown, 500–600 m, *Sheridan, P.* 1A (K).

78. **Pholidota clemensii**: SABAH: Mt. Kinabalu, Mesilau Cave Trail, 1700–1900 m, *Beaman* 7999 (K). Mt. Kinabalu, Kinunut River Head, 1200 m, *Carr* 3336, SFN 27398 (SING). Mt. Kinabalu, Tenompok, 1500 m, *J. & M.S. Clemens* 27398 (BM, K). Mt. Kinabalu, Gurulau Spur, 1500 m, *J. & M.S. Clemens* 50349 (BM, K). Tambunan District, Mt. Trus Madi, above Kidukarok, 1560 m, *Wood* 873 (K). SARAWAK: Limbang District, Batu Lawi, 1400 m, *Awa & Lee* S. 50799 (K, L, SAR). Marudi District, Tama Abu Range, 2100m, *Moulton* SFN 6656 (AMES). Marudi District, Mt. Api, 800–1000 m, *Nielsen* 491 (K). Kapit District, Hose Mountains, Batu Hill, 1200 m, *de Vogel* LC 913979 (L), & *de Vogel* LC 914024 (SAR).

79. **Pholidota schweinfurthiana**: SABAH: Sipitang District, near Long Pa Sia, *Phillipps et al.* SNP 2951 (K, Sabah Parks Herbarium, Kinabalu Park). Sipitang District, Ulu Long Pa Sia, 1400 m, *Wood* 647 (K). SARAWAK: Marudi District, Tama Abu Range, 1200 m, Moulton SFN 6678 (AMES, K, SING).

80. **Pholidota sigmatochilus**: SABAH: Mt. Kinabalu, Mesilau Cave, 2000–2100 m, *Beaman* 8154 (K). Mt. Kinabalu, Kemburongoh, 2900 m, *Carr* 3524, SFN 27533 (K). Mt. Kinabalu, Paka-paka Cave, 3500 m, *J. & M.S. Clemens* 27153 (BM). Mt. Kinabalu, Kemburongoh, 2900 m, *Fuchs* 21072 (K, L). Mt. Kinabalu, Pig Hill, *Lamb* AL 734/87 (K).

81. **Pholidota ventricosa**: SABAH: Mt. Kinabalu, Kaung, 500 m, *J. & M.S. Clemens* 26022 (BM). Mt. Kinabalu, Penataran Basin, 1200 m, *J. & M.S. Clemens* 34445 (BM). Mt.

Kinabalu, Penibukan, 1700 m, *J. & M.S. Clemens* 50364 (BM, K). Sipitang District, near Long Pa Sia to Long Miau trail, 1100 m, *Vermeulen & Duistermaat* 1144 (K, L). SARAWAK: Limbang District, Mt. Pagon, 1000 m, *Awa & Lee* S. 47848 (K, L, SAR). Kuching District, Matang River, *Beccari* 3494 (FI). Belaga District, Dema Hill, 900 m, *Burtt* 11337 (E, SAR). Marudi District, Mt. Mulu, *Kerby* RBGE 773433 (E). Marudi District, Mt. Mulu/Batu Pajang, *Lai & Jugah* S. 44172 (K, L, SAR). Marudi District, Mt. Laiun, Upper Tinjar River, *Richards* 2422 (K, SING). Kapit District, Hose Mountains, Batu Hill, 1200 m, *de Vogel* LC 914071 (L).

82. **Podochilus marsupialis**: KALIMANTAN BARAT: Serawai, Merah River, 700 m, *Church et al.* 2095 (AMES, BO, K, L). KALIMANTAN TIMUR: West Kutai, Long Hut, 130–200 m, *Endert* 2723 (BO, L). SABAH: Taman Tawau, *Lim, W.H.* 1.26 (K). SARAWAK: Kapit District, Teneong, 500 m, *Brooke, W.* 9182 (BM, SAR). Julau District, Lanjak Entimau Protected Forest/Entimau Hill, 800 m, *Schuiteman, Mulder & Vogel* LC 932914 (L).

83. **Pomatocalpa kunstleri**: KALIMANTAN TIMUR: Lempake, Tanah merah, 40 m, *Maskuri* 164 (BO, K, L). SABAH: Nabawan, 400–500 m, *Lamb* AL 1153/89 (K). SARAWAK: Lundu District, Lundu, *Anderson, J.W.* 156 (SING). Kuching District, Kuching, *Brooke, W.* 10783 (BM). Kuching District, Semenggoh Forest Reserve, *George* 954 (K). Bau District, Bau/Jagoi, 200 m, *Roelfsema, Schuiteman & Vogel* LC 970005 (L).

84. **Porphyrodesme sarcanthoides**: KALIMANTAN BARAT: Sintang, Posang River, *Church et al.* 1141 (AMES, BO, K). SABAH: Pensiangan District, Sapulut, Batu Ponggol, 400 m, *Lamb* AL 1200/89 (K). Danum Valley, 160–250 m, *O'Byrne* (sight record). Sapulot Forest Reserve, *Suali & Sumbing* SAN 101310 (K, SAN).

85. **Sarcoglyphis potamophila**: SABAH: Danum Valley, *O'Byrne* (sight record). SARAWAK: Kuching District, Kuching River, 200 m, *O'Byrne* (sight record).

86. **Schoenorchis buddleiflora**: SABAH: Mt. Kinabalu, Marai Parai, 2134 m, *Argent C.* 14844 (E, K). Mt. Kinabalu, Tenompok Orchid Garden, 1500 m, *J. & M.S. Clemens* 51057 (BM, K). Ulu Segama, *Krispinus* SAN 95592 (K, L, SAN). SARAWAK: Kuching District, Mt. Penrissen, 1100 m, *Beaman* 11108 (K, MSC, UNIMAS). Kuching District, Mt. Penrissen, 1000 m, *Beaman, T.E.* 279 (K, UNIMAS).

87. **Schoenorchis endertii**: SABAH: Mt. Kinabalu, Lohan River, 500–600 m, *Clements* 3392 (K). Mt. Kinabalu, Lohan River, 500–600 m, *Lamb* AL 433/85 (K). SARAWAK: Samarahan District, Kuap, *Hewitt* s.n. (K).

88. **Smitinandia micrantha**: KALIMANTAN SELATAN: Martapura, Pleihari Reserve, Muratus Mountains, 250 m, *Lamb* AL 1148/89 (K).

89. **Spathoglottis kimballiana**: SABAH: Mt. Kinabalu, kampung Serinsim, *Bakia* 434 (K, Sabah Parks Herbarium, Kinabalu Park). Labuk & Sugut District, Maliau River, SE of Telupid, 170 m, *Beaman* 10229 (K). Porog River, Labuk, lowlands, *Collenette* 504 (K). Mt. Kinabalu, Penataran Ridge, 600 m, *Lamb* SAN 93370 (K, SAN).

90. **Thrixspermum tortum**: SABAH: Mt. Kinabalu, Tenompok Orchid Garden, *J. & M.S. Clemens* 50181 (BM, K). Crocker Range, Kimanis road, 1000 m, *Vermeulen & Lamb* 305 (K, L). Crocker Range, Kimanis road, 1250 m, *Wood* 818 (K).

91. **Thrixspermum triangulare**: SABAH: Mt. Kinabalu, Mesilau Cave Trail, 1700–1900 m, *Beaman* 9146 (K). Mt. Kinabalu, West Mesilau River, 1600 m, *Beaman* 8999 (K). Mt. Kinabalu, Kemburongoh, *Carr* SFN 27503 (K). Mt. Kinabalu, Kilembun Basin, 2300 m, *J. & M.S. Clemens* 32889 (BM). Mt. Kinabalu, Layang-Layang, 3000 m, *Collenette* 888 (K).

Mt. Kinabalu, Kemburongoh, 2100 m, *Price* 220 (K). Mt. Kinabalu, summit trail, 2700-3400 m, *Wood* 611 (K).

92. **Trichoglottis jiewhoei**: SABAH: Crocker Range, Sinsuron road, 1400 m, *Lamb* AL 1335/91 (K).

93. **Trichoglottis kinabaluensis**: SABAH: Mt. Kinabalu, Bundu Tuhan, *Darnton* 226 (BM). Mt. Kinabalu, Minitinduk, 900 m, *Carr* SFN 27040 (SING). Mt. Kinabalu, Kiau, 900 m, *J. Clemens* 31 (AMES). Mt. Kinabalu, Tenompok, 1200 m, *J. & M.S. Clemens* 26126 (BM, K), & 1500 m, *J. & M.S. Clemens* 28137 (BM).

94. **Trichoglottis magnicallosa**: SABAH: Mt. Kinabalu, Kaung, 500 m, *Carr* SFN 27321 (SING). Mt. Kinabalu, Dallas, 900 m, *J. & M.S. Clemens* 27163 (BM, K). Lamb AL 864/87.

95. **Trichoglottis tinekeae**: SABAH: Pensiangan District, Batu Ponggol, 360–450 m, *Lamb* AL 330/85 (K), & *Lamb* AL 815/87 (K). Sipitang District, Rurun River headwaters, 1700 m, *Vermeulen & Duistermaat* 1089 (K, L). SARAWAK: Marudi District, Bario, 1000 m, *Nooteboom & Chai* 2340 (L). Marudi District, Bario/Pa Berang, 1000 m, *Roelfsema, Schuiteman & Vogel* LC 970164 (L). Marudi District, Bario/Pa Umor, 1100 m, *Yong* 90 (K).

96. **Trichotosia vestita**: BRUNEI: Belait Melilas Paleh Bangawong, 100–200 m, *Thomas* 88 (K, BRUN). KALIMANTAN SELATAN: Banjarmassin, *Motley* 878 (K). SABAH: Keningau District, Pinangah Forest Reserve, *Krispinus* SAN 110257 (K, SAN). Nabawan area, 400–500 m, *Lamb & Lohok* in *Lamb* AL 320/85 (K). Lamag District, Namatoi River, 390 m, *Madani* SAN 88841 (K, SAN). Sandakan District, Wanyang River, Pamol Klagan, *Meijer* 51646 (K, SAN). SARAWAK: Miri District, Niah, *Alphonso & Sansuri* A. 230 (SING). Samarahan District, Samarahan, *Brooke, W.* 10878 (L). Kapit District, Hose Mountains, Simpurai River, 500 m, *Burtt & Martin* B. 4929 (E, SAR). Marudi District, Lio Matoh, 200 m, *Moulton* SFN 6727 (AMES, K, SING). Belaga District, Sepakang Panjang Waterfalls, *Yong* 216 (K).

97. **Vanilla havilandii**: SABAH: Mt. Kinabalu, near Nalumad, *Andau* 440 (K, Sabah Park Herbarium, Kinabalu Park). Tenom District, Mendalom Forest Reserve, *Krispinus* SAN 116658 (K, SAN). Tenom District, Kallang Waterfall, Tenom Valley, 270–300 m, *Lamb* SAN 89632 (K, SAN). Tenom District, Kallang Waterfall area, 470 m, *Sands* 3702 (K). SARAWAK: Bau District, Bau Hills, *Anderson, J.A.R. & Ashton,* S. 20293 (SAR). Miri District, Baram, *Hewitt* s.n. (SAR). Lundu district, Datu Protected Forest Reserve, *Lee* S. 41919 (K, L, SAR). Kuching District, Santubong, *Paie* S. 8309 (K, L, SAR).

98. **Vanilla kinabaluensis**: SABAH: Crocker Range, Sinsuron road, *Joseph & Donggop* SAN 128637 (K, SAN). Tenom District, Paling Paling Hills, 800 m, *Lamb* AL 1230/90 (K). Mt. Kinabalu, Melangkap Tomis, *Lugas* 1724 (K, Sabah Parks Herbarium, Kinabalu Park). Tambunan District, Mt. Trus Madi, trail from Toboban to Kaingaran, 750 m, *Wood* 909 (K).

99. **Ventricularia borneensis**: SABAH: Sipitang District, above Maga River, 1350 m, *Wood* 664 (K). *Lamb* s.n. Heath forest at Long Pa Sia. Tawau District, Luasong, 200 m, *Postar et al.* SAN 143533 (SAN). SARAWAK: Marudi District, Bario/Pa Berang, 1000 m, *Roelfsema, Schuiteman & Vogel* LC 970153 (L).

100. **Vrydagzynea grandis**: SABAH: Mt. Kinabalu, Lubang, 1300 m, *Carr* 3195 (BM, K, SING). Mt. Kinabalu, Kiau, *J. Clemens* 355 (AMES). Mt. Kinabalu, Tinekuk River, 1100 m, *J. & M.S. Clemens* 51721 (BM). Mt. Kinabalu, Haye-Haye/Tinekuk Rivers, 900–1100 m, *Lamb* AL 51/83 (K).

HERBARIUM ABBREVIATIONS

A: Herbarium, Arnold Arboretum, Harvard University, 22 Divinity Avenue, Cambridge, Massachusetts 02138, USA.

AAU: Herbarium Jutlandicum, Department of Systematic Botany, University of Aarhus, Building 137, Universitetsparken, DK-8000, Aarhus C, Denmark.

AMES: Orchid Herbarium of Oakes Ames, Botanical Museum, Harvard University, Cambridge, Massachusetts 02138, USA.

B: Herbarium, Botanischer Garten und Botanisches Museum Berlin-Dahlem, Königin-Luise-Strasse 6-8, D-14195, Berlin, Germany.

BM: Herbarium, Botany Department, The Natural History Museum, Cromwell Road, London SW7 5BD, England, UK.

BO: Herbarium Bogoriense, Jalan Raya Juanda 22-24, Bogor, Java, Indonesia.

BR: Herbarium, Jardin Botanique National de Belgique, Domein van Bouchout, B-1860, Meise, Belgium.

BRUN: Herbarium, Brunei Forestry Centre, Sungai Liang 7490, Belait, Bandar Seri Begawan, Brunei Darussalam.

CAL: Central National Herbarium, Botanical Survey of India, P.O. Botanic Garden, Howrah, Calcutta 711 103, West Bengal, India.

E: Herbarium, Royal Botanic Garden, Edinburgh EH3 5LR, Scotland, UK.

FI: Herbarium Universitatis Florentinae, Museo Botanico, Via G. La Pira 4, I-50121, Firenze, Italy.

HBG: Herbarium, Institut für Allgemeine Botanik und Botanischer Garten Hamburg, Ohnhorststrasse 18, D-2000, Hamburg 52, Germany.

K: Herbarium, Royal Botanic Gardens, Kew, Richmond, Surrey, TW9 3AB, England, U.K. (LINDL refers to the John Lindley herbarium housed in the Orchid Herbarium; WALL refers to the Nathaniel Wallich herbarium housed at Kew).

KEP: Herbarium, Forest Research Institute of Malaysia, PO Box 201, Kepong, 52109 Kuala Lumpur, Malaysia.

L: Rijksherbarium, Van Steenis Gebouw, Einsteinweg 2, P.O. Box 9514, 2300 RA Leiden, The Netherlands.

LAE: Papua New Guinea National Herbarium, Forest Research Institute, PO Box 314, Lae, Papua New Guinea.

MEL: National Herbarium of Victoria, Royal Botanic Gardens, Birdwood Avenue, South Yarra, Victoria 3141, Australia.

MSC: Herbarium, Botany & Plant Pathology Department, 166 Plant Biology Building, Michigan State University, East Lansing, Michigan 48824–1312, USA.

NY: Herbarium, New York Botanical Garden, Bronx, New York 10458-5126, U.S.A.

P: Herbier, Laboratoire de Phanérogamie, Muséum National d'Histoire Naturelle, 16 rue Buffon, F-75005, France.

SAN: Herbarium, Forest Research Centre, Forest Department, PO Box 1407, 90715, Sandakan, Sabah, Malaysia.

SAR: Sarawak Herbarium, Department of Forestry, Mile 6, Penrissen Road, 93250 Kuching, Sarawak, Malaysia.

SING: Herbarium, Parks and Recreation Department, Botanic Gardens, Cluny Road, Singapore 259569, Singapore.

SNP: Herbarium, Sabah National Parks, Mt. Kinabalu National Park, Sabah, Malaysia.

TI: Herbarium, University of Tokyo, 7-3-1 Hongo, Bunkyo-ku, Tokyo, Tokyo 113-0033, Japan.

TNS: Herbarium, Botany Department, National Science Museum, Amakubo 4-1-1, Tsukuba, Ibaraki 305–0005, Japan.

UKMS: Herbarium Jabatan Biologi, Universiti Kebangsaan Malaysia, Kampus Sabah, Locked Bag No. 62, 88996 Kota Kinabalu, Sabah, Malaysia.

UNIMAS: Universiti Malaysia Sarawak, 94300, Kota Samarahan, Sarawak, Malaysia.

W: Herbarium, Department of Botany, Naturhistorisches Museum Wien, Burgring 7, A-1014 Wien, Austria.

COLOUR PLATES

(The number in brackets refers to the page in the text where the species appears)

PLATE 1

A. *Abdominea minimiflora* (Hook.f.). J.J. Sm. Sabah. Cult. Tenom Orchid Centre. *(Photo: A. Lamb)* (21)

B. *Agrostophyllum glumaceum* Hook.f. Sabah, Mount Kinabalu. *(Photo: K. Barrett)* (22)

C. *Aphyllorchis montana* Rchb.f. Sabah, Tenom District, Paling-Paling Hills. *(Photo: A. Lamb)* (28)

D. *Ania borneensis* (Rolfe) Senghas. Sabah, Tambunan. *(Photo: A. Lamb)* (24)

E. *Ania ponggolensis* A. Lamb in H. Turner. Sabah, Batu Ponggol. *(Photo: A. Lamb)* (26)

PLATE 2

A. *Appendicula congesta* Ridl. Sabah, Mount Kinabalu. *(Photo: K. Barrett)* (35)

B. *Appendicula congesta* Ridl. Sabah, Mount Kinabalu. *(Photo: K. Barrett)* (35)

C. *Appendicula foliosa* Ames & C. Schweinf. Sabah. *(Photo: A. Schuiteman)* (37)

D. *Appendicula foliosa* Ames & C. Schweinf. Sabah. *(Photo: A. Schuiteman)* (37)

PLATE 3

A. *Appendicula calcarata* Ridl. Sabah, Long Pa Sia. *(Photo: A. Lamb)* (33)

B. *Appendicula rostellata* J.J. Sm. Sabah, Mount Kinabalu. *(Photo: K. Barrett)* (41)

C. *Ascidieria longifolia* (Hook.f.) Seidenf. Sabah, Long Pa Sia. *(Photo: A. Lamb)* (43)

D. *Brachypeza indusiata* (Rchb.f.) Garay. Sabah. *(Photo: K. Barrett)* (45)

PLATE 4

A. *Bromheadia divaricata* Ames & C. Schweinf. Sabah, Mount Kinabalu. *(Photo: K. Barrett)* (50)

B. *Bulbophyllum lemniscatoides* Rolfe. Sabah, cult. Tenom Orchid Centre. *(Photo: A. Lamb)* (56)

C. *Bulbophyllum coriaceum* Ridl. Sabah, Mount Kinabalu. *(Photo: K. Barrett)* (54)

D. *Calanthe otuhanica* C.L. Chan & T.J. Barkman. Sabah, Mount Kinabalu. *(Photo: C.L. Chan)* (58)

E. *Cheirostylis spathulata* J.J. Sm. Sabah, Tenom District, Kallang Waterfall. *(Photo: A. Lamb)* (60)

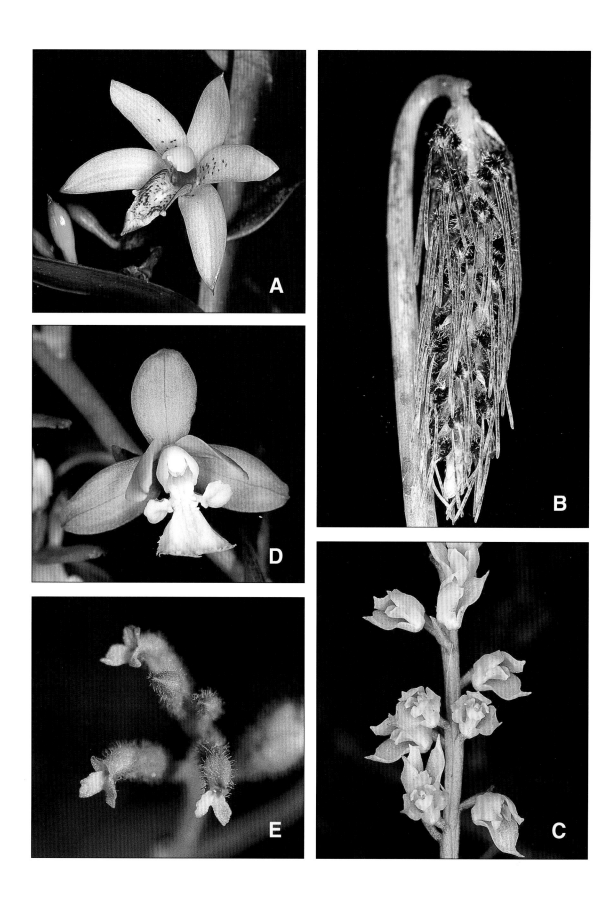

PLATE 5

A. *Cleisostoma suaveolens* Blume. Sabah. *(Photo: A. Schuiteman)* (62)

B. *Cordiglottis fulgens* (Ridl.) Garay. Sabah, Nabawan. *(Photo: A. Lamb)* (64)

C. *Corybas serpentinus* J. Dransf. Sabah, Lahad Datu District, Mount Silam.
(Photo: J. Dransfield) (68)

D. *Corybas serpentinus* J. Dransf. Sabah, Telupid, Mount Tawai.
(Photo: A. Lamb) (68)

E. *Cymbidium kinabaluense* K.M. Wong & C.L. Chan. Sabah, Mount Kinabalu.
(Photo: C.L. Chan) (70)

F. *Cymbidium kinabaluense* K.M. Wong & C.L. Chan. Sabah, Mount Kinabalu.
(Photo: C.L. Chan) (70)

PLATE 6

A. *Cystorchis salmoneus* J.J. Wood. Sabah, Sipitang District, Long Pa Sia.
(Photo: A. Lamb) (74)

B. *Cystorchis saprophytica* J.J. Sm. Sabah, Sipitang District, Long Pa Sia.
(Photo: A. Lamb) (76)

C. *Dendrobium bifarium* Lindl. Sabah, cult. Tenom Orchid Centre.
(Photo: C.L. Chan) (81)

D. *Dendrobium aurantiflammeum* J.J. Wood. Sabah, Crocker Range, Kimanis road.
(Photo: A. Lamb) (78)

PLATE 7

A. *Dendrobium cinnabarinum* Rchb.f. Sabah, Kinabatangan District, Maliau Basin. *(Photo: A. Lamb)* (83)

B. *Dendrobium cinnabarinum* Rchb.f. Sabah, Sipitang District, Long Pa Sia. *(Photo: A. Lamb)* (83)

C. *Dendrobium cinnabarinum* Rchb.f. Sabah, Sipitang District, Ulu Meligan. *(Photo: A. Lamb)* (83)

D. *Dendrobium cinnabarinum* Rchb.f. Sabah, Sipitang District, Long Pa Sia. *(Photo: A. Lamb)* (83)

E. *Dendrobium dearei* Rchb.f. Sabah, Sandakan. cult. Tenom Orchid Centre. *(Photo: A. Lamb)* (86)

PLATE 8

A. *Dendrobium derryi* Ridl. Sabah. *(Photo: K. Barrett)* (89)

B. *Dendrobium derryi* Ridl. Sabah. *(Photo: K. Barrett)* (89)

C. *Dendrobium derryi* Ridl. Cult. Hortus Botanicus, Leiden University, The Netherlands. *(Photo: A. Vogel)* (89)

D. *Dendrobium hendersonii* A.D. Hawkes & A.H. Heller. Sabah. *(Photo: A. Schuiteman)* (93)

E. *Dendrobium kurashigei* Yukawa. Sabah, Mount Kinabalu. *(Photo: K. Barrett)* (95)

PLATE 9

A. *Dendrobium lamrianum* C.L. Chan. Sabah, Mount Kinabalu.
(Photo: A. Lamb) (97)

B. *Dendrobium lamrianum* C.L. Chan. Sabah, Kinabalu Park Mountain Garden.
(Photo: C.L. Chan) (97)

C. *Dendrobium metachilinum* Rchb.f. Sabah, Telupid area. *(Photo: A. Lamb)* (101)

D. *Dendrobium parthenium* Rchb.f. Sabah, Kinabalu, Hempuen Hill. *(Photo: A. Lamb)*
(103)

E. *Dendrochilum alpinum* Carr. Sabah, Mount Kinabalu, 3200 m.
(Photo: T.J. Barkman) (108)

PLATE 10

A. *Dendrochilum alpinum* Carr. Sabah, Mount Kinabalu, 3200 m. *(Photo: A. Lamb)* (108)

B. *Dendrochilum joclemensii* Ames. Sabah, Mount Kinabalu. *(Photo: K. Barrett)* (116)

C. *Dendrochilum joclemensii* Ames. Sabah, Mount Kinabalu. *(Photo: K. Barrett)* (116)

D. *Dilochia rigida* (Ridl.) J.J. Wood. Sabah, Crocker Range. *(Photo: J. Stone)* (133)

E. *Diplocaulobium brevicolle* (J.J. Sm.) Kraenzl. Cult. Hortus Botanicus, Leiden University, The Netherlands. *(Photo: A. Vogel)* (135)

F. *Diplocaulobium longicolle* (Lindl.) Kraenzl. Brunei, Brunei Horticultural Show. *(Photo: A. Lamb)* (137)

PLATE 11

A. *Dossinia marmorata* E. Morren. Sarawak, Gunung Mulu National Park. *(Photo: A. Lamb)* (140)

B. *Dossinia marmorata* E. Morren. Sarawak, Gunung Mulu National Park. *(Photo: A. Lamb)* (140)

C. *Entomophobia kinabaluensis* (Ames) de Vogel. Cult. Hortus Botanicus, Leiden University, The Netherlands. *(Photo: A. Schuiteman)* (142)

D. *Entomophobia kinabaluensis* (Ames) de Vogel. Cult. Hortus Botanicus, Leiden University, The Netherlands. *(Photo: A. Schuiteman)* (142)

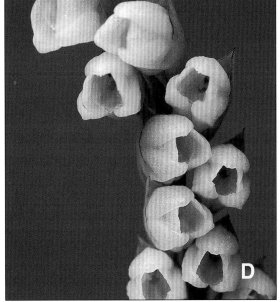

PLATE 12

A. *Epigeneium zebrinum* (J.J. Sm.) Summerh. Sarawak, Kuching District. *(Photo: J. Dransfield)* (144)

B. *Eria cymbidifolia* Ridl. var. *cymbidifolia.* Sabah, ex Nabawan, cult. Tenom Orchid Centre. *(Photo: K. Barrett)* (149)

C. *Eria saccifera* Hook.f. Sabah, Mount Kinabalu. *(Photo: K. Barrett)* (154)

PLATE 13

A. *Geesinkorchis alaticallosa* de Vogel. Cult. Hortus Botanicus, Leiden University, The Netherlands. *(Photo: A. Schuiteman)* (157)

B. *Geesinkorchis alaticallosa* de Vogel. Sabah, Sipitang District, Meligan Range. *(Photo: A. Lamb)* (157)

C. *Habenaria lobbii* Rchb.f. Cult. Hortus Botanicus, Leiden University, The Netherlands. *(Photo: A. Schuiteman)* (163)

D. *Habenaria lobbii* Rchb.f. Sarawak, Gunung Mulu National Park. *(Photo: A. Lamb)* (163)

PLATE 14

A. *Goodyera condensata* Ormerod & J.J. Wood. Sabah, Mount Kinabalu.
(Photo: A. Lamb) (161)

B. *Hetaeria anomala* (Lindl.) Rchb.f. Sabah, Mount Kinabalu, Mesilau Valley.
(Photo: A. Lamb) (166)

C. *Liparis anopheles* J.J. Wood. Cult. Hortus Botanicus, Leiden University, The
Netherlands. *(Photo: A. Schuiteman)* (168)

D. *Liparis grandiflora* Ridl. Sabah, Kinabatangan District, Maliau Basin.
(Photo: A. Lamb) (170)

E. *Liparis grandiflora* Ridl. Sabah. *(Photo: A. Schuiteman)* (170)

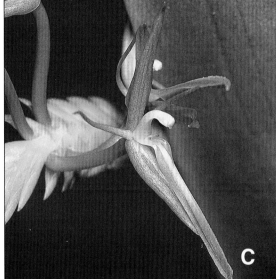
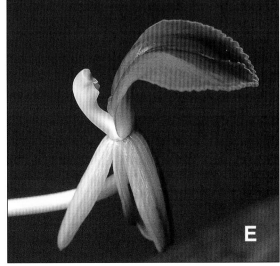

PLATE 15

A. *Liparis lacerata* Ridl. Sabah, Crocker Range. *(Photo: A. Lamb)* (172)

B. *Liparis lacerata* Ridl. Cult. Hortus Botanicus, Leiden University, The Netherlands. *(Photo: A. Schuiteman)* (172)

C. *Pennilabium struthio* Carr. Sabah. *(Photo: J. Stone)* (180)

D. *Oberonia patentifolia* Ames & C. Schweinf. Sabah, Crocker Range. *(Photo: A. Lamb)* (176)

PLATE 16

A. *Peristylus hallieri* J.J. Sm. Sabah. *(Photo: K. Barrett)* (182)

B. *Peristylus hallieri* J.J. Sm. Sabah. *(Photo: J. Stone)* (182)

C. *Phalaenopsis cochlearis* Holttum. Cult. Hortus Botanicus, Leiden University, The Netherlands. *(Photo: A. Schuiteman)* (184)

D. *Pholidota clemensii* Ames. Sabah, Mount Kinabalu. *(Photo: K. Barrett)* (186)

E. *Pholidota clemensii* Ames. Sabah, Mount Kinabalu. *(Photo: K. Barrett)* (186)

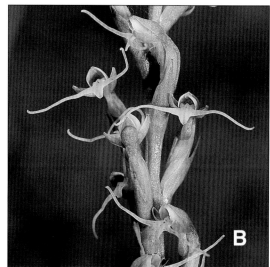

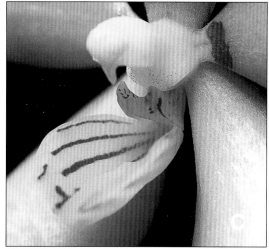

PLATE 17

A. *Pholidota sigmatochilus* (Rolfe) J.J. Sm. Sabah, Mount Kinabalu. *(Photo: K. Barrett)* (191)

B. *Pholidota sigmatochilus* (Rolfe) J.J. Sm. Sabah, Mount Kinabalu. *(Photo: K. Barrett)* (191)

C. *Pholidota ventricosa* (Blume) Rchb.f. Sabah, Kimanis road. *(Photo: A. Lamb)* (193)

D. *Pholidota ventricosa* (Blume) Rchb.f. Sabah, Mount Kinabalu. *(Photo: K. Barrett)* (193)

PLATE 18

A. *Pholidota ventricosa* (Blume) Rchb.f. Sabah, Mount Kinabalu. *(Photo: K. Barrett)* (193)

B. *Pholidota ventricosa* (Blume) Rchb.f. White-flowered form. Sabah, Mount Kinabalu. *(Photo: K. Barrett)* (193)

C. *Podochilus marsupialis* Schuit. Cult. Hortus Botanicus, Leiden University, The Netherlands. *(Photo: A. Schuiteman)* (196)

D. *Pomatocalpa kunstleri* (Hook.f.) J.J. Sm. Sabah, Crocker Range. *(Photo: A. Lamb)* (198)

PLATE 19

A. *Pomatocalpa kunstleri* (Hook.f.) J.J. Sm. Cult. Hortus Botanicus, Leiden University, The Netherlands. *(Photo: A. Schuiteman)* (198)

B. *Porphyrodesme sarcanthoides* (J.J. Sm.) U.W. Mahyar. Sabah, Sapulut, Labang Basin. *(Photo: A. Lamb)* (200)

C. *Porphyrodesme sarcanthoides* (J.J. Sm.) U.W. Mahyar. Sabah, Sapulut, Labang Basin. *(Photo: A. Lamb)* (200)

D. *Sarcoglyphis potamophila* (Schltr.) Garay & W. Kittr. Sabah, Nabawan. *(Photo: A. Lamb)* (202)

PLATE 20

A. *Schoenorchis buddleiflora* (Schltr. & J.J. Sm.) J.J. Sm. Sabah, Mount Kinabalu. *(Photo: K. Barrett)* (204)

B. *Schoenorchis endertii* (J.J. Sm.) E.A. Christenson & J.J. Wood. Sabah, Hempuen Hill. *(Photo: A. Lamb)* (207)

C. *Schoenorchis endertii* (J.J. Sm.) E.A. Christenson & J.J. Wood. Sabah, Hempuen Hill. *(Photo: A. Lamb)* (207)

D. *Smitinandia micrantha* (Lindl.) Holttum. South Kalimantan, cult. Tenom Orchid Centre. *(Photo: A. Lamb)* (209)

E. *Smitinandia micrantha* (Lindl.) Holttum. Cult. Hortus Botanicus, Leiden University, The Netherlands. *(Photo: A. Schuiteman)* (209)

PLATE 21

A. *Spathoglottis kimballiana* Hook.f. Sabah, Mount Kinabalu, Penataran River. *(Photo: A. Lamb)* (211)

B. *Thrixspermum triangulare* Ames & C. Schweinf. Sabah, Mount Kinabalu. *(Photo: K. Barrett)* (216)

C. *Thrixspermum tortum* J.J. Sm. Sabah, cult. Tenom Orchid Centre. *(Photo: A. Lamb)* (213)

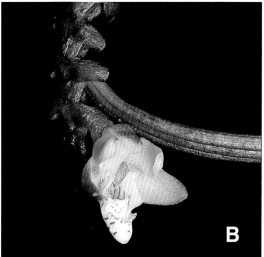

PLATE 22

A. *Trichoglottis jiewhoei* J.J. Wood, A. Lamb & C.L. Chan. Sabah, Kimanis road, cult. Tenom Orchid Centre. *(Photo: A. Lamb)* (218).

B. *Trichoglottis jiewhoei* J.J. Wood, A. Lamb & C.L. Chan. Sabah, cult. Tenom Orchid Centre. *(Photo: A. Lamb)* (218)

PLATE 23

A. *Trichoglottis kinabaluensis* Rolfe. Sabah, Mount Kinabalu. *(Photo: A. Lamb)* (220)

B. *Trichoglottis kinabaluensis* Rolfe. Sabah, Mount Kinabalu. *(Photo: A. Lamb)* (220)

C. *Trichoglottis magnicallosa* Ames & C. Schweinf. Sabah, cult. Tenom Orchid Centre. *(Photo: A. Lamb)* (222)

PLATE 24

A. *Trichoglottis tinekeae* Schuit. Sabah, Sipitang District, Long Pa Sia. *(Photo: A. Lamb)* (224)

B. *Trichoglottis tinekeae* Schuit. Cult. Hortus Botanicus, Leiden University, The Netherlands. *(Photo: A. Schuiteman)* (224)

C. *Trichotosia vestita* (Lindl.) Kraenzl. Sabah, Nabawan. *(Photo: A. Lamb)* (226)

D. *Vanilla havilandii* Rolfe. Sabah, Tenom District, Kallang Waterfall. *(Photo: A. Lamb)* (228)

E. *Vanilla kinabaluensis* Carr. Sabah, cult. Tenom Orchid Centre. *(Photo: A. Lamb)* (231)

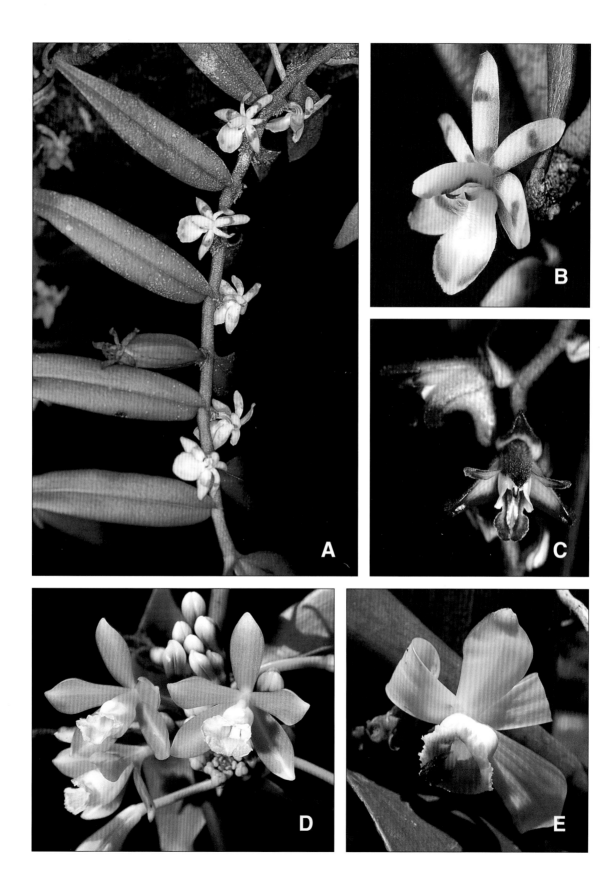

PLATE 25

A. *Ventricularia borneensis* J.J. Wood. Cult. Hortus Botanicus, Leiden University, The Netherlands. *(Photo: A. Schuiteman)* (233)

B. *Ventricularia borneensis* J.J. Wood. Sabah, Sipitang District, Long Pa Sia. *(Photo: A. Lamb)* (233)

C. *Vrydagzynea grandis* Ames & C. Schweinf. Sabah, Mount Kinabalu. *(Photo: A. Lamb)* (235)

INDEX TO ORCHID SCIENTIFIC NAMES

(Accepted names appear in roman type. Synonyms appear in *italics*. Numbers in **bold** type indicate pages with detailed treatment. Numbers within brackets indicate pages with illustrations).

A

Abdominea 2, 15, 22
  micrantha 21
  minimiflora (20), **21,** 243
Acanthephippium 2, 9
Acanthoglossum (section of Pholidota) 190, 193
Acianthinae (subtribe) 1
Acoridium
  corrugatum 110
Acriopsidinae (subtribe) 2
Acriopsis 2, 13
Adenoncos 2, 15
Aerides 2, 17
Aeridinae (subtribe) 2, 3, 4, 14, 239, 240
Aetheria
  anomala 166
Agrostophyllum 2, 11, 149
  glumaceum **22,** (23), 243
  ? khasiyanum 22
Ania 2, 8, 26, 28, 241
  borneensis **24,** (25), 26, 243
  malayana 24
  ponggolensis **26,** (27), 28, 243
Anoectochilus 1, 5
  lowii 140
Apaturia
  montana 28
Aphanobulbon (section of Bulbophyllum) 54
Aphyllorchis 1, 14, 30
  benguetensis 28
  borneensis 28
  kemulensis 30
  montana **28,** (29), 30, 243
  ? odoardi 28
  pallida 30
  prainii 28
  purpurea 28
  spiculaea 30
  striata 28
  tanegashimensis 28
  unguiculata 28
Aporodes (section of Bromheadia) 52
Aporum (genus)
  hendersonii 93
Aporum (section of Dendrobium) 9
Apostasia 1, 4

Apostasioideae (subfamily) 1
Appendicula 2, 7, 11, 31, 33, 35, 147
  bilobulata **31,** (32), 33, 243
  calcarata **33,** (34), 35, 43, 243
  congesta **35,** (36), 37, 243
  foliosa **37,** (38), 39, 243
  fractiflexa 33
  kinabaluensis 35
  latibracteata 41
  merrillii 37
  niahensis 39, 41
  pendula 39, 41
  pilosa **39,** (40), 41, 244
  rostellata **41,** (42), 43, 244
  torta 31
  undulata 33
Arachnis 2, 15
Arethusae (tribe) 2, 131
Arundina 2, 7, 133
  gracilis 133
  parviflora 131
Arundinae (subtribe) 2
Ascidieria 2, 11, 45
  longifolia **43,** (44), 244
Ascocentrum 2, 18
  micranthum 209
Ascochilopsis 2, 17
Ascochilus 2, 4, 17
  emarginatus 4
Ascotainia
  borneensis 24

B

Biermannia 2, 17
Bletiinae (subtribe) 2
Bogoria 2, 16
Brachypeza 2, 17, 47
  indusiata **45,** (46), 244
  stenoglottis 45
  zamboangensis 47
Bracisepalum 241
Bromheadia (genus) 2, 7, 11, 52, 240
Bromheadia (section of Bromheadia) 49, 52, 135
  borneensis var. borneensis 7, **47,** (48), 49, 50, 244
  borneensis var. longiflora **49,** 244

crassifolia 7
divaricata 49, **50,** (51), 52, 244
finlaysoniana 7, 49, 126
graminea **52,** (53), 245
rigida 133, 135
Bromheadiinae (subtribe) 2
Broughtonia
  sanguinea 79
Bulbophyllinae (subtribe) 2, 3, 10
Bulbophyllum 2, 3, 10, 58, 239, 241
  anceps frontispiece
  apodum 56
  comosum 56
  coriaceum **54,** (55), 56, 245
  disjunctum xi
  dracunculus xii
  farinulentum subsp. farinulentum xi
  goebelianum xi
  kinabaluense 54
  lemniscatoides **56,** (57), 58, 245
  lemniscatum 58
  longiflorum 3
  macranthum xi
  microglossum xi
  nabawanense 126
  polygaliflorum xii
  unguiculatum 56
  venustum 54

C
Calanthe 2, 8, 60
  otuhanica **58,** (59), 60, 245
  transiens 60
Calcarifera (section of Dendrobium) 91
Callista
  bifaria 81
  metachilina 101
Ceratochilus 4
  jiewhoei xi, 4
Ceratostylis 2, 11
Cestichis (section of Liparis) 170
Chamaeanthus 2, 18
Cheirorchis 66
  fulgens 64, 66
Cheirostylis 1, 6
  marmorata 140, 142
  montana 62
  spathulata **60,** (61), 245
Chelonanthera
  ventricosa 193
Chelonistele 2, 12, 193, 241
  kinabaluensis 191
  sulphurea var. sulphurea 135
Chilopogon
  kinabaluensis 35, 37

merrillii 37
Chroniochilus 2, 18
Chrysoglossum 2, 8
Cirrhopetalum (genus) 3, 10, 239
Cirrhopetalum (section of Bulbophyllum) 3
Claderia 2, 6
Cleisocentron 2, 16
Cleisomeria 2, 15
Cleisostoma 2, 15, 16, 64, 202, 204
  borneense 62
  discolor 64
  kunstleri 198
  longifolium 62
  potamophilum 202
  micranthum 209
  poilanei 209
  petitiana 209
  suaveolens **62,** (63), 245
  tixieri 209
Coelogyne 2, 12
  papillosa 60
  phaiostele 159
  ridleyana 159
  rupicola 60
  ventricosa 193
Coelogyneae (tribe) 2
Coelogyninae (subtribe) 2, 159, 241, 242
Collabiinae 2
Collabium 2, 8
Conostalix (section of Dendrobium) 7, 103
Cordiglottis 2, 16, 66
  fulgens **64,** (65), 68, 245
  major 66
  multicolor 66
  pulverulenta 66
  westenenkii **66,** (67), 68, 245
Coriifoliae (section of Liparis) 172
Corybas 1, 6
  pictus 68
  serpentinus **68,** (69), 245
Corymborkis 1, 5
Cranichideae (tribe) 1
Crumenata (section of Dendrobium) 79, 95, 97
Cryptostylidinae (subtribe) 1
Cryptostylis 1, 6
Cylindrolobus (section of Eria) 11, 147, 228
Cymbidieae (tribe) 2, 135
Cymbidium 2, 7, 8, 13, 149, 151
  bicolor subsp. pubescens 151
  borneense 8
  elongatum 7
  ensifolium subsp. haematodes 8
  kinabaluense 12, **70,** (71), 245
  lancifolium 8, 13

sigmoideum 70, 72
Cymboglossum (genus)
   longifolium 43
Cymboglossum (section of Eria) 151, 241
Cyperorchis (section of Cymbidium) 70
Cypripedioideae (subfamily) 1
Cypripedium 213
Cyrtopodiinae (subtribe) 2
Cyrtosia 1, 14
Cystorchis 1, 5, 14, 72, 74, 78
  aphylla 14, 74, 76
  macrophysa **72**, (73), 245
  peliocaulos 74, 76
  salmoneus 14, **74**, (75), 245
  saprophytica 14, 74, **76**, (77), 78, 245

D
Dendrobiinae (subtribe) 2, 240
Dendrobium 2, 7, 9, 137
  aurantiflammeum **78,** 79, (80), 81, 83, 85,
    245
  bifarium **81**, (82), 83, 245
  brevicolle 135
  calicopis 91
  cinnabarinum 79, 81, **83**, (84), 101, 246
  cinnabarinum var. *angustitepalum* auct., *non*
    Carr 78
  cinnabarinum var. *angustitepalum* Carr 79,
    83
  cinnabarinum var. *lamelliferum* 99, 101
  cinereum 89, 91
  citrinocastaneum 144, 240
  connatum 83
  crumenatum 95
  dearei **86**, (87), 88, 105, 240, 246
  derryi **89**, (90), 91, 246
  erythropogon 93
  excisum 81
  fugax 93, 95
  groeneveldtii 89, 91
  hallieri **91**, (92), 93, 246
  hendersonii **93**, (94), 95, 246
  holttumianum 79, 83
  hymenopterum 91
  kurashigei **95**, (96), 246
  lamelluliferum 97
  lamrianum **97**, (98), 246
  lawiense **99**, (100), 101, 246
  longicolle 139
  lowii 93
  metachilinum **101**, (102), 103, 246
  metachilinum var crenulatum 103
  olivaceum 99
  ovipostoriferum J.J. Sm. 88
  oviposteriferum auct., *non* J.J. Sm. 86

parthenium **103**, (104), 105, 240, 246
piranha 99
radians **105**, 106, (107), 246
revolutum 105
ridleyanum 93
rorulentum 101
rudolphii 93
sanderae 88
sanderianum 103, 105
sandsii 99
sanguineum 79, 81, 83
schuetzei 88
sculptum 106
singkawangense xi
sulphuratum 144
takahashii 88
vestitum 226
zebrinum 144
Dendrochilum (genus) 2, 12, 120, 239, 242
  alatum 110
  alpinum **108**, (109), 246
  angustipetalum 120
  corrugatum **110**, (111), 247
  crassum 114
  fimbriatum 110
  gravenhorstii **112**, (113), 114, 247
  jiewhoei **114**, (115), 116, 247
  joclemensii **116**, 117, 118, 247
  kamborangense 116
  kelabitense **118**, (119), 120, 247
  longipes **120**, (121), 122, 247
  longirachis **122**, (123), 124, 247
  magaense 127
  mantis 120, 121, 122
  pallidiflavens 114
  papillilabium xii
  remotum 124, 126
  simplex **124**, (125), 126, 247
  subintegrum 122
  subulibrachium **127**, (128), 247
  tenompokense xii
  transversum **129**, (130), 248
Dendrochilum (subgenus) 114
Dendrocolla (genus) 66
  fulgens 64
Dendrocolla (section of Thrixspermum) 215
Didymoplexiella 1, 13
Didymoplexis 1, 13
Dilochia 2, 7, 131, 147
  gracilis 133
  parviflora **131**, (132), 248
  rigida **133**, (134), 248
Dimorphorchis 2, 16
Diplocaulobium 2, 9, 137
  brevicolle **135**, (136), 248

310

longicolle **137**, (138), 248
malayanum 139
vanleeuwenii (J.J. Sm.) P.F. Hunt &
Summerh. 137
vanleeuwenii sensu Wood & Cribb 135
Dipodium 2, 6
Distichae (sction of Liparis) 168
Distichophyllum (section of Dendrobium) 7,
83, 99, 103
Diurideae (tribe) 1
Doritis 2, 14
Dossinia 1, 5, 142
marmorata **140**, (141), 142, 248
Dyakia 2, 18

E
Entomophobia 2, 12, 144, 241
kinabaluensis **142**, (143), 248
Epicrianthes (genus) 3, 10
Epicrianthes (section of Bulbophyllum) 3
Epidendreae (tribe) 2
Epidendroideae (subfamily) 1, 3, 4, 239
Epigeneium 2, 7, 9, 146
kinabaluense 7, 146
longerepens 146
treacherianum 146
tricallosum 135, 146
zebrinum **144**, (145), 146, 249
Epipogiinae (subtribe) 1
Epipogium 1, 14
Eria 2, 11, 13, 147, 228, 241
apertiflora 156
aurantia **147**, (148), 149, 249
aurantiaca 147
bipunctata 156
brookesii 154, 156
cymbidifolia var. cymbidifolia **149**, (150),
151, 152, 249
cymbidifolia var. *longipes* 152
cymbidifolia var. pandanifolia **151,** 249
cymbiformis 149
densa 156
longifolia 43, 45
pseudocymbiformis var. hirsuta 151, **154,**
249
pseudocymbiformis var. pseudocymbiformis
152, (153), 154, 249
saccata 154, 156
saccifera **154**, (155), 156, 157, 249
suaveolens 156
verticillaris 43, 45
vestita 226
Eriinae (subtribe) 2
Erythrodes 1, 5
glandulosa 72

Erythrorchis 1, 14
Eulophia 2, 8, 14
graminea 8
spectabilis 8
zollingeri 14
Eulophiinae (subtribe) 2
Eurybrachium (section of Dendrochilum)
108, 118, 129

F
Flickingeria 2, 9, 95
Formosae (section of Dendrobium) 88, 93
Fuscatae (section of Phalaenopsis) 186

G
Galeola 1, 14
Galeolinae (subtribe) 1
Gastrochilus 2, 15, 18
parviflorus 209
patinatus, 15
Gastrodia 1, 3, 13
Gastrodieae (tribe) 1, 3
Gastrodiinae (subtribe) 1, 3
Geesinkorchis 2, 8, 12, 159, 241
alaticallosa **157**, (158), 249
phaiostele **159**, (160), 161, 250
Geodorum 2, 8
Glomera 24
Glomerinae (subtribe) 2
Goodyera 1, 5, 166
condensata **161**, (162), 163, 250
rostellata 163
ustulata 163
Goodyerinae (subtribe) 1, 60
Grammatophyllum 2, 12
Grastidium (section of Dendrobium) 147
Grosourdya 2, 18

H
Habenaria 1, 6, 165, 184, 242
havilandii 163
lobbii **163**, (164), 165, 250
marmorophylla 163, 165
setifolia 165
Habenariinae (subtribe) 1
Hancockia 241
Hapalochilus 3, 10
lohokii 3
Hetaeria 1, 6, 166
angustifolia 166
anomala **166**, (167), 250
biloba 166
grandiflora 166
hylophiloides 166
rotundiloba 166

Hippeophyllum 1, 9
Hylophila 1, 5
Hymeneria (section of Eria) 157

J
Jejewoodia xi, 2, 4, 15

K
Katherinea
  citrinocastanea 144
  zebrina 144
Katocolla (section of Thrixspermum) 215
Kingidium 2, 16
Kuhlhasseltia 1, 6

L
Lecanorchidinae (subtribe) 1
Lecanorchis 1, 13
Lepidogyne 1, 5
Leptorchis
  lacerata 172
Limodorinae (subtribe) 1
Liparis 1, 7, 170
  anopheles **168**, (169), 250
  crenulata 170
  grandiflora **170**, (171), 250
  lacerata **172**, (173), 250
  latifolia 170
  lobongensis 168
  pandurata 168
Lockhartia 196
Luisia 2, 18

M
Macodes 1, 6, 142
Macropodanthus 2, 17
Malaxideae (tribe) 1
Malaxis 1, 7
Malleola 2, 18, 235
Mastigion 3, 10
  putidum 3
Micropera 2, 15
Microsaccus 2, 14
Microtatorchis 2, 18
Mischobulbum 2, 9, 241
Mycaranthes (section of Eria) 11
Myrmechis 1, 6

N
Nabaluia 2, 12, 174, 241
  exaltata **174**, (175), 250
Neoclemensia 1, 3, 13
Neottieae (tribe) 1

Nephelaphyllum 2, 8
Nervilia 1, 6
Nervilieae (tribe) 1
Neuwiedia 1, 4
Nigrohirsutae (section of Dendrobium) 93

O
Oberonia 1, 9, 176, 241
  hispidula 178
  patentifolia **176**, (177), 178, 250
Octarrhena 2, 11, 180
  amesiana 178
  angraecoides 180
  condensata 180
  lorentzii 180
  nana 178
  parvula **178**, (179), 251
Oeceoclades 2, 8
Orchideae (tribe) 1
Orchidinae (subtribe) 1
Orchidoideae (subfamily) 1
Orcidice (section of Thrixspermum) 215
Ornithochilus 2, 15
Osyricera (genus) 3, 10
  crassifolia 3
  osyriceroides 3
Osyricera (section of Bulbophyllum) 3
Oxygenianthe (section of Dendrobium) 93
Oxystophyllum (section of Dendrobium) 9

P
Pachystoma 2, 8
Pantlingia 1, 6
Paphiopedilum 1, 4
Papilionanthe 2, 17
Paraphalaenopsis 2, 17
Pennilabium 2, 18, 182
  angraecoides 182
  lampongense 182
  struthio **180**, (181), 182, 251
Peristylus 1, 6, 184, 242
  hallieri **182**, (183), 251
Phaius 2, 8
Phalaenopsis xii, 2, 14, 17, 239, 241
  bellina xii, 239
  cochlearis **184**, (185), 186, 251
  corningiana xi
  fuscata 186
  violacea xii, 239
Pholidota 2, 12, 174, 193, 242
  camelostalix 193
  clemensii **186**, (187), 188, 193, 251
  ? *dentiloba* 186, 188

grandis 193
kinabaluensis 142, 144, 193
pectinata 190
schweinfurthiana **188**, (189), 190, 251
sesquitorta 193
sigmatochilus 60, **191**, (192), 193, 239, 251
sororia var. *sororia* 193
sororia var. *djamuensis* 193
triloba 159
ventricosa **193**, (194), 251
Phreatia 2, 11, 13
amesiana 11
densiflora 11
listrophora 13
monticola 11
nana 178
parvula 178
secunda 11
sulcata 13
Physurus
glandulosus 72, 74
Pilophyllum 2, 7
Pinalia (genus)
longifolia 43
vestita 226
Pinalia (section of Eria) 156
Platanthera 1, 6, 14
saprophytica 14
Platyclinis (genus)
corrugata 110
Platyclinis (section of Dendrochilum) 124, 126, 127
Pleiophyllus (section of Bulbophyllum) 58
Plocoglottis 2, 8
Poaephyllum 2, 11
Podochileae (tribe) 2, 3
Podochilinae (subtribe) 2
Podochilus 2, 11, 31, 196
calcaratus 33
congestus 35
marsupialis **196**, (197), 252
Polychilos
cochlearis 184
Polystachya 2, 12, 149
Polystachyinae (subtribe) 2
Pomatocalpa 2, 16, 198
kunstleri **198**, (199), 252
merrillii 198
poilanei 209
Porpax 2, 13
Porphyrodesme 2, 17, 200, 202
hewittii 200
papuana 200
sarcanthoides **200**, (201), 202, 252
Porphyroglottis 2, 12

Porrorhachis 2, 18
Pristiglottis 1, 6
hasseltii xi
hydrocephala xi
Pteroceras 2, 17, 47, 182
stenoglottis 45

R
Renanthera 2, 4, 16
auyongii xi
histrionica xi
moluccana 202
sarcanthoides 200
Renantherella xi, 4, 16
histrionica subsp. *auyongii* xi
Rhomboda 1, 5, 240
Rhopalanthe (section of Dendrobium) 79
Rhynchostylis 2, 17
Rhytionanthos 3, 4, 10
mirum 4
Robiquetia 2, 17, 207
endertii 207

S
Saccolabium
buddleiflorum 204
cortinatum 21
fissum 209
micranthum 209
minimiflorum 21
pubescens 198
Sarcanthus 202
potamophilus 202
robustus 62
suaveolens 62
Sarcochilus
keyensis 45
indusiatus 45
sigmoideus 45
stenoglottis 45
Sarcoglyphis 2, 15, 204, 235
fimbriatus 204
potamophila **202**, (203), 204, 252
Sarcopodium
citrinocastaneum 144
sulphuratum 144
zebrinum 144
Sarcostoma 2, 11
Sarganella (section of Podochilus) 196
Schoenorchis 2, 16, 206, 207
buddleiflora **204**, (205), 206, 252
endertii **207**, (208), 252
juncifolia 206
minimiflora 21
philippinensis 21

Sigmatochilus 191
  kinabaluensis 191
Smitinandia 2, 15
  micrantha **209**, (210), 211, 252
Spathoglottis 2, 9, 213, 240
  aurea Lindl. 213
  aurea Rchb.f., *non* Lindl. 213
  confusa 213
  gracilis 213
  kimballiana var.angustifolia 213
  kimballiana var. kimballiana **211**, (212), 213, 252
  microchilina 213
  plicata 213
Spiranthes 1, 6
Spiranthinae (subtribe) 1
Spongiola 2, 18
Staurochilus 2, 16
Stereosandra 1, 14
Strongyle (section of Dendrobium) 9
Synarmosepalum 3, 4, 10
  heldiorum 4

T
Taeniophyllum 2, 14
Tainia 2, 9, 26, 28, 241
  malayana 24
  paucifolia 26
  penangiana 24
  rolfei 24
  steenisii 24
Thecopus 2, 13
Thecostele 2, 12
Thecostelinae (subtribe) 2
Thelasiinae (subtribe) 2
Thelasis 2, 11, 13
  capitata 13
  carinata 11
  carnosa 13
  micrantha 11
  variabilis 13
Trias 2, 10
  tothastes 10
Thrixspermum 2, 16, 66, 200, 215, 216
  amplexicaule 215
  arachnites 215
  centipeda 215
  indusiatum 45
  longicauda 215
  pensile 215
  platyphyllum 45
  scopa 215
  tortum **213**, (214), 215, 252
  triangulare **216**, (217), 252

Trichoglottis 2, 16, 233
  jiewhoei **218**, (219), 253
  kinabaluensis **220**, (221), 253
  lanceolaria 233
  magnicallosa **222**, (223), 253
  scapigera xi
  tinekeae **224**, (225), 226, 253
  uexkulliana 222
  zollingeriana xi
Trichotosia 2, 7, 11, 228
  brevipedunculata 228
  ferox 228
  vestita **226**, (227), 253
Tropidia 1, 5, 14
  connata 14
  saprophytica 14
Tropidieae (tribe) 1
Tuberolabium 2, 18

U
Uncifera
  albiflora 209
Urostachya (section of Eria) 156

V
Vanda 2, 17
Vandeae (tribe) 2, 3, 4, 14
Vanilla 1, 6, 230, 231, 241
  abundiflora 231, 233
  albida 230
  borneensis 231
  havilandii **228**, (229), 230, 231, 253
  kinabaluensis **231**, (232), 233, 253
Vanilleae (tribe) 1
Vanillinae (subtribe) 1
Vanilloideae (subfamily) 1
Ventricularia 2, 4, 16, 233
  borneensis 4, **233**, (234), 235, 253
  tenuicaulis 235
Vrydagzynea 1, 5, 74, 235
  grandis **235**, (236), 253

Z
Zeuxine 1, 6, 78
  biloba 166